# Human-Technology Shifts
## In 2026 & Beyond
### Point/Counterpoint
Series #1

No AIs were injured in the making of this book.

I did wrangle research with the help of several AI toolsets including Anthropic's Claude, Open AI's ChatGPT as well as my own. The analysis is mine as are any errors, omissions or gaffs.

I also had ChatGPT generate concept art that I used for many of the graphics in this book.

The content of this book is intended for entertainment and informational purposes, and I don't assume responsibility for any actions taken based on this content.

## Human-Technology Shifts in 2026 & Beyond


*Intro* ____ *4*

*The Approach* ____ *7*

*Conversation Is the Thing* ____ *8*

*Quantum Computing breaks out of the lab!* ____ *9*

*The Edge Is Real. The Economics Are Still Being Figured Out.* ____ *20*

*Open-Source Flames Out!* ____ *31*

*Neuromorphic Computing Breakthrough* ____ *43*

*6G Standards Fragmentation* ____ *54*

*Multimodal Model Commoditization* ____ *65*

*AI-Generated Code Quality Ceiling* ____ *76*

*Solid-State Battery Commercialization Delay* ____ *87*

*Satellite Internet Consolidation* ____ *98*

*Social Media Regulation's Defining Year* ____ *110*

*Smartphone Upgrade Cycle Death* ____ *121*

*Gaming Subscription Fatigue* ____ *132*

*Foldable Phones Breakthrough. Or Bust?* ____ *143*

*Streaming Service Consolidation Wave* ____ *154*

*Cashless Society Backlash* ____ *166*

*AI Voice Clone Regulation Emergency* ____ *178*

*AI-Powered Personal Finance Revolution* ____ *190*

*Voice Assistant Platform Abandonment* ____ *202*

*AI Customer Service Consumer Revolt* ____ *213*

*Wearable Health Monitoring Medical Revolution* ____ *225*

*Wireless Charging Infrastructure Reality Check* ____ *237*

*AI-Generated Influencer Content Destroys Marketing Trust* ____ *248*

*Smart TV Advertising Overreach Sparks Consumer Backlash* ____ *260*

*Subscription Box Model Collapses From Market Saturation* ____ *272*

*AI Search Engines Disrupt Google's Two-Decade Dominance* ____ *284*

*AI Tutoring Revolutionizes Personalized Education* ____ *295*

*Biometric Authentication Replaces Passwords Universally* ____ *305*

*The Analog Backlash Among Gen Z. Rebellion or Just a Rebrand?* ____ *317*

*Agentic AI: The 40% Reckoning Is Already Here* ____ *329*

*Voice as the New Browser: Transformation or Another False Start?* ____ *340*

*Smart Glasses Reach a True Tipping Point?* ____ *352*

*The PC Replacement Cycle (Innovation or Engineered Obsolescence?)* ____ *367*

*AI Erodes Consumer Trust in Digital Media (and Brands Feel It)* ____ *378*

*The Smart Home Interoperability Promise (Still Waiting After 20 Years)* ____ *389*

*Q-Day: The Security Crisis That Started Before Anyone Was Ready* ____ *401*

*The Captive Repair Economy Is Breaking* ____ *413*

*The Most Personal Data Relationship You've Ever Had* ____ *424*

*The Roundup* ____ *435*

*Your Turn* ____ *438*

*Attribution* ____ *439*

## Intro

Technology doesn't announce its most consequential shifts. It whispers them.

Quietly it whispers in a gaming headset that reads your brain signals, in a quadruped robot running for 72 hours on a single charge, in a nation-state quietly copying your encrypted data while it waits for the quantum computer that can unlock it.

The shifts that reshape how people live, work, communicate, and trust arrive not as dramatic ruptures but as accumulating pressures, until the day they don't feel like shifts at all. The shifts are just the new normal nobody quite remembers choosing.

This series exists to name them before that happens.

I launched Point/Counterpoint in 2025 with a specific frustration about the dominant mode of technology forecasting: confident predictions delivered without accountability, trend decks that call everything a revolution, analysis that tells you what to think rather than showing you the evidence and asking what you see.

The world has enough of that.

What it has less of is honest, evidence-grounded engagement with genuine uncertainty underneath the headlines.

In this series I examine more than thirty technology inflection points across more than ten domains of daily life; from the quantum computing lab to your living room and the standards bodies shaping digital technology to the teenager quietly buying a record player.

Each topic gets the same treatment: a serious case for the prediction, a serious argument against it, and a likelihood percentage that has to earn its precision with evidence rather than instinct.

What you won't find here is false certainty in either direction. The most interesting position in technology analysis is almost never "this will definitely happen" or "this is obviously hype." It's the harder, more useful claim: here is what the evidence actually shows, here is what would have to be true for the enthusiasts or optimists to be right, and here is what would have to fail for the pragmatists or pessimists to win.

The infographic that accompanies this series maps something important that a topic-by-topic reading can obscure: these trends don't operate in isolation.

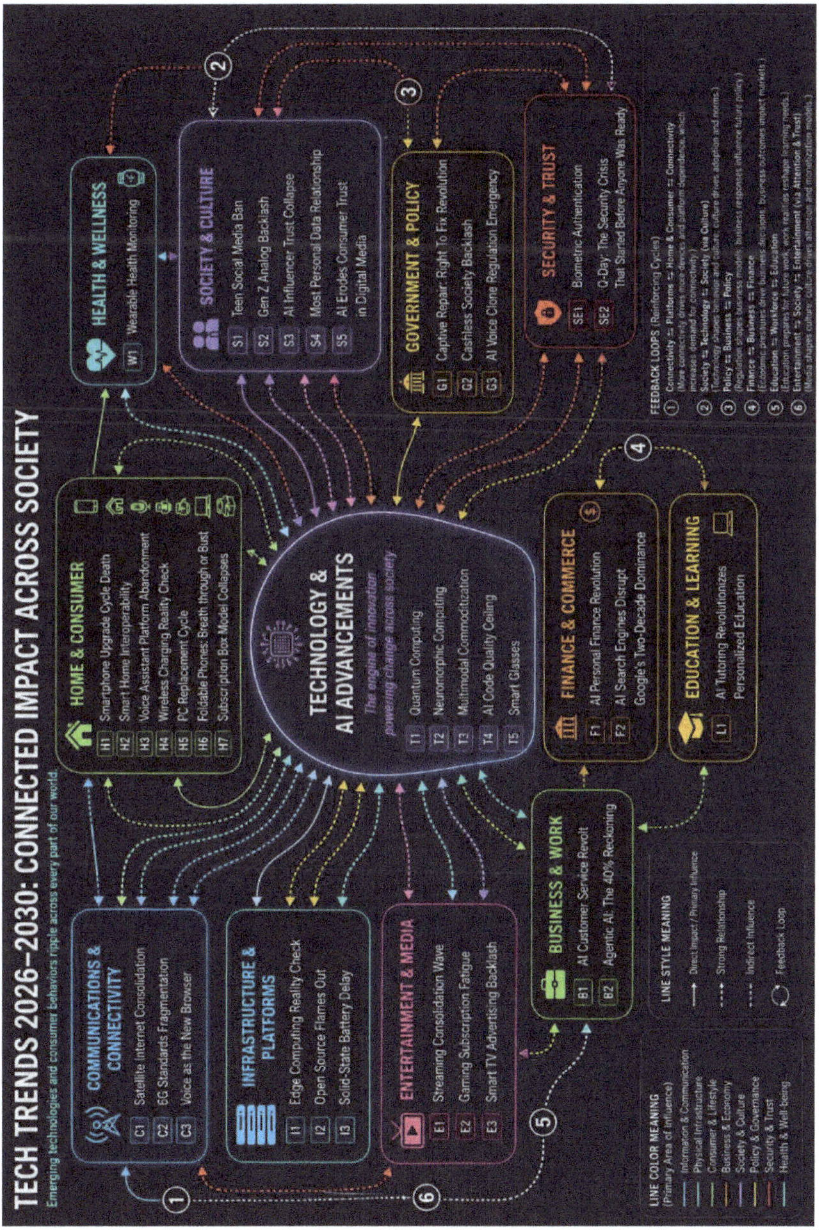
TECH TRENDS 2026–2030: CONNECTED IMPACT ACROSS SOCIETY
Emerging technologies and consumer behaviors ripple across every part of our world
TECHNOLOGY & AI ADVANCEMENTS
The engine of innovation powering change across society
T1 Quantum Computing
T2 Neuromorphic Computing
T3 Multimodal Commoditization
T4 AI Code Quality Ceiling
T5 Smart Glasses
COMMUNICATIONS & CONNECTIVITY
C1 Satellite Internet Consolidation
C2 6G Standards Fragmentation
C3 Voice as the New Browser
INFRASTRUCTURE & PLATFORMS
I1 Edge Computing Reality Check
I2 Open Source Flames Out
I3 Solid-State Battery Delay
ENTERTAINMENT & MEDIA
E1 Streaming Consolidation Wave
E2 Gaming Subscription Fatigue
E3 Smart TV Advertising Backlash
HOME & CONSUMER
H1 Smartphone Upgrade Cycle Death
H2 Smart Home Interoperability
H3 Voice Assistant Platform Abandonment
H4 Wireless Charging Reality Check
H5 PC Replacement Cycle
H6 Foldable Phones: Breakthrough or Bust
H7 Subscription Box Model Collapses
HEALTH & WELLNESS
W1 Wearable Health Monitoring
SOCIETY & CULTURE
S1 Teen Social Media Ban
S2 Gen Z Analog Backlash
S3 AI Influencer Trust Collapse
S4 Most Personal Data Relationship
S5 AI Erodes Consumer Trust in Digital Media
GOVERNMENT & POLICY
G1 Captive Repair: Right To Fix Revolution
G2 Cashless Society Backlash
G3 AI Voice Clone Regulation Emergency
SECURITY & TRUST
SE1 Biometric Authentication
SE2 Q-Day: The Security Crisis That Started Before Anyone Was Ready
BUSINESS & WORK
B1 AI Customer Service Revolt
B2 Agentic AI: The 40% Reckoning
FINANCE & COMMERCE
F1 AI Personal Finance Revolution
F2 AI Search Engines Disrupt Google's Two-Decade Dominance
EDUCATION & LEARNING
L1 AI Tutoring Revolutionizes Personalized Education
LINE COLOR MEANING (Primary Area of Influence)
Information & Communication
Physical Infrastructure
Consumer & Lifestyle
Business & Economy
Society & Culture
Policy & Governance
Security & Trust
Health & Well-being
LINE STYLE MEANING
Direct Impact / Primary Influence
Strong Relationship
Indirect Influence
Feedback Loop
FEEDBACK LOOPS (Reinforcing Cycles)
1 Connectivity ⇄ Platforms ⇄ Home & Consumer ⇄ Connectivity
2 Society ⇄ Technology ⇄ Society (via Culture)
3 Policy ⇄ Business ⇄ Policy
4 Finance ⇄ Business ⇄ Finance
5 Education ⇄ Workforce ⇄ Education
6 Entertainment ⇄ Society ⇄ Entertainment (via Attention & Trust)

Q-Day, the quantum computing threat to consumer encryption, connects directly to Biometric Authentication replacing passwords and to AI Voice Clone Regulation, all three converging on the Security and Trust cluster that underpins every other domain in the map.

The Smart Home Interoperability failure connects to the PC Replacement Cycle and the Smart Glasses tipping point. All three are stories about who controls the hardware layer of consumer life and on whose terms.

The Analog Backlash Among Gen Z, AI Erodes Consumer Trust, and the AI Influencer Content crisis are not separate trends. These topics are about the same crisis of authenticity arriving simultaneously in three different channels.

The infographic's arrows tell you which topics are causes, which are consequences, and which are in feedback loops that amplify each other in ways neither the optimists nor the pessimists in any individual debate are fully accounting for.

The organizing question running through every topic in this series is deceptively simple:

***Is the human-to-technology relationship improving or worsening?***

Not for the industry. Not for the investors.

For the people on the receiving end of decisions made in boardrooms, standards bodies, regulatory agencies, and research labs. The people whose attention, data, identity, health, security, and increasingly their neural signals are the raw material the technology economy runs on.

Some of the trends examined here represent genuine, evidence-backed improvements in human capability and access.

AI tutoring that delivers personalized instruction at scale.

Wearable health monitoring approaching clinical diagnostic capability.

Neuromorphic computing enabling always-on AI in devices that couldn't previously run it.

Others represent structural deteriorations in human agency, privacy, and trust that are moving faster than the governance frameworks designed to constrain them.

Most are both simultaneously, depending on which end of the access and power divide you occupy.

This is not a prediction analysis for the market. It's not a technology investment thesis. It's an ongoing, evidence-anchored argument about the direction of change that is explicitly designed to be argued with.

Every topic adds a piece to a picture that no single topic can show alone. The picture, when complete, is a map of the choices being made right now that will determine what kind of digital world the near decade produces, and whether the people living in it had any meaningful say in its design.

Several of the topics were originally written in 2025 with a "by end of 2026" analytical window.

Shifting that window to 2030 as a revision is material for those topics. In some cases, new research validated the original analysis but provided important aspects. In a few cases, the broader window provided updated research that adjusted the original take on the topic.

I've changed the graphics of the earlier pieces. On the one hand, the initial graphics template was short and to the point. Simple and economical.

With the aim of being more "infographic" and providing 3 focused graphics per topic, the adjusted graphics I introduce may provide a better summary for when you just want to scan.

Let me know which you prefer so I can revise for the Series 2 publication (yes, technology continues to shift so I have more to analyze).

Going the other direction, I've also created a research guide for each topic so you can dig into the material yourself. It's not included in this book to keep the size of the book down, but subscribe to digitalcowboystudio.substack.com (either as a free or paid subscription) and you will be able to get access to the guide. I am also posting continued analysis of Human-Technology shifts so you can keep up with all that newness as well.

# The Approach

I've spent a lot of time cross-referencing the research cited for each and every article. I've done my best to eliminate aggregator-laundered content and cite primary sources rather than repeat second-hand gossip. If you find something that's not verifiable or correct, feel free to let me know and I will make corrections in future editions.

Developer ecosystem metrics throughout this series (registered users, SDK downloads, platform enrollments) reflect cumulative historical participation rather than active developer counts. Vendors have no commercial incentive to distinguish between the two, and independent verification is rarely available. These figures should be read as indicators of platform reach and historical interest, not as measures of current active development activity.

With all that said, I hope you enjoy it and find the content interesting. More than anything I hope you engage in conversations around the topics that directly impact your life.

All these Human-Technology shifts have significant impact on the human condition. We need to be involved in how and to what degree that impact materializes.

## Conversation Is the Thing

The conversation I want to invite is not the debate between the prediction of The Point and the argument of the Counterpoint. It's the conversation between the analysis and your life.

These shifts are not abstractions. They are the customer service call that left you furious. The wearable that told you something your doctor hadn't. The AI tutor your kid is using without the school having a policy. The smart glasses someone is wearing in the restaurant. The subscription you forgot you had. The password you've been meaning to replace.

Every topic in this series touches something you are already navigating, and the people who navigate it most effectively are the ones who understand both what is genuinely changing and what is genuinely contested.

The system changes.

The hype changes.

The tools change.

Reality remains stubbornly real.

Two Sides. One Reality. You Decide.

That is what this series is for. Come for the data. Stay for the questions. The most interesting ones are yours.

Let's get to it!

## Quantum Computing breaks out of the lab!

*The "perpetual five years away" narrative is dead.*

*HSBC achieved a 34% improvement in bond trading predictions using IBM's Heron quantum processor. Google's Willow chip cracked a 30-year challenge in error correction. PsiQuantum raised $1 billion at a $7 billion valuation. IBM has committed to a fault-tolerant quantum computer by 2029, with the manufacturing facility already under construction in Poughkeepsie.*

*The question is no longer whether quantum computing works. It's whether it works for your problem, at your scale, at a cost your business can justify, and that question has a more complicated answer than the headline numbers suggest.*

OVERVIEW
QUANTUM COMPUTING BREAKS OUT OF THE LAB!
THE "PERPETUAL FIVE YEARS AWAY" NARRATIVE IS DEAD.
HSBC achieved a 34% improvement in bond trading predictions using IBM's Heron quantum processor.
Google's Willow chip cracked a 30-year challenge in error correction.
PsiQuantum raised $1 billion at a $7 billion valuation.
IBM has committed to a fault-tolerant quantum computer by 2029, with the manufacturing facility already under construction in Poughkeepsie.
THE EVIDENCE IS REAL
HSBC: 34% BETTER
Improvement in bond trading predictions using IBM's Heron quantum processor.
GOOGLE WILLOW
Cracked a 30-year challenge in quantum error correction.
PSIQUANTUM: $1B RAISE
Raised $1 billion at a $7 billion valuation.
IBM: FAULT TOLERANCE
Committed to a fault-tolerant quantum computer by 2029 with a new facility under construction in Poughkeepsie, NY.
The question is no longer whether quantum computing works. It is whether it works for your problem, at your scale, at a cost your business can justify, and that question has a more complicated answer than the headline numbers suggest.
BUT THE CHALLENGE REMAINS
NOT GENERAL (YET)
Advantage proven in specific domains, not across all problem types.
ERRORS STILL HIGH
Fault tolerance is coming, but today's systems are still noisy and costly.
COST IS A BARRIER
Quantum resources are scarce and error correction is exponentially expensive.
ENTERPRISE GAP
Tools, talent, and integration pipelines aren't ready for broad enterprise deployment.
REAL COMMERCIAL RESULTS
Live trading. Measurable improvement. Not a lab benchmark.
TECHNICAL BREAKTHROUGHS
Error correction scaling solved. The physics is catching up.
INFRASTRUCTURE COMMITMENT
Manufacturing facilities. Public roadmaps. Billions in funding.
COMMERCIAL ADOLESCENCE
Real value in specific domains today. Broader utility coming.
THE CHALLENGE REMAINS
From advantage in controlled settings to enterprise scale.
THE QUANTUM ERA HAS BEGUN. THE QUESTION IS HOW YOU PREPARE FOR WHAT COMES NEXT.
WEEK 1 | GOVERNANCE TRILOGY

## Opening

The original prediction that quantum computing crossing a commercial utility threshold by end of 2026 was right about the direction and more conservative than events warranted about the pace.

What 2026 actually looks like is a technology in genuine commercial adolescence: demonstrating real and reproducible value in specific, well-defined problem classes, with funded roadmaps to fault tolerance now backed by manufacturing infrastructure, not just research projections.

The 2027–2030 window is where the story becomes genuinely compelling and genuinely uncertain in equal measure.

The hardware trajectory is credible and primary-sourced. The fault-tolerance roadmaps are funded and moving. The application domains where quantum advantage is clearest (molecular simulation, specific optimization classes, post-quantum cryptography) are large enough to justify the investment levels the industry is attracting.

The gap between "quantum advantage demonstrated in a controlled setting" and "quantum advantage reliably reproduced in enterprise production environments" remains the central challenge. But for the first time in the technology's history, the tools to close that gap are being built on schedule.

## Intro

The quantum computing industry has long suffered from the perception that practical applications remain perpetually five years away.

The 2025-2026 evidence forces a revision of that narrative, not into triumphalism, but into something more analytically interesting: domain-specific commercial utility, demonstrably real, accompanied by a credible and funded path to the broader fault-tolerant era that will expand that utility to a wider range of enterprise problems.

The HSBC bond trading result is the single most important commercial data point in the quantum record to date. A real financial institution, real production-scale trading data, a peer-reviewed pre-print, and a 34% measurable improvement that ended the "still a research curiosity" dismissal.

What it did not end, and what the Counterpoint correctly preserves, is the question of generalizability.

That question defines the 2027-2030 window.

## Background

The quantum computing ecosystem entered 2026 with the strongest evidence base in the technology's history.

National governments had invested more than $10 billion in quantum technologies by mid-2025, up from $1.8 billion in all of 2024.

Quantum computing companies raised $3.77 billion in equity funding in the first nine months of 2025 alone, nearly triple the $1.3 billion raised in all of 2024.

IBM's formal roadmap commits to near-term quantum advantage by end of 2026, with IBM Quantum Kookaburra targeting 2026 and the Starling fault-tolerant system targeting 2029, to be housed in a new 511,000-square-foot IBM Quantum Data Center under construction in Poughkeepsie, New York.

IonQ reported $64.7 million in revenue for Q1 2026, representing 755% year-over-year growth, with full-year 2026 guidance raised to $260–270 million and remaining performance obligations growing 554% year-over-year.

These are not speculative projections.

They are audited financial results and funded engineering commitments from public companies with manufacturing infrastructure and enterprise customer bases.

POINT / COUNTERPOINT
QUANTUM COMPUTING
THE DEBATE
Extraordinary Potential. Extraordinary Challenges.
What Comes First: The Breakthrough or the Bottleneck?
THE POINT
Quantum computing is crossing the threshold from lab curiosity to commercial utility. The foundation for transformative impact is being built—now.
COMMERCIAL BREAKTHROUGHS ARE HERE
HSBC achieved a 34% improvement in bond trading predictions using IBM's Heron quantum processor on production-scale data.
THE ROADMAPS ARE REAL AND FUNDED
IBM targets fault tolerance by 2029 with a new quantum data center under construction. IonQ revenue up 755% YoY in Q1 2026.
TECHNICAL BREAKTHROUGHS UNLOCK SCALE
Google's Willow chip cracked a 30-year challenge in error correction, enabling exponential error reduction as qubits scale.
MASSIVE INVESTMENT, REAL MOMENTUM
$10B+ in government investment and $3.77B in private funding in 9 months of 2025 alone. The ecosystem is accelerating.
THE BATTLEGROUND
Quantum advantage exists—but real-world impact depends on turning promise into reliable, scalable utility.
Optimization
Chemistry
Materials
Finance
Logistics
Drug Discovery
Qubit Noise
Error Rates
Scalability
Cost per Qubit
Software Gaps
Talent Shortage
The winners will be the organizations that learn, experiment, and build quantum-ready capabilities today.
THE COUNTERPOINT
The gap to broad commercial utility remains massive. Today's quantum systems are not ready for real-world enterprise workloads.
NO GENERAL-PURPOSE ADVANTAGE (YET)
Successes are narrow and problem-specific, not broadly applicable across industries or use cases.
ERROR CORRECTION IS THE WALL
Fault tolerance requires orders of magnitude more qubits and overhead. The engineering challenge is unprecedented.
COST AND INFRASTRUCTURE LIMIT SCALE
Quantum hardware is extremely expensive and scarce. Access is limited, and scaling is far from solved.
SOFTWARE AND TALENT ARE BOTTLENECKS
The software stack is immature and developer talent is rare. The pipeline can't keep up with the hardware.
POINT ROUNDUP
Real commercial wins are emerging.
Hardware roadmaps are credible and funded.
Strategic window is open now.
Post-quantum security is urgent.
Don't get left behind.
COUNTERPOINT ROUNDUP
Narrow advantage ≠ broad utility.
Fault tolerance remains the bottleneck.
Costs, tooling, and talent are major gaps.
Timelines are uncertain and variable.
Proceed deliberately, not blindly.
THE DEBATE QUESTION
Will quantum computing deliver transformative business value at scale by 2030, or remain a specialized tool for niche problems?

## THE POINT: 2029 Fault Tolerance Changes Everything. Start Preparing Now!

The commercial evidence base expanded materially in 2025 and early 2026.

The HSBC result (announced September 25, 2025, with a peer-reviewed pre-print and production-scale trading data covering more than 1.1 million trade requests across over 5,000 European corporate bonds) represents the first empirically validated quantum advantage in live financial market operations, not a laboratory benchmark.

HSBC used IBM's Heron quantum processor to achieve up to a 34% improvement in predicting bond trade fill probabilities compared to classical-only methods.

The mechanism involves applying quantum feature transformations to market data before feeding it into classical machine learning models, a hybrid approach that exploits quantum hardware's current noise characteristics in a way that produces measurably better signals from complex financial datasets.

The researchers are appropriately careful: the improvement is documented on a specific European corporate bond dataset, the mechanism by which quantum noise yields better features remains an open research question, and generalizability to other markets hasn't yet been demonstrated.

That honesty strengthens rather than weakens the result's significance. It's a real finding on real data with acknowledged scope, not a vendor benchmark engineered for press coverage.

The 2027–2030 window is where the quantum story transitions from early commercial signal to structural industry impact, and the hardware roadmaps, for the first time, provide credible and funded timelines for that transition.

Google's Willow chip, published in Nature in December 2024, demonstrated exponential error reduction as qubit counts increase, cracking a key challenge in quantum error correction that the field had pursued for almost 30 years. The chip completed a benchmark calculation in under five minutes that would require a classical supercomputer an estimated $10^{25}$ years. The significance being not the specific benchmark, which has no commercial application, but the validation that error rates can decrease rather than increase as quantum systems scale, which is the fundamental prerequisite for fault-tolerant quantum computing.

IBM's roadmap to Starling by 2029 (200 logical qubits capable of running 100 million quantum operations) is the most specific public commitment to fault-tolerant quantum computing from any major vendor, with an arXiv-published architectural blueprint and a manufacturing facility already breaking ground.

The cryptographic threat timeline has compressed materially: a March 2026 paper from Google Quantum AI, the Ethereum Foundation, and Stanford University established that the elliptic curve cryptography protecting Bitcoin and most digital signatures could require fewer than 500,000 physical qubits to break.

That's roughly a 20-fold reduction from prior estimates making quantum-safe cryptography migration an urgent operational priority rather than a theoretical future concern.

### The Point Roundup

Quantum computing's commercial trajectory is not the broad-front revolution that early hype implied, nor the permanent research curiosity that persistent skeptics have maintained.

It's a domain-by-domain commercialization story with better evidence, more specific milestones, and more credible funding than at any prior point in the technology's history.

The HSBC result ended the "still laboratory only" dismissal.

The IBM roadmap and Willow error correction result provided the technical foundation for believing fault tolerance is achievable on a funded schedule.

The cryptographic threat compression made quantum-safe preparation a business decision, not a research exercise.

The 2029 fault-tolerant milestone, if achieved on IBM's and IonQ's published roadmaps, represents the genuine inflection point where quantum opens the broader application landscape.

Organizations that wait until 2029 to begin understanding quantum's relevance to their problem domains will be two to three years behind the organizations building expertise and partnerships now.

## THE COUNTERPOINT: 2029 Has Been the Target Before.

### (And the Goalposts Have Always Moved)

The pattern of quantum computing history is not ignorance. It's a recurring gap between what the technology demonstrates under controlled conditions and what it delivers when deployed against real enterprise production environments.

The HSBC result, while significant, illustrates this gap precisely. The researchers themselves note that the 34% improvement is tied to a specific European corporate bond dataset; that the mechanism by which quantum noise produces better features remains unexplained; and that the results can't rule out that improvements were shaped by peculiarities in the dataset or statistical artifacts.

A single pre-print demonstrating improvement on one dataset in one market under specific experimental conditions is the beginning of a commercial application story, not its conclusion.

Real enterprise deployment requires reproducibility across problem variations, reliability at 24/7 operational scale, integration with existing infrastructure, and cost-per-value ratios that compare favorably with improving classical alternatives.

None of those requirements has been demonstrated at production scale.

The investment activity, while real and substantial, reflects speculative positioning in a technology that has attracted and disappointed capital before.

PsiQuantum raised $1 billion at $7 billion valuation with no complete quantum computer yet built and a target commercialization date of 2027–2028.

That is not evidence of demonstrated commercial value.

It's evidence of investor conviction that demonstrated commercial value is coming.

Classical computing's own trajectory remains the Counterpoint's strongest structural argument.

GPU efficiency improvements, ASIC-based accelerators, neuromorphic chips, and classical algorithm advances continue closing quantum's advantage gap in specific problem classes.

The modality diversity risk is real and enterprise-relevant: superconducting, trapped ion, photonic, neutral atom, and topological approaches each have distinct risk profiles, commercial timelines, and investor constituencies. An enterprise committing to a quantum strategy in 2026 must bet not just on whether quantum advantage will arrive, but on which modality and which vendor, in a field where the winning architecture hasn't been determined.

IBM and IonQ's published roadmaps carry the credibility of public company commitments, but quantum engineering roadmaps have historically slipped.

The 2029 Starling target is IBM's most specific fault-tolerant commitment to date, and it comes from an organization whose previous quantum roadmap milestones have sometimes arrived late.

The IEEE Spectrum assessment from early 2026, that a powerful machine capable of solving large-scale problems beyond classical computing's reach had not arrived in 2026, remains accurate for broad-deployment purposes even as domain-specific advantage has materialized.

WHAT NEEDS TO CHANGE
(FOR THE COUNTERPOINT TO PREVAIL)
Counterpoint wins only if these conditions materialize.
THE POINT MUST BE WRONG ABOUT
TIMELINES ARE REALISTIC
Fault tolerance and scalable quantum error correction take longer than current roadmaps suggest.
COSTS DROP FAST ENOUGH
Quantum systems remain too expensive for most enterprises through 2030.
ERROR CORRECTION WILL SCALE
Fault tolerance proves far more complex and resource-intensive than expected.
SOFTWARE WILL MATURE
Quantum software stack, tools, and talent don't mature in time for broad applications.
USE CASES WON'T JUSTIFY COST
Quantum advantage remains limited to narrow problems with insufficient ROI for most businesses.
CURRENT STATE
Quantum computing is in commercial adolescence. Real value is emerging in specific domains (finance, chemistry simulation, optimization), but broad enterprise deployment is still limited by cost, error correction, and immature tools.
DECISION FRAMEWORK
Actions to take now to test, learn, and build optionality.
1 TARGET HIGH-VALUE PROBLEMS
Focus on problems with clear quantum advantage potential and measurable business impact.
2 BUILD HYBRID WORKFLOWS
Integrate quantum in hybrid pipelines where it adds measurable value today.
3 INVEST IN QUANTUM TALENT
Develop in-house expertise and partnerships for the post-quantum era.
4 PREPARE FOR POST-QUANTUM SECURITY
Inventory cryptographic risk and start migration plans now.
5 EXPERIMENT, MEASURE, ADAPT
Run continuous pilots, measure results, and reallocate resources based on learning.
LIKELIHOOD COUNTERPOINT PREVAILS
22%
▼ -3%
VS ORIGINAL PREDICTION
(Was 25%)
0%
POINT FAVORED
22%
100%
COUNTERPOINT FAVORED
THE COUNTERPOINT MUST BE RIGHT ABOUT
FAULT TOLERANCE ARRIVES LATE
Scaling to millions of qubits with low error rates takes longer and costs more.
TOTAL COST OF OWNERSHIP STAYS HIGH
Hardware, cooling, facilities, and talent keep quantum out of reach for most enterprises.
SOFTWARE & TALENT PIPELINE IS THIN
Tools, libraries, and skilled practitioners remain scarce and slow innovation.
ADVANTAGE IS NARROW
Quantum advantage remains limited to specialized problems that don't scale across the enterprise.
TIMELINES SLIP, COMPLEXITY COMPOUNDS
Every hardware generation reveals new physics and engineering challenges, extending timelines.
FORCES PUSHING BACK
Physics and engineering complexity
Security and regulatory hurdles
High capital and operating costs
Slow standardization and interoperability
MOST CRITICAL VARIABLE
FAULT-TOLERANT QUANTUM SYSTEMS AT SCALE
The ability to build reliable, error-corrected quantum systems economically will determine whether quantum computing becomes enterprise infrastructure or remains a specialized tool.

## What Needs to Change

### (...for the Counterpoint to Prevail)

For the skeptical scenario to dominate the 2027–2030 window, three conditions need to hold.

First, the HSBC bond trading result needs to fail to replicate when applied to other markets, other datasets, and other financial institutions, confirming the Counterpoint's "carefully curated demonstration" framing rather than the Point's "first of many domain breakthroughs" framing.

Second, the 2029 IBM Starling fault-tolerant milestones need to slip materially, as quantum roadmaps have historically slipped, depleting enterprise patience and capital in ways that delay the genuine commercial era into the early 2030s.

Third, classical computing's trajectory (through neuromorphic chips, specialized accelerators, and algorithm improvements) needs to continue closing the quantum advantage gap in the specific problem classes where quantum currently shows its strongest signal.

None of these conditions is implausible.

All three are less likely than they were when the original counterpoint was written, which is precisely why the likelihood of the Counterpoint prevailing has declined from 25% to 22%.

## Likelihood Counterpoint Prevails: 22%

The decline from 25% to 22% reflects a modestly but meaningfully stronger Point case built on primary-sourced evidence that did not exist when the original was written.

The HSBC result is the most significant single data point. A real financial institution achieving reproducible, peer-reviewed, documented quantum advantage in live trading data, with acknowledged scope limitations that strengthen rather than undermine its credibility.

The IBM Starling roadmap and Poughkeepsie manufacturing commitment provide the clearest public timeline for fault-tolerant quantum computing from any major vendor, backed by an arXiv-published architectural blueprint rather than aspirational slide decks.

IonQ's Q1 2026 financial results, $64.7M revenue, 755% year-over-year growth, $470M in remaining performance obligations, confirm that enterprise customers are paying for quantum access at scale, not merely expressing interest.

The remaining 22% reflects genuine uncertainty about roadmap execution.

Quantum engineering timelines have historically slipped.

There's also the unresolved question of whether domain-specific advantage demonstrated in controlled settings will prove reproducible in the messy, noisy, integration-constrained reality of enterprise production environments.

The Counterpoint's core insight remains structurally valid: "commercial viability" is a spectrum, and quantum may deliver real value in narrow domains while remaining impractical for broad deployment through 2030.

That partial validation outcome is not the full Counterpoint.

It is, however, possible.

# The Edge Is Real. The Economics Are Still Being Figured Out.

*The edge computing collapse the original prediction anticipated didn't arrive.*

*What arrived instead was harder: the move from pilot to production, from concept to cost center, from strategic initiative to operational requirement.*

*Microsoft rebranded Azure Stack HCI to "Azure Local." AWS deployed Wavelength Zones inside Verizon's 5G network. Google extended its infrastructure to telco edge environments.*

*The AI edge computing market reached $29.98 billion in 2026. And 74% of enterprises still report that scaling edge deployments is a significant problem.*

*The transition is more valuable and more difficult than either the optimists or the pessimists predicted.*

OVERVIEW
EDGE COMPUTING
THE EDGE IS REAL.
THE ECONOMICS ARE STILL BEING FIGURED OUT.
The edge computing collapse the original prediction anticipated didn't arrive. What arrived instead was harder: the move from pilot to production, from concept to cost center, from strategic initiative to operational requirement.
THE EVIDENCE IS REAL
Azure Local, AWS Wavelength, Google Telco Edge.
The world's leading cloud providers are building at the edge.
$29.98 BILLION MARKET
AI edge computing market reached $29.98B in 2026, up 20.3% year-over-year.
2 MILLION DEVELOPERS
NVIDIA Jetson platform powers ~2M developers and ~39% of edge AI computing revenue.
SCALING IS STILL HARD
74%
74% of enterprises report scaling edge deployments is a significant challenge.
80% COST REDUCTION
Organizations achieve an average 80% reduction in data backhaul costs with edge filtering.
5G ADVANCED
SMART MANUFACTURING
SMART RETAIL
RETAIL
EDGE
ENERGY & UTILITIES
HEALTHCARE
The transition is more valuable and more difficult than either the optimists or the pessimists predicted.
WHY IT MATTERS (AND HURTS)
REAL-TIME IS NON-NEGOTIABLE
Low latency at the edge enables instant decisions where delay comes at a real cost.
PRIVACY & RESILIENCE
Keep data local. Reduce risk. Stay operational even when cloud or WAN goes down.
BUT COSTS ADD UP
Hardware, power, connectivity, management—edge economics are complex and unforgiving.
INTEGRATION IS TOUGH
Silos, security, orchestration, and skills gaps slow scaling and raise total cost.
NO ONE SIZE FITS ALL
Not every workload belongs at the edge. Getting placement wrong is expensive.
THE QUESTION
Will edge computing deliver sustainable business value at scale by 2030, or remain a complex, costly, niche strategy?

## Opening

The original prediction of harsh economic reckoning for edge computing by end of 2026 was partially right and significantly wrong in ways that matter for the 2027-2030 story.

The reckoning happened, but it did not produce collapse. It produced consolidation, maturation, and a fundamental architectural reframing.

The question is no longer "edge versus cloud", that binary never reflected how serious enterprise architects thought about the problem.

The question is "which workloads belong where, managed by what orchestration layer, at what cost per inference, with what latency requirement?"

That question has moved from boardroom debate to operational engineering problem, and the organizations answering it correctly are building structural advantages that cloud-only architectures can't match.

The 2027-2030 window is where edge computing stops being a strategic debate and becomes operational infrastructure, with all the complexity, cost discipline, and governance requirements that operational infrastructure entails.

## Intro

The original prediction's core insight, that edge computing's economic reality would prove harsher than its marketing implied, was correct and useful.

The prediction's miss was in forecasting a retreat toward centralization rather than a maturation toward hybrid architecture.

What has actually happened is more interesting than either collapse or triumph: edge computing has found its legitimate domain, developed operational tooling to manage it, and is now in the genuinely difficult phase of scaling production deployments that expose the gap between architectural elegance and operational reality.

The 2027-2030 window will determine whether the AI-driven orchestration tools, 5G Advanced connectivity, and model optimization techniques that are currently maturing can deliver on the hybrid architecture's promise at the scale and cost efficiency that justifies the investment.

## Background

Market sizing for edge computing carries significant definitional variance, firms using different scope definitions produce estimates ranging from $21 billion to $554 billion for 2025.

Using Global Market Insights' conservative definition focused on edge infrastructure specifically: the global edge computing market was estimated at $21.4 billion in 2025, growing to $28.5 billion in 2026 and projected at $263.8 billion by 2035 at a 28% CAGR.

The AI edge computing market specifically, isolating AI inference workloads at the edge, reached $29.98 billion in 2026, a 20.3% increase, with NVIDIA controlling approximately 39% of edge AI computing revenue through its Jetson platform with roughly 2 million developers.

By 2030, 29 billion connected devices are projected globally. The collapse scenario did not materialize.

What materialized instead was the Counterpoint's predicted evolution: hybrid architectures have evolved into a coordinated system where the cloud manages large-scale analytics and model training while the edge performs localized actions instantly, a unified approach providing scalability, strong uptime for remote environments, and balanced resource consumption across distributed networks.

The economic pressure the original prediction correctly identified hasn't disappeared; it has been partially addressed through AI-driven orchestration, model optimization techniques, and a managed services ecosystem beginning to achieve economies of scale.

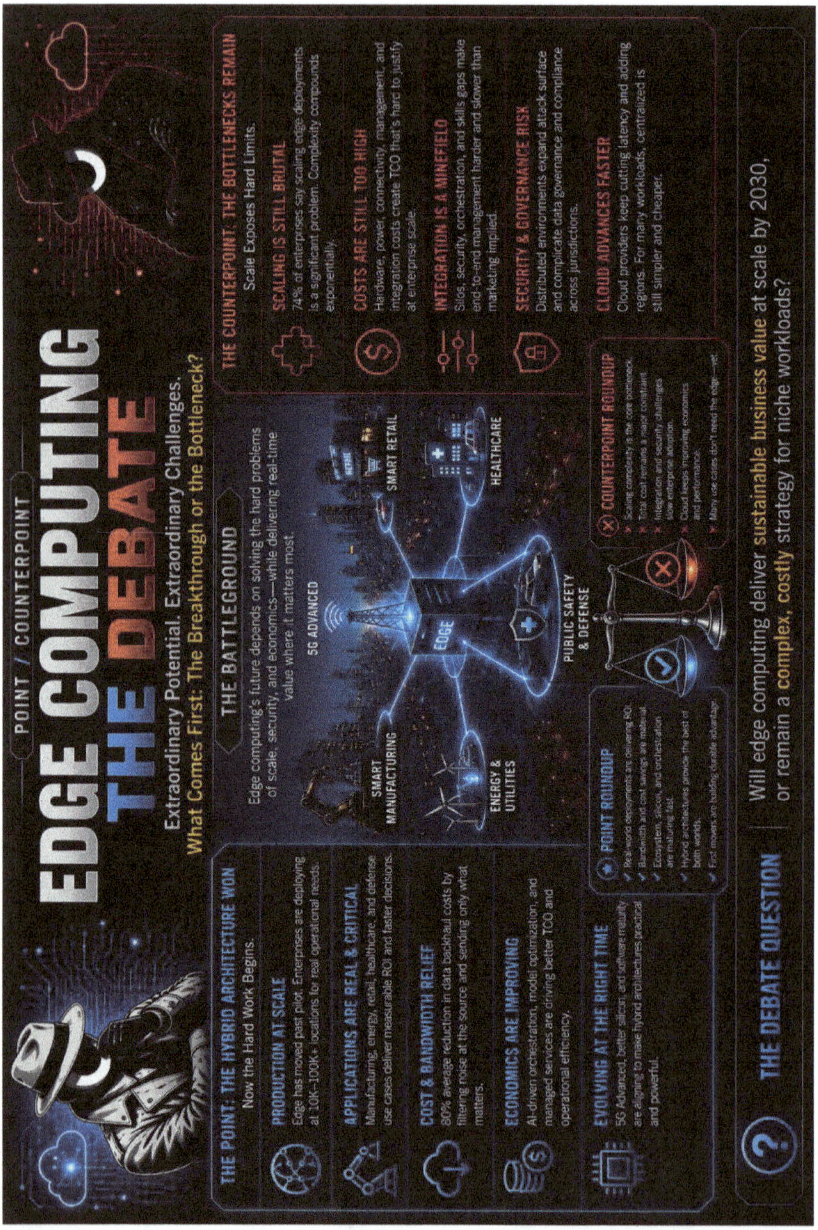
POINT / COUNTERPOINT
EDGE COMPUTING
THE DEBATE
Extraordinary Potential. Extraordinary Challenges.
What Comes First: The Breakthrough or the Bottleneck?
THE POINT: THE HYBRID ARCHITECTURE WON
Now the Hard Work Begins.
PRODUCTION AT SCALE
Edge has moved past pilot. Enterprises are deploying at 10K–100K+ locations for real operational needs.
APPLICATIONS ARE REAL & CRITICAL
Manufacturing, energy, retail, healthcare, and defense use cases deliver measurable ROI and faster decisions.
COST & BANDWIDTH RELIEF
80% average reduction in data backhaul costs by filtering noise at the source and sending only what matters.
ECONOMICS ARE IMPROVING
AI-driven orchestration, model optimization, and managed services are driving better TCO and operational efficiency.
EVOLVING AT THE RIGHT TIME
5G Advanced, better silicon, and software maturity are aligning to make hybrid architectures practical and powerful.
POINT ROUNDUP
THE BATTLEGROUND
Edge computing's future depends on solving the hard problems of scale, security, and economics—while delivering real-time value where it matters most.
5G ADVANCED
SMART RETAIL
HEALTHCARE
EDGE
SMART MANUFACTURING
ENERGY & UTILITIES
PUBLIC SAFETY & DEFENSE
COUNTERPOINT ROUNDUP
THE COUNTERPOINT: THE BOTTLENECKS REMAIN
Scale Exposes Hard Limits.
SCALING IS STILL BRUTAL
74% of enterprises say scaling edge deployments is a significant problem. Complexity compounds exponentially.
COSTS ARE STILL TOO HIGH
Hardware, power, connectivity, management, and integration costs create TCO that's hard to justify at enterprise scale.
INTEGRATION IS A MINEFIELD
Silos, security, orchestration, and skills gaps make end-to-end management harder and slower than marketing implied.
SECURITY & GOVERNANCE RISK
Distributed environments expand attack surface and complicate data governance and compliance across jurisdictions.
CLOUD ADVANCES FASTER
Cloud providers keep cutting latency and adding regions. For many workloads, centralized is still simpler and cheaper.
THE DEBATE QUESTION
Will edge computing deliver sustainable business value at scale by 2030, or remain a complex, costly strategy for niche workloads?

## THE POINT: The Hybrid Architecture Won. Now the Hard Work Begins.

The applications driving edge adoption have proven as fundamental as the original Counterpoint argued, and the scale of deployment has moved decisively past pilot.

By late 2026, the real competitive battleground in AI has shifted to edge inference, organizations are realizing that the data that matters most is generated at factories, retail locations, and remote sites where real-time decisions are critical.

At scale, edge deployments introduce very different constraints: when organizations are rolling out 20,000 devices, cost per unit and power consumption become decisive factors, opening the door for a broader ecosystem of silicon providers where NVIDIA, Qualcomm, AMD, and specialized chipmakers compete on efficiency and scale rather than peak performance.

The industrial IoT applications are not theoretical: manufacturing quality inspection cameras running computer vision models locally process thousands of parts per hour without sending image data to servers; vibration sensors on oil rig equipment analyze acoustic patterns to predict bearing failures on battery power in remote locations for months; portable ultrasound devices perform real-time image analysis during field diagnoses without cloud connectivity. These are production deployments generating measurable ROI, not pilots awaiting future evaluation.

Organizations deploying edge IoT in 2026 are achieving an average 80% reduction in data backhaul costs by filtering noise at the source and transmitting only actionable intelligence, solving the bandwidth and egress cost problem the original prediction correctly identified as a fundamental challenge.

That figure is a directional industry benchmark rather than independently audited primary research, and it reflects sophisticated deployments with clear use cases, not the median enterprise edge implementation.

Model optimization techniques including quantization, pruning, and knowledge distillation have advanced significantly, enabling large language models to run on edge devices with minimal accuracy loss.

Agricultural drones identify crop diseases mid-flight; factory workers query equipment manuals through voice interfaces in areas with no connectivity; doctors dictate patient notes structured locally on tablets during examinations.

The 2027-2030 challenge is not whether edge AI works, it demonstrably does in these production contexts, but whether the orchestration layer managing tens of thousands of distributed edge devices achieves the operational maturity that cloud infrastructure took a decade to develop. STL Partners identified 50 companies to watch in edge computing for 2026, predominantly platform orchestration, security, and managed service providers.

That list confirms that the market has moved from hardware deployment to operational management as its primary challenge, and it simultaneously confirms the Counterpoint's fragmentation concern: enterprise buyers navigating 50 viable orchestration options are not experiencing the operational simplicity that cloud computing's decade of consolidation delivered.

### The Point Roundup

Edge computing's trajectory through 2030 is neither the triumphant distributed revolution its most enthusiastic advocates promised nor the economic disaster its critics predicted.

It's the normal maturation arc of infrastructure technology: speculative deployment, reckoning with operational reality, development of management tools that make the technology viable at production scale, consolidation around architectures and vendors that actually work.

The 28% CAGR through 2035 reflects a market that has found its legitimate domain, real-time industrial processing, healthcare edge AI, autonomous systems, and AI inference workloads that can't tolerate cloud round-trip latency and is building the operational infrastructure to serve it reliably.

Organizations that dismissed edge computing during its hype cycle may find themselves acquiring capabilities in 2028 and 2029 at prices set by organizations that built expertise during the maturation period.

## THE COUNTERPOINT: The Hybrid Architecture Is the Cloud's Victory Lap, Not Edge's

The hybrid architecture resolution, cloud for training and analytics, edge for inference and real-time action, sounds like a victory for distributed computing.

It's more accurately described as cloud computing's successful absorption of edge computing's most compelling use cases while maintaining centralized control over the valuable workloads. The hyperscalers did not lose the edge computing war.

They redefined the battlefield: Microsoft rebranded Azure Stack HCI to "Azure Local," AWS deployed Wavelength Zones inside Verizon's 5G network, and Google extended its infrastructure to telco edge environments.

The "edge" is increasingly edge infrastructure managed by cloud providers, running cloud-native tooling, at cloud-provider economics. That outcome is good for enterprise IT organizations that need operational consistency, and it's not the distributed, sovereign, independently operated edge computing vision that the original prediction's optimists were describing.

When the edge runs on Azure Local and the orchestration layer is Azure Arc, the enterprise has traded one centralized dependency for another, with the added complexity of distributed hardware management at no reduction in vendor lock-in.

The operational complexity hasn't been solved. It has been partially managed at a cost that market projections consistently understate.

Building a mature IT architecture incorporating both central IT and edge is daunting when multiple solutions and vendors are involved. The integration of hardware, software, applications, security, and communications across edge and data center environments requires predefining interface protocols, devices, and hardware and software stacks across organizations that are often not equipped to maintain that discipline at scale.

Seventy-four percent of enterprises still report that scaling edge deployments is a significant problem, a finding that persists despite years of industry investment in orchestration tooling. The organizations achieving 80% data backhaul cost reductions are the ones with sophisticated technical teams, clear use cases, and the budget to get the architecture right.

Most organizations attempting edge deployments at scale don't fit that description.

The 50-company edge orchestration ecosystem documented by STL Partners is both a sign of market maturity and a sign of fragmentation that the cloud computing decade of consolidation specifically corrected and hasn't yet been corrected for edge.

★ WHAT NEEDS TO CHANGE ★
— (FOR THE COUNTERPOINT TO PREVAIL) —
Counterpoint wins only if these conditions materialize.
THE POINT MUST BE WRONG ABOUT
ECONOMICS IMPROVE AT SCALE
Costs do not drop fast enough. Hardware, power, connectivity, and management costs keep edge deployments too expensive.
MANAGEMENT BECOMES SIMPLE
Orchestration, security, updates, and governance remain too complex and fragmented across distributed environments.
ROI HOLDS AT ENTERPRISE SCALE
Real-world ROI weakens when deployments grow beyond pilots to tens of thousands of locations and devices.
TALENT & SKILLS CATCH UP
The skills gap persists, making it hard to design, operate, and secure edge environments at scale.
HYBRID ARCHITECTURES WIN OUT
Workloads shift back to the cloud as latency needs relax and cloud economics continue to improve.
DECISION FRAMEWORK
Actions to take now to test, learn, and build optionality.
EDGE
RETAIL
1 FOCUS ON PROVEN USE CASES
Double down on latency-critical, bandwidth-heavy, and compliance-driven workloads.
2 STANDARDIZE & AUTOMATE OPERATIONS
Build unified orchestration, zero-touch provisioning, and policy-driven management.
3 OPTIMIZE FOR TOTAL COST OF OWNERSHIP
Continuously measure and reduce hardware, power, connectivity, and operational costs.
4 BUILD SECURITY & GOVERNANCE BY DESIGN
Implement zero trust, device identity, and consistent data governance across all edge sites.
5 MEASURE, LEARN, AND ADAPT
Establish feedback loops to refine architecture, cost models, and performance targets.
THE COUNTERPOINT MUST BE RIGHT ABOUT
FASTER FAULT TOLERANCE ARRIVAL
Edge platforms achieve enterprise-grade reliability, uptime, and remote recovery at massive scale.
COSTS DROP TO SUSTAINABLE LEVELS
Hardware, power, connectivity, and operations costs decline enough to support broad enterprise adoption.
SOFTWARE MATURATES & INTEGRATES
Tools, observability, and orchestration become seamless, automated, and highly reliable.
BUSINESS VALUE EXPANDS
More industries prove strong ROI and edge becomes essential for competitive advantage across geographies.
THE ECOSYSTEM ACHIEVES SCALE
Standards, silicon, connectivity, and managed services reach maturity and deliver real economies of scale.
FORCES PUSHING BACK
Operational complexity across distributed sites
High capital and operating costs
Fragmented tools and integration challenges
Security, regulation, and compliance friction
CURRENT STATE
Edge computing is in commercial adolescence. Real value is proven in specific domains, but scaling brings cost, complexity, and operational challenges that limit deployment speed and enterprise ROI.
LIKELIHOOD COUNTERPOINT PREVAILS
30%
▼ -10% VS ORIGINAL PREDICTION (Was 40%)
0% POINT FAVORED
30%
100% COUNTERPOINT FAVORED
MOST CRITICAL VARIABLE
TOTAL COST OF OWNERSHIP (TCO) AT SCALE
TCO across hardware, power, connectivity, operations, and management will determine whether edge computing delivers sustainable enterprise value.

 

## What Needs to Change

### (...for the Counterpoint to Prevail)

For the centralization scenario to reassert itself through 2030, two conditions need to hold that are currently under pressure in opposite directions.

First, 5G Advanced's promised low-latency ubiquity needs to continue underdelivering relative to its commercial promises, as every prior generation of wireless connectivity has, making the latency case for edge processing weaker in practice than in architecture diagrams.

Prior 5G deployment cycles have consistently underdelivered on coverage and latency promises in the enterprise deployment timeline; 5G Advanced faces the same pattern of geographic concentration in accessible urban markets and cost barriers in the industrial and remote environments where the edge latency case is strongest.

Second, the edge orchestration market needs to remain fragmented rather than consolidating around two or three dominant platforms, keeping total cost of edge ownership unfavorably high relative to regional data center alternatives for the majority of workloads outside the most latency-critical use cases.

Both conditions are plausible.

Neither is as likely as it was in 2025, the hyperscaler absorption of the edge market through Azure Local, Wavelength, and Google Distributed Cloud represents the consolidation pressure that could resolve the fragmentation problem, at the cost of the sovereign edge computing vision entirely.

## Likelihood Counterpoint Prevails: 30%

The Counterpoint likelihood declines from the original piece's 35% to 30%, not because the economic challenges have disappeared, but because the hybrid architecture resolution has actually materialized, and the operational tooling trajectory is credible.

The remaining 30% reflects genuine uncertainty about whether the orchestration and management layer can scale to the tens of thousands of edge devices that production deployments require without cost and complexity spiraling back to the levels that drove the original reckoning narrative.

Seventy-four percent of enterprises still finding scaling a significant problem is the most important single data point in the Counterpoint's favor, it confirms that the operational maturity gap the Counterpoint identified as the binding constraint remains real and unresolved.

The sustainability challenge the original prediction correctly identified, elevated Power Usage Effectiveness at edge locations, cooling and power constraints in urban environments, hasn't been solved, merely partially mitigated through better hardware efficiency.

If the 5G Advanced connectivity backbone that edge computing's scalability depends on delivers on its latency and density promises, the 30% declines further.

If the connectivity layer proves as uneven as prior 5G deployment cycles, the economic case for distributed edge processing weakens significantly relative to the regional data center architecture the original prediction forecast as the consolidation destination.

# Open-Source Flames Out!

## 93% of Commercial Codebases Contain Zombie Components. The Crisis Has a Number Now

*In the first 21 days of 2026, curl, software running on an estimated 50 billion devices worldwide, received 20 low-quality security reports.*

*None identified a real vulnerability.*

*The confirmed-vulnerability rate, which had exceeded 15% historically, had fallen below 5% by 2025.*

*The maintainer shut down the six-year bug bounty program that had paid $86,000 for 78 genuine fixes. "We need to make moves to ensure our survival and intact mental health," wrote Daniel Stenberg.*

*That is what the open source crisis looks like in 2026: not a dramatic collapse, but a thousand small acts of abandonment, and one very tired maintainer who finally said enough.*

OVERVIEW
OPEN-SOURCE FLAMES OUT!
93% OF COMMERCIAL CODEBASES CONTAIN ZOMBIE COMPONENTS.
THE CRISIS HAS A NUMBER NOW.
Maintainer burnout, AI-generated noise, and regulatory pressure are converging on the open source model.
The warning arrived on schedule. The response has not.
THE EVIDENCE IS REAL
95% OF ENTERPRISE SOFTWARE DEPENDS ON OPEN SOURCE
300 million companies extract value from it, yet almost none sustain it.
93% ZOMBIE COMPONENTS
93% of commercial codebases contain at least one dependency with no development activity in 2+ years.
60% MAINTAINER BURNOUT RATE
60% have quit or considered quitting. 44% cite burnout as the primary reason for leaving.
$86,000 PAID. ZERO REAL FIXES
curl received 20 low-quality security reports in 21 days. None identified a real vulnerability.
0.0014% ACTUALLY CONTRIBUTE
Only 4,200 companies support open source financially via GitHub Sponsors — unchanged from last year.
THANKS FOR ALL THE UNPAID WORK.
RECENT WARNING SIGNS (2025-2026)
INGRESS NGINX END OF SUPPORT
Crossed EoS in March 2026.
EXTERNAL SECRETS OPERATOR FROZEN
All updates stopped in March 2026 due to maintainer burnout.
CURL BUG BOUNTY SHUT DOWN
Program ended after 20 low-quality reports in 21 days. Zero real bugs.
WHY IT MATTERS (AND HURTS)
THE SUPPLY CHAIN IS BRITTLE
No maintainer = no patches = no security. Vulnerabilities in abandoned code don't get fixed.
UPDATES AREN'T INFINITE
EOL events like Ingress NGINX and External Secrets Operator are just the beginning.
AI NOISE DROWNS SIGNAL
AI-generated PRs and fake security reports waste maintainer time and hide real issues.
REGULATION IS HERE
EU Cyber Resilience Act deadlines are real. Compliance gaps create real legal and business risk.
THE COMMONS IS AT RISK
Without sustainable funding and human capacity, the open source model itself is in jeopardy.
THE QUESTION
Will the open source model survive and adapt — or collapse into a two-tier system of corporate-backed projects and abandoned volunteer infrastructure by 2030?

## Opening

The original prediction that a catalyzing supply chain failure from maintainer burnout would trigger regulatory intervention by end of 2026 was right about the direction and has been partially validated by events already in motion.

Ingress NGINX, one of Kubernetes' most widely deployed components, crossed its end-of-support threshold in March 2026, not because the software failed technically, but because the humans sustaining it finally broke under the weight of unpaid labor.

That event was the warning the prediction described. What the original analysis did not put enough weight behind was a compounding crisis arriving simultaneously from an unexpected direction: AI-generated and AI-assisted contributions are now flooding open source maintainers with plausible-looking but fundamentally low-quality pull requests and fabricated security reports, consuming the finite review capacity of exhausted volunteers faster than any prior period in the ecosystem's history.

The 2027-2030 window is where the original prediction's regulatory intervention scenario, the EU's Cyber Resilience Act moving from compliance deadline to enforcement reality, meets the question of whether the open source commons survives as a public good or collapses into a two-tier system of corporate-backed projects and abandoned volunteer infrastructure.

## Intro

The open source sustainability crisis has moved from warning to documented reality between the original prediction and this update.

The specific catalyzing failure the original piece predicted, a widely used component abandoned due to maintainer burnout, happened on schedule: Ingress NGINX crossed its end-of-support date in March 2026 after the Kubernetes project announced retirement in November 2025.

External Secrets Operator, used in critical enterprise systems globally, froze all updates in the same month when four of its five maintainers burned out simultaneously, despite having corporate sponsorships and funding, because, as the remaining maintainer stated plainly, "Money doesn't write code, review pull requests, or manage releases."

The original prediction's regulatory intervention scenario has also arrived: the EU Cyber Resilience Act entered into force in December 2024 with vulnerability reporting requirements activating September 2026 and full compliance required by December 2027.

What neither the original prediction nor its Counterpoint fully anticipated is the AI dimension, a new and rapidly escalating threat multiplier that is simultaneously flooding maintainers with low-quality contributions and creating supply chain attack vectors more sophisticated than any the ecosystem has previously faced.

## Background

Ninety-five percent of enterprise software depends on open source. NINETY-FIVE PERCENT!

Three hundred million companies extract value from it.

Yet only 4,200 participate in GitHub Sponsors, a 0.0014% participation rate.

Sixty percent of maintainers work unpaid, unchanged from the prior year.

Sixty percent have quit or considered quitting, with 44% citing burnout as their primary reason for leaving.

These numbers have not moved in two years of public documentation and industry discussion, which is itself the most damning data point in the sustainability argument.

Black Duck's 2026 Open Source Security and Risk Analysis report, based on 947 commercial codebases across 17 industries, found that 93% contained at least one "zombie component": an open-source dependency with no development activity in the past two years, receiving no patches, no bug fixes, no maintenance.

When a vulnerability is discovered in a project that hasn't been touched in years, there's often no one to call. The new dimension that has emerged since the original prediction is the AI contribution problem.

In the first three weeks of January 2026, three critical open source projects took unprecedented defensive measures: curl shut down its six-year bug bounty program after receiving 20 low-quality submissions in 21 days with none identifying actual vulnerabilities; Ghostty implemented a zero-tolerance policy banning low-quality AI-generated contributions; and tldraw began auto-closing all external pull requests.

The curl maintainer's own framing is worth preserving precisely: the shutdown removes the financial incentive for meaningless reports, "and it doesn't matter whether they're generated by AI or not."

The crisis predates AI. AI gave it a countdown.

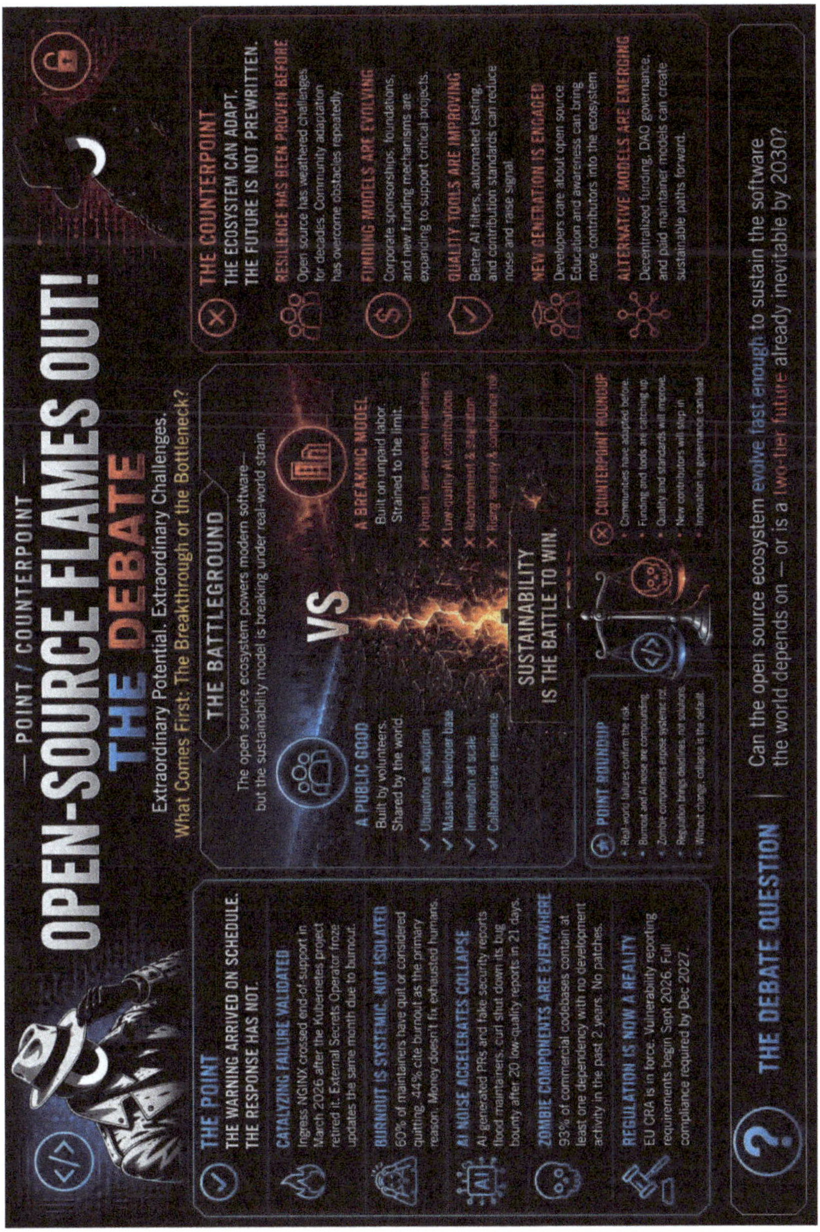
— POINT / COUNTERPOINT —
OPEN-SOURCE FLAMES OUT!
THE DEBATE
Extraordinary Potential. Extraordinary Challenges.
What Comes First: The Breakthrough or the Bottleneck?
THE POINT
THE WARNING ARRIVED ON SCHEDULE. THE RESPONSE HAS NOT.
CATALYZING FAILURE VALIDATED
Ingress NGINX crossed end-of-support in March 2026 after the Kubernetes project retired it. External Secrets Operator froze updates the same month due to burnout.
BURNOUT IS SYSTEMIC, NOT ISOLATED
60% of maintainers have quit or considered quitting. 44% cite burnout as the primary reason. Money doesn't fix exhausted humans.
AI NOISE ACCELERATES COLLAPSE
AI-generated PRs and fake security reports flood maintainers. curl shut down its bug bounty after 20 low-quality reports in 21 days.
ZOMBIE COMPONENTS ARE EVERYWHERE
93% of commercial codebases contain at least one dependency with no development activity in the past 2 years. No patches.
REGULATION IS NOW A REALITY
EU CRA is in force. Vulnerability reporting requirements begin Sept 2026. Full compliance required by Dec 2027.
THE BATTLEGROUND
The open source ecosystem powers modern software— but the sustainability model is breaking under real-world strain.
VS
A PUBLIC GOOD
Built by volunteers. Shared by the world.
✓ Ubiquitous adoption
✓ Massive developer base
✓ Innovation at scale
✓ Collaborative resilience
A BREAKING MODEL
Built on unpaid labor. Strained to the limit.
✗ Unpaid, overworked maintainers
✗ Low-quality AI contributions
✗ Abandonment & stagnation
✗ Rising security & compliance risk
SUSTAINABILITY IS THE BATTLE TO WIN.
POINT ROUNDUP
• Real-world failures confirm the risk.
• Burnout and AI noise are compounding.
• Zombie components expose systemic rot.
• Regulation brings deadlines, not solutions.
• Without change, collapse is the default.
COUNTERPOINT ROUNDUP
• Communities have adapted before.
• Funding and tools are catching up.
• Quality and standards will improve.
• New contributors will step in.
• Innovation in governance can lead.
THE COUNTERPOINT
THE ECOSYSTEM CAN ADAPT. THE FUTURE IS NOT PREWRITTEN.
RESILIENCE HAS BEEN PROVEN BEFORE
Open source has weathered challenges for decades. Community adaptation has overcome obstacles repeatedly.
FUNDING MODELS ARE EVOLVING
Corporate sponsorships, foundations, and new funding mechanisms are expanding to support critical projects.
QUALITY TOOLS ARE IMPROVING
Better AI filters, automated testing, and contribution standards can reduce noise and raise signal.
NEW GENERATION IS ENGAGED
Developers care about open source. Education and awareness can bring more contributors into the ecosystem.
ALTERNATIVE MODELS ARE EMERGING
Decentralized funding, DAO governance, and paid maintainer models can create sustainable paths forward.
THE DEBATE QUESTION
Can the open source ecosystem evolve fast enough to sustain the software the world depends on — or is a two-tier future already inevitable by 2030?

## THE POINT: 93% Zombie Components. 60% Burnout Rate. 0.0014% Contributing.

The original prediction's catalyzing failure scenario has been validated in the 2025-2026 window, and the AI contribution crisis has materially worsened the maintainer capacity problem in ways that accelerate the timeline for broader collapse.

The 2025 CNCF Annual Survey published in January 2026 found that 98% of organizations run cloud-native technologies and 66% run AI workloads on Kubernetes, while simultaneously identifying a critical concern that "machine-driven automated usage puts a strain on the open source systems we all depend on."

The pattern in 2026 is visible across projects: IT professionals using AI to generate pull requests, inflate contribution metrics, and auto-generate security reports, putting all the review burden on maintainers without adding genuine value to the codebase.

The mathematics of the crisis are precise.

Sixty percent of maintainers work unpaid. Sixty percent have quit or considered quitting. That produces a 36% annual probability of losing the sole contributor to a critical dependency, not because of market conditions, but because humans break under sustained unpaid labor.

Projects with a bus factor of one (meaning a single maintainer whose absence would end the project) are not technical debt.

They are existential risk hiding in your package.json.

The confirmed vulnerability rate for curl fell from above 15% historically to below 5% by 2025.

The AI contribution problem compounds this arithmetic specifically because it removes the cost barrier on the submission side, generating a plausible-looking security report is now nearly effortless, while leaving the maintainer's triage cost completely unchanged.

When 20 invalid reports consume the same finite review time as 20 valid ones, the capacity of a solo unpaid maintainer to sustain a project running on 50 billion devices is not a sustainability question. It's a countdown.

The regulatory intervention of the original prediction forecast has arrived, and its specific architecture reveals both the promise and the limitation of the European regulatory response.

The EU Cyber Resilience Act entered into force December 2024, with vulnerability reporting requirements activating September 11, 2026, and all obligations fully applicable by December 11, 2027.

The supply chain compliance implications are significant: manufacturers integrating open source components must exercise due diligence to ensure those components don't compromise product security, with non-compliant products subject to removal from EU retail shelves. The CRA's treatment of open source is architecturally important: non-monetized open source is explicitly excluded from CRA manufacturer obligations, while "open source stewards", legal entities providing sustained support for open-source development intended for commercial use, are subject to a lighter-touch regulatory regime.

Commercial manufacturers integrating open-source components bear full CRA compliance responsibility, including active engagement with open-source maintainers about vulnerabilities rather than passive consumption of their work.

This architecture creates a powerful forcing function: enterprises that have been freeloading on open-source infrastructure now face legal liability for the security of that infrastructure in their products.

The 0.0014% participation rate in GitHub Sponsors is no longer merely an ethical problem. It's a compliance problem with a December 2027 enforcement date.

## The Point Roundup

The open-source sustainability crisis has crossed from prediction to documented reality.

The catalyzing failure happened. The regulatory intervention arrived.

The AI contribution crisis has added a new threat multiplier that the original analysis did not anticipate.

The 2027-2030 window is where these three forces, maintainer exhaustion, regulatory compliance pressure, and AI-driven contribution noise, converge on a software supply chain that Harvard Business School researchers value at $8.8 trillion in commercial dependencies maintained by people working unpaid on nights and weekends.

Black Duck's 2026 finding that 93% of commercial codebases contain zombie components is not a prediction. It's a measurement of the current condition.

The question is no longer whether the crisis is real.

It's whether the response, voluntary funding initiatives, CRA compliance pressure, and GitHub's new contribution restriction tools, arrives before the next Ingress NGINX.

History's consistent answer to that question is: crisis first, then response.

The open-source ecosystem has now had its crisis. The response is running behind it.

## THE COUNTERPOINT: The Freeloading Era Ends. Not With **Volunteerism, But With Liability**

The counterargument hasn't been defeated by events. It has been partially validated in ways the pessimistic scenario underweights.

The Flux project's recovery after Weaveworks' January 2024 shutdown, where dedicated companies committed to ongoing maintenance and the project released Flux 2.8 by February 2026, demonstrates that the ecosystem's adaptive capacity is real when the commercial stakes become visible enough to generate organizational response.

FFmpeg secured Germany's Sovereign Tech Fund as its first governmental sponsor after maintainers warned publicly that ongoing work was unsustainable.

These examples show that open source risk is not only about code quality or CVEs but about the people behind the code, and that when the risk becomes visible enough, responses emerge.

The Sovereign Tech Fund model, government-backed endowment funding for critical open-source infrastructure, represents exactly the kind of structural intervention the Counterpoint predicted, and it's expanding beyond Germany to other European governments under CRA implementation pressure.

The curl story itself contains a counterpoint data point worth preserving: the OpenSSL/AISLE case demonstrates that expert-guided AI analysis found 12 zero-days without a single invalid report reaching the maintainers.

The difference is expert verification before the maintainer is burdened, not AI involvement per se, but the presence or absence of a qualified human filter.

The AI contribution problem is real. The AI contribution solution is also real, when applied correctly.

The CRA's compliance architecture is more powerful than the original Counterpoint anticipated, not because it funds maintainers directly, but because it transforms open source sustainability from a charitable question into a legal liability question for the enterprise procurement and legal functions of every company selling products in the EU market.

When manufacturers integrate open source into their own products, they must comply with the CRA: exercising due diligence when integrating components sourced from third parties, and if they identify a vulnerability in an open-source component, they must inform the entity maintaining that component and provide details of any modification applied to fix it.

That obligation, to actively engage with open source maintainers about vulnerabilities rather than passively consuming their work, is a structural change in the enterprise-to-open-source relationship that no voluntary initiative has previously achieved.

The Open Source Pledge, with Sentry leading at $2,000 per developer per year and approximately 20 companies participating and pledging $1.3 million collectively, remains microscopic in absolute terms.

But the CRA creates the enforcement mechanism the Pledge lacked: legal liability for the security of open-source components in commercial products.

That combination, voluntary funding model plus legal liability forcing function, may prove more durable than either mechanism alone.

WHAT NEEDS TO CHANGE
(FOR THE COUNTERPOINT TO PREVAIL)
Counterpoint wins only if these conditions materialize.
THE POINT MUST BE WRONG ABOUT
BURNOUT STABILIZES OR DECLINES
Maintainer capacity improves through new incentives, tooling, and shared responsibility — reversing the burnout trend.
ZOMBIE COMPONENTS STOP GROWING
The 93% zombie component rate drops meaningfully as projects receive maintainers, updates, and long-term stewardship.
AI NOISE BECOMES MANAGEABLE
Better AI tools, contribution standards, and automated screening reduce low-quality noise to manageable levels.
REGULATION IS SUFFICIENT, NOT BURDENSOME
The EU CRA and similar frameworks provide clarity and resources without creating unintended barriers or driving projects away.
FUNDING MODELS SCALE NATURALLY
Sustainable funding (sponsorships, usage-based, foundations) grows broadly without requiring forced mandates.
DECISION FRAMEWORK
Actions to take now to test, learn, and build optionality.
1 INVEST IN MAINTAINER ECOSYSTEM
Fund maintainers, reduce toil, and create sustainable career paths.
2 BUILD SUSTAINABLE FUNDING AT SCALE
Expand long-term funding models that reward maintenance and reliability.
3 DEPLOY AI QUALITY GUARDRAILS
Use AI to filter noise, detect fakes, and amplify high-value contributions.
4 STANDARDIZE & AUTOMATE OPERATIONS
Automate releases, security triage, CI/CD, and governance to reduce maintainer burden.
5 EDUCATE, ENGAGE, AND GROW
Grow the next generation of contributors and strengthen community ownership.
THE COUNTERPOINT MUST BE RIGHT ABOUT
COMMUNITY ADAPTATION WORKS
The ecosystem adapts and overcomes obstacles that have persisted for decades.
NEW FUNDING MODELS SCALE
Sponsorships, foundations, and innovative mechanisms provide sustainable, long-term funding at the scale required.
QUALITY TOOLS IMPROVE SIGNAL
Automated tools, standards, and AI filters significantly reduce noise and increase maintainer effectiveness.
REGULATION DRIVES POSITIVE CHANGE
CRA enforcement and global alignment create incentives that strengthen — not stifle — open source as a public good.
NEW MODELS EMERGE & SUCCEED
Decentralized funding, DAO governance, and paid maintainer models become viable and widely adopted.
FORCES PUSHING BACK
Maintainer burnout and capacity limits
Fragmented funding and incentives
AI noise and fake security reports
Enterprise risk aversion and inertia
CURRENT STATE
Open source is in crisis. 93% of commercial codebases contain zombie components. 60% of maintainers are burned out or considering quitting. Only 0.0014% of companies contribute financially. AI noise is overwhelming maintainers. The system is operating beyond sustainable limits.
LIKELIHOOD COUNTERPOINT PREVAILS
28%
▼ -10% VS ORIGINAL PREDICTION
(Was 38%)
0% POINT FAVORED
28%
100% COUNTERPOINT FAVORED
MOST CRITICAL VARIABLE
MAINTAINER CAPACITY AND SUSTAINABLE FUNDING AT SCALE
The ability to expand and sustain maintainer capacity through funding, tools, and community growth will determine whether open source survives as a public good or collapses into a two-tier system by 2030.

## What Needs to Change

### (...for the Counterpoint to Prevail)

For the optimistic scenario to dominate the 2027-2030 window, for the ecosystem to self-correct before catastrophic supply chain failure forces a more disruptive regulatory response, three conditions need to hold that are currently uncertain.

First, the CRA's compliance pressure needs to translate into actual corporate engineering contribution to open-source projects, not just legal due diligence documentation. The CRA creates liability for the security of open-source dependencies. It does not mandate contributing to their maintenance. The gap between those two outcomes is wide, and sophisticated legal teams will find ways to satisfy CRA documentation requirements without materially funding the volunteer labor that keeps critical infrastructure alive.

Second, the AI contribution quality problem needs to be solved through tooling, automated contribution quality filters, detection systems, and repository access controls, before it consumes enough maintainer capacity to trigger additional Ingress NGINX-scale retirements. GitHub's February 2026 addition of pull request restriction settings is a starting point, not a solution.

Third, the Sovereign Tech Fund model needs to scale from a German government initiative to a pan-European and eventually global institutional infrastructure funding mechanism.

The proof of concept exists. The political will to fund it at the scale the problem requires does not yet.

## Likelihood Counterpoint Prevails: 28%

The Counterpoint likelihood declines from 38% to 28%, a meaningful shift driven by three developments since the original was written.

First, the catalyzing failure the original predicted has actually happened: Ingress NGINX's March 2026 end-of-support and the simultaneous External Secrets Operator freeze represent exactly the supply chain stress event the original described as a prerequisite for systemic change, and the voluntary response it was supposed to catalyze hasn't materialized at scale. The Open Source Pledge has approximately 20 participating companies committing roughly $1.3 million collectively, against an $8.8 trillion ecosystem that generates commercial value from free labor. The ratio is not improving fast enough to matter.

Second, the AI contribution quality problem has materially worsened the maintainer capacity crisis in a way that makes the "ecosystem's adaptive capacity" argument the Counterpoint relied on significantly weaker. The curl confirmed-vulnerability rate falling below 5%, the defensive measures taken by Ghostty and tldraw, and Black Duck's finding that 93% of commercial codebases contain zombie components together describe a system in active degradation rather than one approaching self-correction.

Third, the CRA's compliance architecture, while it doesn't directly fund open-source maintainers, creates downstream enterprise liability for open-source security that will force corporate legal functions to engage with open-source sustainability as a compliance risk rather than a charitable gesture.

That forcing function is more powerful than any voluntary initiative, but it operates on a December 2027 timeline that leaves a 12-18 month window during which the gap between the crisis's current severity and the regulatory response's eventual effect remains unaddressed.

The 28% reflects genuine remaining probability that the combination of CRA compliance pressure, the visibility of the Ingress NGINX failure, and the AI contribution crisis produces a faster-than-expected voluntary corporate response.

But the history of this specific problem, publicly documented and widely discussed for a decade without producing meaningful change in the 0.0014% participation rate, argues strongly against optimism about voluntary reform arriving before the next catalyzing failure.

# Neuromorphic Computing Breakthrough

## Brain-Inspired Chips Are Real. Now Try Programming One

*Intel's Loihi 3 is in commercial availability targeting Q4 2026. IBM's NorthPole has moved into full-scale production.*

*Reported deployments in quadruped inspection robots claim battery life improvements of roughly ninefold over GPU-powered predecessors. The 1,000x power efficiency advantage for specific event-driven workloads has been validated in controlled testing.*

*And: as of May 2026, Intel hasn't published an official Loihi 3 product sheet with verified specifications. A developer trained in PyTorch can't program a neuromorphic chip without learning an entirely different computational paradigm.*

*There's still no unified framework making that transition accessible at enterprise deployment speed.*

*The hardware has arrived, in the narrow applications where physics makes it structurally decisive.*

*The software to deploy it broadly hasn't.*

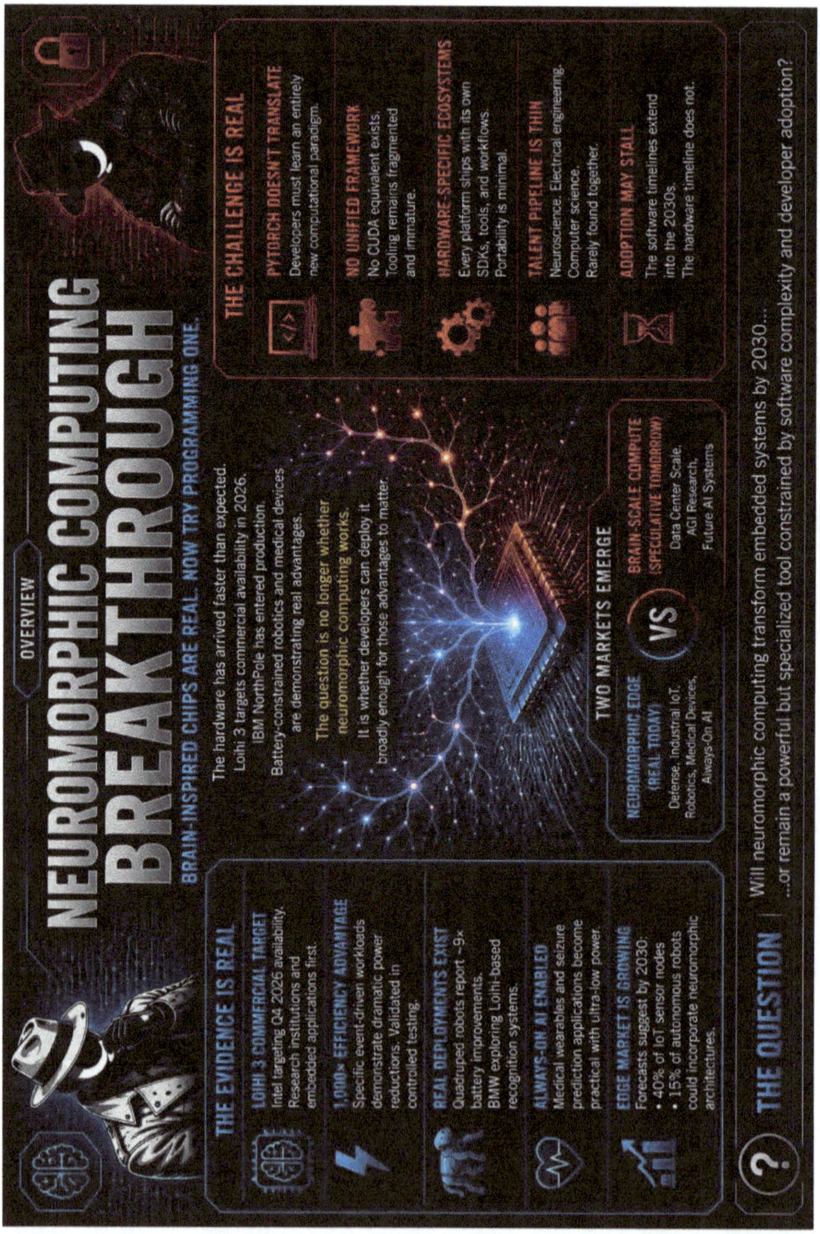
OVERVIEW
NEUROMORPHIC COMPUTING
BREAKTHROUGH
BRAIN-INSPIRED CHIPS ARE REAL. NOW TRY PROGRAMMING ONE.
The hardware has arrived faster than expected.
Loihi 3 targets commercial availability in 2026.
IBM NorthPole has entered production.
Battery-constrained robotics and medical devices are demonstrating real advantages.
The question is no longer whether neuromorphic computing works.
It is whether developers can deploy it broadly enough for those advantages to matter.
THE EVIDENCE IS REAL
LOIHI 3 COMMERCIAL TARGET
Intel targeting Q4 2026 availability. Research institutions and embedded applications first.
1,000× EFFICIENCY ADVANTAGE
Specific event-driven workloads demonstrate dramatic power reductions. Validated in controlled testing.
REAL DEPLOYMENTS EXIST
Quadruped robots report ~9x battery improvements. BMW exploring Loihi-based recognition systems.
ALWAYS-ON AI ENABLED
Medical wearables and seizure prediction applications become practical with ultra-low power.
EDGE MARKET IS GROWING
Forecasts suggest by 2030:
• 40% of IoT sensor nodes
• 15% of autonomous robots
could incorporate neuromorphic architectures.
TWO MARKETS EMERGE
NEUROMORPHIC EDGE (REAL TODAY)
Defense, Industrial IoT, Robotics, Medical Devices, Always-On AI
VS
BRAIN-SCALE COMPUTE (SPECULATIVE TOMORROW)
Data Center Scale, AGI Research, Future AI Systems
THE CHALLENGE IS REAL
PYTORCH DOESN'T TRANSLATE
Developers must learn an entirely new computational paradigm.
NO UNIFIED FRAMEWORK
No CUDA equivalent exists. Tooling remains fragmented and immature.
HARDWARE-SPECIFIC ECOSYSTEMS
Every platform ships with its own SDKs, tools, and workflows. Portability is minimal.
TALENT PIPELINE IS THIN
Neuroscience. Electrical engineering. Computer science. Rarely found together.
ADOPTION MAY STALL
The software timelines extend into the 2030s. The hardware timeline does not.
THE QUESTION
Will neuromorphic computing transform embedded systems by 2030...
...or remain a powerful but specialized tool constrained by software complexity and developer adoption?

## Opening

The original prediction correctly identified neuromorphic computing's genuine architectural advantages and the trajectory from research deployment toward commercial production.

The 2026 update reveals a more nuanced situation than either the optimistic or pessimistic scenario anticipated: the hardware has arrived faster than the original Counterpoint expected, and the software barrier remains more stubborn than the original Point expected.

The result is a market that has bifurcated along exactly the lines the original Counterpoint's workload-specificity argument predicted, but at a scale and with commercial deployment evidence that the Counterpoint's timeline may have dismissed too quickly.

A market bifurcation framing has emerged from independent analyst work as the most useful analytical lens: neuromorphic computing is not one market but two, a small, real "neuromorphic edge" market shipping today for defense and industrial IoT applications, and a much larger speculative "brain-scale compute" market funded on the theory that GPU scaling will hit fundamental limits.

Understanding which market the original prediction was actually about is the 2027-2030 analytical work this update initiates.

### Intro

The neuromorphic prediction requires separating two claims that the original piece treated as a single thesis.

The first claim, that neuromorphic computing's architectural advantages in power efficiency and event-driven processing are genuine and would drive deployment in specific domains, has been validated by the 2026 evidence.

The second claim, that those advantages would translate into broad embedded systems redesign within the prediction's window, hasn't been validated, and the market bifurcation analysis that has emerged since the original writing explains precisely why.

The "neuromorphic edge" market is real, shipping, and commercially meaningful in the domains where physics makes the efficiency advantage structurally decisive: battery-constrained robotics, always-on medical wearables, defense-sector autonomous navigation.

The "brain-scale compute" market, the speculative thesis that neuromorphic architectures will eventually displace GPU clusters for large-scale AI workloads, remains funded by the theory that GPU scaling will hit fundamental limits before neuromorphic software tooling catches up. Whether that theory proves correct is the 2027-2030 analytical question.

The 2026 evidence answers the narrower question definitively: the hardware has arrived in the domains where the physics is most compelling.

The software ecosystem that would expand neuromorphic's reach beyond those domains hasn't.

## Background

Intel's Loihi 3 processor, fabricated on a 4nm process and reported to pack 8 million neurons and 64 billion synapses per chip, an eightfold density increase over its predecessor, has been announced for commercial availability targeting Q4 2026, initially aimed at research institutions and specialized embedded applications.

N.B: as of May 2026, Intel hasn't published an official Loihi 3 product sheet with verified specifications; figures reported in industry coverage should be treated as pre-release claims rather than confirmed specifications.

IBM's NorthPole architecture has moved into full-scale production in 2026. The Hala Point research system, built from 1,152 Loihi 2 processors, has been publicly confirmed by Intel with 1.15 billion neurons and 128 billion synapses, the confirmed research-scale architecture that Loihi 3 would extend toward commercial deployment.

The competitive landscape has sharpened: an April 2026 market analysis argues neuromorphic computing is not one market but two, a real "neuromorphic edge" market shipping today for defense and industrial IoT, and a speculative "brain-scale compute" market, with four competing architectures: digital neuromorphic, analog/mixed-signal, superconducting optoelectronic, and photonic neural compute, each with distinct risk profiles and timelines.

Industry forecasts project neuromorphic chips will be present in 40% of IoT sensor nodes and 15% of autonomous robots by 2030, driven by power efficiency requirements that conventional processors can't meet in battery-operated edge applications.

GRAPHIC 2
POINT / COUNTERPOINT
NEUROMORPHIC COMPUTING
BREAKTHROUGH
BRAIN-INSPIRED CHIPS ARE REAL. NOW TRY PROGRAMMING ONE.
Extraordinary Potential. Extraordinary Challenges.
What Comes First: The Breakthrough or the Bottleneck?
THE POINT: THE INFLECTION IS HERE
The hardware is real. The advantages are proven. The early markets are commercial.
1. HARDWARE HAS ARRIVED
Loihi 3 targets commercial availability in Q4 2026. IBM NorthPole is in production. The next generation is here.
2. PHYSICS DELIVERS UNMATCHED EFFICIENCY
Up to 1,000× better power efficiency for specific event-driven workloads. Validated in controlled testing.
3. REAL DEPLOYMENTS ARE HAPPENING
Quadruped robots report ~9× battery life improvements. BMW exploring Loihi-based systems. Defense and industrial IoT are early but real.
4. NEW APPLICATIONS ARE POSSIBLE
Always-on medical wearables, seizure prediction, and continuous monitoring become practical with ultra-low power.
5. THE EDGE MARKET IS TAKING OFF
Forecasts suggest by 2030:
• 40% of IoT sensor nodes
• 15% of autonomous robots
could incorporate neuromorphic chips.
THE BATTLEFIELD
Real hardware advantages today.
Real software barriers ahead.
The winners will be the organizations that
LEARN FAST, EXPERIMENT BOLDLY,
and BUILD NEUROMORPHIC-READY CAPABILITIES TODAY.
THE COUNTERPOINT: THE BARRIERS ARE REAL
The barriers are structural, not temporary. Broad impact is years away.
1. NEW PARADIGM, NEW MENTAL MODEL
PyTorch skills don't transfer. Developers must learn spike-based, event-driven computing from the ground up. The cognitive load is enormous.
2. NO CUDA MOMENT (YET)
No unified framework or abstraction layer. The tooling ecosystem is fragmented across incompatible platforms.
3. HARDWARE-LOCKED ECOSYSTEMS
Every chip architecture has its own SDKs, workflows, and constraints. Portability is minimal to none.
4. TALENT PIPELINE IS EXTREMELY LIMITED
Requires rare combination of neuroscience, electrical engineering, and computer science. The talent bottleneck is real.
5. ADOPTION TIMELINES SLIP
Even optimistic estimates show broad developer adoption lagging hardware readiness by multiple years into the 2030s.
THE POINT ROUNDUP
• Real hardware with real advantages is shipping.
• Physics doesn't negotiate—efficiency is decisive.
• The edge is large enough to build real businesses.
• The inflection is here—momentum is building.
THE DEBATE IN ONE QUESTION
Will neuromorphic computing transform embedded systems by 2030...
...or remain a powerful but specialized tool constrained by software complexity and developer adoption?

## THE POINT: When Energy Is the Constraint, Physics Always Wins. Eventually.

The production deployment evidence that the original prediction identified as the inflection point has arrived, and in specific domains, it has arrived with efficiency numbers that make the architectural argument impossible to dismiss.

Reported deployments of quadruped inspection robots utilizing Intel Loihi 3 architecture have demonstrated dramatic battery life improvements over GPU-powered predecessors, with a ninefold improvement figure appearing in multiple industry reports, though these claims trace to promotional industry coverage rather than peer-reviewed validation and should be understood as directional rather than independently confirmed.

NVIDIA's competing Jetson Thor module offers raw processing power but carries a 10W to 60W power draw that limits battery life for untethered robots, the direct competitive comparison that makes the neuromorphic efficiency case most visible.

BMW Research's implementation of Loihi 2 clusters for real-time traffic sign recognition represents enterprise adoption in a commercially relevant application.

Lockheed Martin's testing of Innatera SNP processors for autonomous drone navigation reflects defense-sector adoption where power constraints are operationally decisive.

Medical device companies are pursuing FDA approvals for neuromorphic-powered seizure prediction and continuous health monitoring wearables, with ultra-low power consumption enabling always-on monitoring without frequent battery replacement.

The original Counterpoint's argument that "applications like industrial IoT and robotics will prove adequately served by improving conventional low-power chips" is being tested by these deployments.

The 2027-2030 market trajectory reflects the dynamic bifurcation rather than a single adoption curve, and the energy cost argument has become more urgent rather than less.

Data centers are projected to consume 134 TWh annually from AI workloads by 2026, roughly equivalent to Sweden's total energy consumption, creating structural tailwind for any architecture offering credible alternatives to the GPU's power profile.

Critical 2026 milestones include standardized neuromorphic benchmarks through IEEE P2800, mass production of neuromorphic microcontrollers at 100,000+ units per year scale, and first medical device approvals for neuromorphic-powered seizure prediction wearables.

The energy cost argument that the original piece identified as the long-term driver has been validated by data center consumption projections that make efficiency gains genuinely financially material rather than merely technically interesting.

Intel plans commercial Loihi 3 availability in Q4 2026, initially targeting research institutions and specialized embedded applications, the narrow but real market where the physics is most decisive and the software complexity is most manageable within current tooling constraints.

### The Point Roundup

Neuromorphic computing's 2026 reality is the best version of the original prediction's specific claims about the "neuromorphic edge" market and the most contested version of its "broad embedded systems redesign" ambition.

The new category of always-on, energy-efficient AI transforming battery-constrained robotics, always-on medical wearables, and autonomous drone navigation is not a forecast, it's a description of shipping products in late 2026.

What the prediction overestimated is the breadth of that transformation: it's happening in the domains where physics is most decisive and not yet spreading to the broader embedded systems redesign the original piece described.

The 2027-2030 window tests whether the software ecosystem matures fast enough to expand neuromorphic's reach beyond the domains where its energy advantage is structurally unavoidable, or whether the bifurcation hardens into permanent structural limitation.

## THE COUNTERPOINT: Great Hardware. No Developers. We've Seen This Movie.

The original Counterpoint's software ecosystem argument has been validated by every independent assessment of the neuromorphic landscape in 2026, and the CUDA analogy deserves direct acknowledgment as the most precise framing of why hardware milestones don't translate to deployment milestones on the timeframes the optimistic scenario requires.

Programming a neuromorphic system requires learning to "think in spikes" rather than traditional code, each neuromorphic platform includes its own SDK or research-oriented tools, there's virtually no unified mature framework comparable to PyTorch, and the entire deep learning ecosystem built on gradient backpropagation has no direct equivalent for spiking neural networks.

The Counterpoint's prediction that "developers trained in PyTorch and TensorFlow can't translate arbitrary neural networks to neuromorphic hardware without massive performance degradation" has been confirmed by every independent technical assessment published in 2025 and 2026.

The programming paradigm gap, manufacturing scalability, software ecosystem fragmentation, and developer talent shortages all have resolution timelines projected through 2036, not 2030. Those are not analyst pessimism; they are the timelines that the organizations building the software ecosystem themselves are projecting.

The market bifurcation dynamic contains within it the most compelling counterargument to broad neuromorphic adoption. The "neuromorphic edge" market is real, battery-constrained robotics deployments and BMW traffic sign recognition are genuine.

But "real" and "transformative of embedded system design broadly" are not the same claim.

The 40% of IoT sensor nodes by 2030 forecast assumes both that software tooling challenges are resolved and that use cases driving that adoption require neuromorphic's specific energy profile rather than being adequately served by improving conventional embedded processor efficiency.

The "brain-scale compute" speculative thesis faces NVIDIA's Vera Rubin GPU roadmap as a direct competitive threat, and NVIDIA's track record of defending its market position through architectural efficiency improvements is the strongest single counterargument to the long-term neuromorphic adoption scenario.

The software ecosystem development that made GPU computing accessible took a decade of sustained investment, a massive developer community, and the commercial pressure of gaming before it became the infrastructure AI training depends on.

The neuromorphic software ecosystem is earlier on that journey than the hardware milestones suggest.

GRAPHIC 3
NEUROMORPHIC COMPUTING
BREAKTHROUGH
BRAIN-INSPIRED CHIPS ARE REAL. NOW TRY PROGRAMING ONE.
What Would Have to Be True for the Point to Prevail?
What Would Have to Be True for the Counterpoint to Prevail?
CURRENT STATE (MID-2026)
SPECIALIZED HARDWARE IS HERE
Intel Loihi 3 targeting Q4 2026 commercial availability. IBM NorthPole in production.
PROVEN ADVANTAGES IN NARROW DOMAINS
Up to 1,000x power efficiency for specific event-driven workloads validated in testing. Real deployments in robotics and medical.
EARLY ECOSYSTEM, EARLY ADOPTERS
Defense, industrial IoT, and medical device companies leading. Tooling fragmented and platform-specific.
SOFTWARE GAP IS THE PRIMARY LIMITER
No unified programming model. Developers must learn new paradigms and tools. Talent pool remains extremely shallow.
MARKET BIFURCATION HAS BEGUN
"Neuromorphic edge" market is real today. "Brain-scale compute" market is still speculative and long horizon.
MOST CRITICAL VARIABLE
DEVELOPER PRODUCTIVITY INFLECTION
Will software abstractions, frameworks, and tools reach a tipping point where developers can achieve real productivity—*without* needing to become neuroscientists?
LIKELIHOOD:
COUNTERPOINT PREVAILS
35%
The software barrier is the higher-risk constraint. Timeline to broad adoption extends into the 2030s.
DECISION FRAMEWORK
DEFINE THE TARGET WORKLOADS
Focus on event-driven, sparse, or spiking problems where physics delivers a decisive advantage.
MEASURE SOFTWARE PRODUCTIVITY
Benchmark developer velocity and time-to-solution against GPUs/CPUs on relevant tasks.
BUILD ON ABSTRACTIONS, NOT SILICON
Invest in high-level frameworks and compilers that span multiple neuromorphic architectures.
DEVELOP THE TALENT PIPELINE
Cross-train across neuroscience, EE, and CS. Build curriculum and communities.
PROVE TOTAL ECONOMIC VALUE
Beyond power savings—demonstrate system-level ROI in real deployments at scale.
THE POINT MUST BE WRONG ABOUT
The software gap closing quickly enough to enable broad deployment by early 2030s.
Frameworks becoming unified and easy enough for mainstream developers.
Enterprises prioritizing power efficiency over developer velocity and familiarity.
Edge market success scaling naturally into cloud and data center use cases.
THE COUNTERPOINT MUST BE RIGHT ABOUT
Software complexity remaining high and slowing adoption for many years.
Tooling fragmentation persisting across incompatible hardware-specific ecosystems.
Talent scarcity limiting how fast organizations can build internal capabilities.
Hardware advancing faster than software matures.
THE BOTTOM LINE
Neuromorphic computing is real and delivering advantages—but the future depends on whether software can catch up to the silicon.
The next 24–36 months of software progress will determine the next decade of impact.

## What Needs to Change

### (...for the Counterpoint to Prevail)

The Counterpoint's conditions have been confirmed as accurate descriptions of the current state.

For the software barrier to remain the binding constraint through 2030, three conditions exist.

First, the IEEE P2800 neuromorphic benchmarking standard needs to either stall or produce results revealing the hardware efficiency advantages as more workload-specific than production deployments suggest.

Second, Intel's commercial Loihi 3 launch needs to find that research institutions and specialized embedded applications represent the ceiling, not the floor, that the transition from specialist to mainstream developer adoption requires tooling investments the current ecosystem can't fund at the required scale.

Third, NVIDIA's efficiency roadmap needs to continue improving fast enough that the 1,000x advantage in specific edge workloads shrinks to a range that conventional architectures can close through hardware optimization rather than architectural reinvention.

The third condition is the one most likely to prove decisive, and it's the one the original Counterpoint identified first.

## Likelihood Counterpoint Prevails: 35%

The Counterpoint likelihood increases from 30% to 35%, not because the hardware trajectory has weakened, but because the software barrier has proven more persistent than the original 30% implied, and the market bifurcation reveals a more structurally constrained adoption ceiling than the original prediction's broad framing anticipated.

The hardware has delivered ahead of schedule in the narrow domains where it's deployed.

The software ecosystem hasn't closed the gap.

No industry-wide standard exists for neuromorphic instruction sets.

The talent pipeline requires expertise spanning neuroscience, electrical engineering, and computer science, a rare combination that universities are only beginning to produce at scale.

The CUDA development analogy holds: the software ecosystem required to make GPU computing accessible took a decade of sustained investment before it became the infrastructure AI training depends on.

The 35% reflects genuine and growing probability that the "neuromorphic edge" market, shipping today in defense, robotics, and medical wearables, remains a real but narrow commercial category through 2030, while the "brain-scale compute" speculative thesis either waits for software tooling that does not yet exist or gets superseded by GPU efficiency improvements.

The physics remains genuinely advantageous. The software ecosystem gap remains genuinely wide.

The resolution of that tension is the 2027-2030 story.

# 6G Standards Fragmentation

## The Wedge Is Real. The Question Is Whether It Gets Driven Before 2029

*The original prediction's 2026 deadline requires precise interpretation to assess correctly.*

*The formal ITU-R submission window for 6G radio interface proposals opens February 2027, that deadline is not 2026.*

*But the prediction was never really about the formal standards calendar. It was about a political convergence deadline: whether Western vendors (Qualcomm, Ericsson, Nokia, Samsung), and their governments would arrive at sufficiently unified technical positions before the normative 3GPP work began, or whether internal Western fragmentation would give China the leverage to drive incompatible requirements into the standard from the inside.*

*That political window was 2024-2026.*

*The evidence on whether it closed with Western consensus intact, or with enough seams for a wedge, is the analytically meaningful update this piece provides.*

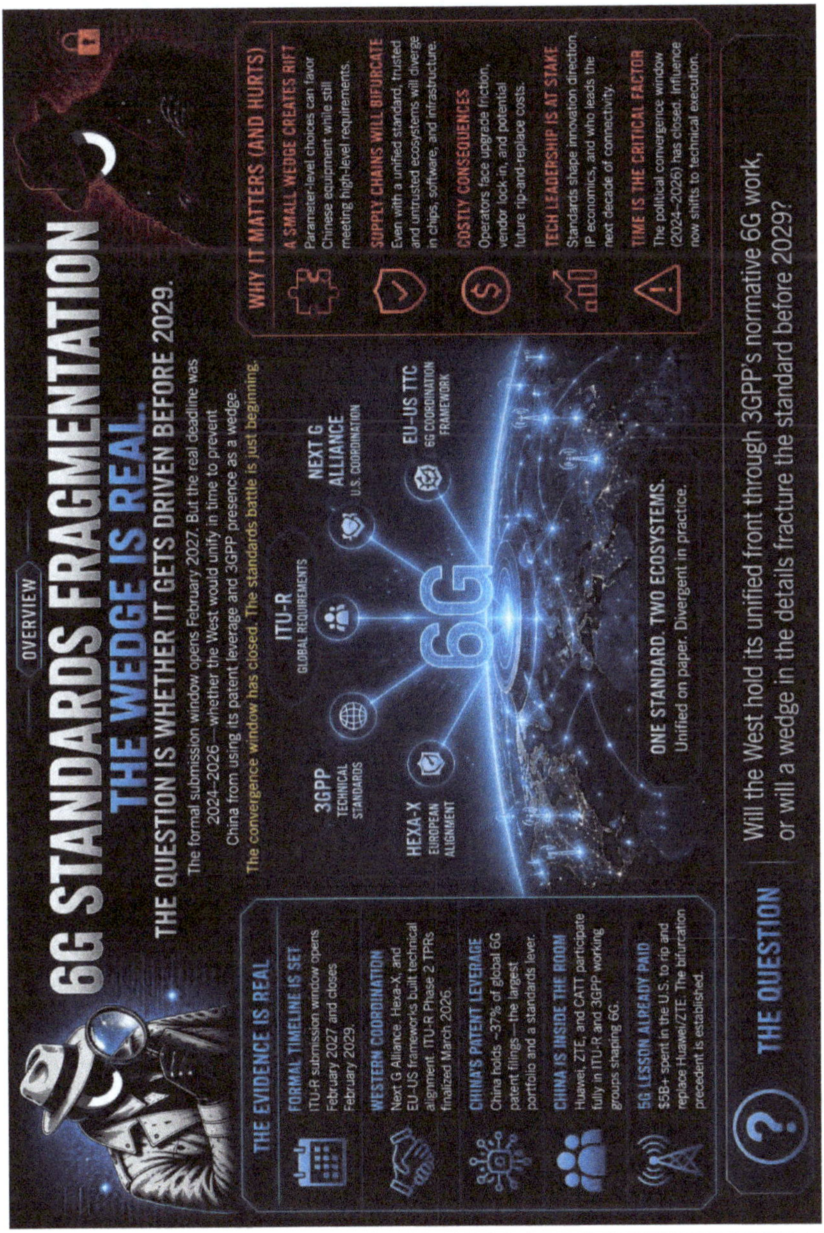
OVERVIEW
6G STANDARDS FRAGMENTATION
THE WEDGE IS REAL.
THE QUESTION IS WHETHER IT GETS DRIVEN BEFORE 2029.
The formal submission window opens February 2027. But the real deadline was 2024–2026—whether the West would unify in time to prevent China from using its patent leverage and 3GPP presence as a wedge.
The convergence window has closed. The standards battle is just beginning.
THE EVIDENCE IS REAL
FORMAL TIMELINE IS SET
ITU-R submission window opens February 2027 and closes February 2029.
WESTERN COORDINATION
Next G Alliance, Hexa-X, and EU-US frameworks built technical alignment. ITU-R Phase 2 TPRs finalized March 2026.
CHINA'S PATENT LEVERAGE
China holds ~37% of global 6G patent filings—the largest portfolio and a standards lever.
CHINA IS INSIDE THE ROOM
Huawei, ZTE, and CATT participate fully in ITU-R and 3GPP working groups shaping 6G.
5G LESSON ALREADY PAID
$5B+ spent in the U.S. to rip and replace Huawei/ZTE. The bifurcation precedent is established.
3GPP
TECHNICAL STANDARDS
ITU-R
GLOBAL REQUIREMENTS
NEXT G ALLIANCE
U.S. COORDINATION
HEXA-X
EUROPEAN ALIGNMENT
EU-US TTC
6G COORDINATION FRAMEWORK
6G
ONE STANDARD. TWO ECOSYSTEMS.
Unified on paper. Divergent in practice.
WHY IT MATTERS (AND HURTS)
A SMALL WEDGE CREATES RIFT
Parameter-level choices can favor Chinese equipment while still meeting high-level requirements.
SUPPLY CHAINS WILL BIFURCATE
Even with a unified standard, trusted and untrusted ecosystems will diverge in chips, software, and infrastructure.
COSTLY CONSEQUENCES
Operators face upgrade friction, vendor lock-in, and potential future rip-and-replace costs.
TECH LEADERSHIP IS AT STAKE
Standards shape innovation direction, IP economics, and who leads the next decade of connectivity.
TIME IS THE CRITICAL FACTOR
The political convergence window (2024–2026) has closed. Influence now shifts to technical execution.
THE QUESTION
Will the West hold its unified front through 3GPP's normative 6G work, or will a wedge in the details fracture the standard before 2029?

## Opening

Understanding what the 6G fragmentation prediction was actually claiming requires distinguishing two different timelines that the original piece treated as one.

The formal standards timeline, ITU-R submission windows, 3GPP release cycles, technology evaluation phases, runs to 2030, and nothing in that process closes or becomes irrecoverable in 2026.

The political convergence timeline, the window during which Western vendors and governments needed to align sufficiently to prevent China from using its 37% patent position and active 3GPP participation as a wedge, ran 2024-2026.

The original prediction's analytical core is the second claim, not the first.

And the 2026 evidence shows that Western consolidation has happened, materially and consequentially, through the Next G Alliance, Hexa-X, the EU-US 6G coordination frameworks, and the ITU-R Phase 2 requirements process that concluded in March 2026 with a unified set of technical performance requirements.

The question the 2027-2030 window answers is not whether that consolidation occurred. It did. The question is whether it holds under the pressure of the normative 3GPP Release 21 work where China's engineers are in the same working groups and the specific parameter choices that determine equipment compatibility will be made line by line.

## Intro

The 6G fragmentation prediction is analytically correct about a real and consequential geopolitical dynamic, and factually imprecise about the timeline on which that dynamic operates.

Correcting the timeline does not weaken the argument, it sharpens it.

The 2026 evidence shows Western consolidation happened, materially and through the specific institutional vehicles the prediction's logic required: the Next G Alliance, Hexa-X, and the EU-US coordination frameworks all produced the unified Phase 2 technical performance requirements the prediction needed to close without a wedge.

That is the good news for the convergence scenario.

The 2027-2030 window is the bad news: the consolidation that succeeded at the high-level TPR layer now has to hold at the parameter level inside 3GPP Release 21, where China's engineers are in the same working groups and the specific waveform choices, beam management procedures, and numerology decisions that determine real-world equipment compatibility will be negotiated line by line.

Winning the political convergence window and losing the parameter negotiation produces the same fragmentation outcome the original prediction described, through a slower and less visible mechanism than "irreconcilable approaches" implied, but with equivalent practical consequences for the multinational corporations and governments that depend on interoperability.

## Background

The 6G standardization process operates in two parallel streams that the fragmentation prediction correctly identified as distinct.

The formal institutional stream:

- ITU-R M.2160 Framework finalized December 2023 through a unified global process;
- ITU-R Technical Performance Requirements for IMT-2030 finalized March 2026 with 20 requirements including 7 new to 6G;
- the submission window for candidate Radio Interface Technologies (RIT/SRITs) formally opens February 2027 and closes February 2029;
- 3GPP Release 20 study phase began June 2025 and Release 21 normative 6G work is the standard.

China participates fully in both bodies. Huawei, ZTE, and CATT engineers sit in the same 3GPP working groups as Qualcomm, Ericsson, Nokia, and Samsung.

The political coordination stream: the US Next G Alliance, the EU Hexa-X flagship program, the EU-US Joint Technology Competition and Trade Council 6G coordination, and the 6G-IA (Infrastructure Association) have been building Western technical consensus specifically to prevent fragmentation through Chinese wedge strategies.

China holds approximately 37% of global 6G patent filings, the largest single-country portfolio, which represents both a standards negotiation lever and a signal of technical investment that will be deployed in the 2027-2029 submission process.

The 5G rip-and-replace program, costing approximately $5 billion to remove Huawei and ZTE equipment from US carrier networks, established the supply chain bifurcation precedent that applies regardless of how the standards process concludes.

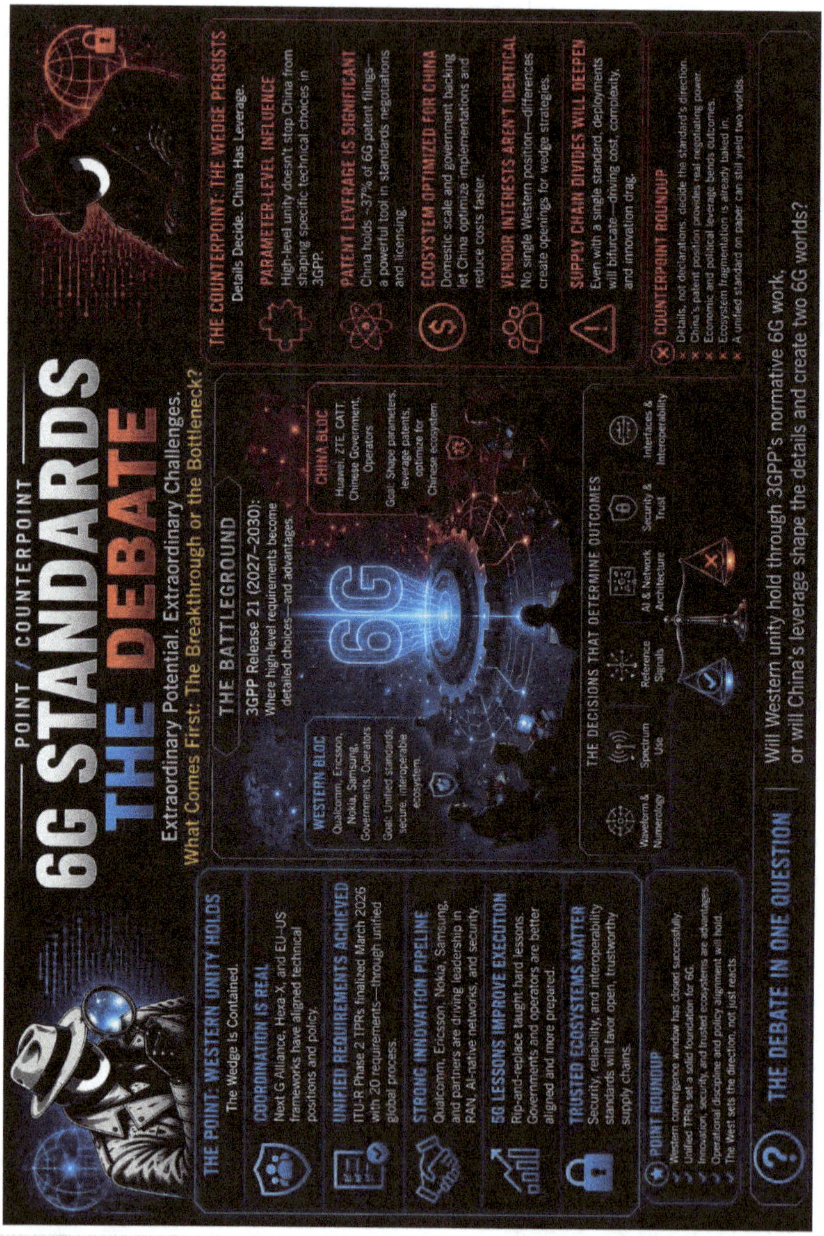
POINT / COUNTERPOINT
6G STANDARDS
THE DEBATE
Extraordinary Potential. Extraordinary Challenges.
What Comes First: The Breakthrough or the Bottleneck?
THE POINT: WESTERN UNITY HOLDS
The Wedge Is Contained.
COORDINATION IS REAL
Next G Alliance, Hexa-X, and EU-US frameworks have aligned technical positions and policy.
UNIFIED REQUIREMENTS ACHIEVED
ITU-R Phase 2 TPRs finalized March 2026 with 20 requirements—through unified global process.
STRONG INNOVATION PIPELINE
Qualcomm, Ericsson, Nokia, Samsung, and partners are driving leadership in RAN, AI-native networks, and security.
5G LESSONS IMPROVE EXECUTION
Rip-and-replace taught hard lessons. Governments and operators are better aligned and more prepared.
TRUSTED ECOSYSTEMS MATTER
Security, reliability, and interoperability standards will favor open, trustworthy supply chains.
POINT ROUNDUP
Western convergence window has closed successfully.
Unified TPRs set a solid foundation for 6G.
Innovation, security, and trusted ecosystems are advantages.
Operational discipline and policy alignment will hold.
The West sets the direction, not just reacts.
THE BATTLEGROUND
3GPP Release 21 (2027–2030):
Where high-level requirements become detailed choices—and advantages.
WESTERN BLOC
Qualcomm, Ericsson, Nokia, Samsung, Governments, Operators
Goal: Unified standards, secure, interoperable ecosystem.
6G
CHINA BLOC
Huawei, ZTE, CATT, Chinese Government, Operators
Goal: Shape parameters, leverage patents, optimize for Chinese ecosystem
THE DECISIONS THAT DETERMINE OUTCOMES
Waveform & Numerology
Spectrum Use
Reference Signals
AI & Network Architecture
Security & Trust
Interfaces & Interoperability
THE COUNTERPOINT: THE WEDGE PERSISTS
Details Decide. China Has Leverage.
PARAMETER-LEVEL INFLUENCE
High-level unity doesn't stop China from shaping specific technical choices in 3GPP.
PATENT LEVERAGE IS SIGNIFICANT
China holds ~37% of 6G patent filings—a powerful tool in standards negotiations and licensing.
ECOSYSTEM OPTIMIZED FOR CHINA
Domestic scale and government backing let China optimize implementations and reduce costs faster.
VENDOR INTERESTS AREN'T IDENTICAL
No single Western position—differences create openings for wedge strategies.
SUPPLY CHAIN DIVIDES WILL DEEPEN
Even with a single standard, deployments will bifurcate—driving cost, complexity, and innovation drag.
COUNTERPOINT ROUNDUP
Details, not declarations, decide the standard's direction.
China's patent position provides real negotiating power.
Economic and political leverage bends outcomes.
Ecosystem fragmentation is already baked in.
A unified standard on paper can still yield two worlds.
THE DEBATE IN ONE QUESTION
Will Western unity hold through 3GPP's normative 6G work, or will China's leverage shape the details and create two 6G worlds?

## THE POINT: Unified Requirements Today, Divergent Parameters Tomorrow: How 6G Fragmentation Actually Works

The original prediction's wedge-mechanism argument is analytically sound and well-supported by the 2026 evidence, even as the formal timeline requires correction.

The argument was never that China would refuse to participate in 3GPP or ITU-R.

It was that China, holding a dominant patent position and active participation in the same working groups as Western vendors, would exploit any internal Western disunity to shape technical requirements in ways that favor Huawei and ZTE implementation while creating friction for Western equipment.

The political window the prediction identified, 2024-2026, during which Western vendors needed to consolidate sufficiently unified positions to hold ground, has produced meaningful coordination.

The ITU-R Phase 2 TPR finalization in March 2026 going through unified reflects that the Western bloc arrived with cohesive enough positions to prevent Chinese requirements from dominating that phase.

But the TPRs are high-level capability targets. The fragmentation the prediction describes occurs at the parameter level: specific waveform choices, numerology specifications, reference signal designs, beam management procedures, the dozens of detailed technical decisions in 3GPP Release 21 where a Chinese company holding essential patents can legitimately propose parameters that, while technically compliant with the TPRs, are optimized for Chinese equipment architectures.

This is precisely how 5G produced unified specifications with Chinese-equipment-incompatible implementations: the core standard was unified, but specific parameter choices created friction in mixed-vendor deployments that the rip-and-replace programs then resolved through exclusion.

The supply chain bifurcation the prediction described is already sufficiently advanced that even a fully unified 6G standard produces functionally incompatible ecosystems in deployment.

The 5G rip-and-replace program, removing Huawei and ZTE from US carrier networks at approximately $5 billion cost, established the policy precedent that security concerns override technical compatibility in critical telecommunications infrastructure.

That precedent now extends to equipment purchase decisions, vendor certification frameworks, and supply chain audits across the UK, Sweden, Australia, Japan, and most EU member states.

When Western carriers can't purchase Chinese-manufactured 6G equipment regardless of its technical compliance with unified ITU-R specifications, the "two incompatible global communication ecosystems" the original prediction described exist de facto through procurement policy rather than de jure through incompatible standards.

Multinational corporations operating in both Western and Chinese-infrastructure markets already face the dual-standard operational friction the prediction described, separate vendor relationships, separate certification processes, different spectrum allocations, and different security audit requirements.

The fragmentation mechanism is already operational.

The 2027-2030 standards process determines whether it hardened into the physical layer or remains confined to the supply chain and security policy layer.

### The Point Roundup

The prediction's core insight, that geopolitical competition produces two functionally incompatible global communication ecosystems regardless of what any standards body produces, is being validated by the 2026 evidence at the supply chain and policy layer even before the formal standards process concludes.

The political convergence deadline the prediction described was real: Western vendors did need to consolidate positions before the normative 3GPP work began, and the evidence suggests they largely did. What the update clarifies is that convergence at the standards body level and fragmentation at the deployment level are not mutually exclusive outcomes.

The 2030 destination is not a single unified standard or two incompatible standards, it's a unified specification with bifurcated implementation that produces the operational friction and cost the original prediction identified, delivered through a more complex mechanism than the piece fully anticipated.

## THE COUNTERPOINT: The Wedge Strategy Requires Western Disunity. Look at March 2026

The Counterpoint's most important 2026 evidence is precisely the evidence the Point cites for its own case: the ITU-R Technical Performance Requirements finalized in March 2026 through a unified global process represent the Western consolidation the original prediction required as a precondition for preventing fragmentation.

The original prediction's wedge-mechanism argument required Western vendors to fail to converge, if they arrived at the Phase 2 requirements process with sufficiently unified technical positions, China's leverage is constrained rather than amplified.

The evidence suggests they did arrive unified: the Next G Alliance, Hexa-X, and EU-US coordination frameworks produced a coherent Western technical block that participated in the TPR finalization without the internal fractures that would have given China a wedge strategy to exploit.

Qualcomm's Vice President of Technical Standards confirmed that the normative 3GPP Release 21 process involves all major players including Chinese companies, but the critical insight is that Western companies are not entering that process as a fragmented bloc.

The wedge mechanism the prediction described requires a seam to drive through. The 2026 consolidation evidence suggests the seam is narrower than the original predicted it would be.

The 5G precedent the prediction relied on as an analog actually demonstrates the Counterpoint's strongest case when read precisely.

5G produced unified global specifications through 3GPP despite significant geopolitical tension during the 2018-2020 period when Huawei bans were being implemented across Western markets.

Those unified specifications enabled genuine global interoperability for the devices and use cases that constitute the majority of 5G commerce (smartphones, IoT, mobile broadband) even while the equipment supply chain bifurcated. International business continued. Roaming worked.

The operational costs of parallel infrastructure were real but manageable for the multinational corporations the original prediction described as facing catastrophic dual-infrastructure requirements.

The economic case against true standards fragmentation has only strengthened since the original was written: 6G is projected to enable $1.5-2 trillion in annual economic value by 2035, and the industries depending on global interoperability, automotive, logistics, healthcare, financial services, have stronger lobbying positions in both Western and Chinese technology policy than they did during 5G.

The Counterpoint's most plausible 2030 outcome is not convergence versus fragmentation but the 5G model at greater scale: unified core specifications, bifurcated supply chains, workable-if-expensive international interoperability, and continued international commerce despite the friction.

WHAT NEEDS TO CHANGE
— (FOR THE COUNTERPOINT TO PREVAIL) —
Counterpoint wins only if these conditions materialize.
THE POINT MUST BE WRONG ABOUT
WESTERN UNITY HOLDS
Internal alignment fractures, creating multiple positions in 3GPP and diluting negotiating power.
UNIFIED IMPLEMENTATIONS SCALE
Interoperability and cost pressures prove too hard to solve, slowing cross-vendor adoption.
POLICY COORDINATION ENDURES
EU-US and alliance coordination weakens under political change, reducing policy and technical alignment.
PATENT POSITION STABILIZES
Western patent filings stagnate relative to China, shrinking leverage in licensing negotiations.
TRUST & SECURITY CONCERNS EASE
Security frameworks fail to hold consensus, reopening trusted vendor qualification to Chinese firms.
DECISION FRAMEWORK
Actions to take now to test, learn, and build optionality.
6G
1 ENGAGE DEEPLY IN 3GPP RELEASE 21
Shape parameter-level decisions, build coalitions, and influence specification details early.
2 INVEST IN VERIFIABLE & INTEROPERABLE TECH
Develop reference implementations and testbeds to prove interoperability across ecosystems.
3 INCREASE PATENTS & LICENSING STRENGTH
Accelerate 6G patent filings and secure cross-licensing to balance China's leverage.
4 BUILD SECURITY & ASSURANCE STANDARDS
Codify security, supply chain assurance, and trusted vendor criteria into technical specs.
5 PREPARE DUAL-ECOSYSTEM OPERATIONS
Design architectures and operations that can support one or two 6G ecosystems if needed.
THE COUNTERPOINT MUST BE RIGHT ABOUT
CHINA DRIVES KEY PARAMETERS
China steers waveform, numerology, and reference signal choices in 3GPP that favor Chinese implementations.
WEDGE STRATEGIES SUCCEED
Subtle differences create ecosystem friction that locks in Chinese equipment advantages and tool chains.
ECONOMIC LEVERAGE EXPANDS
China's scale, pricing power, and government backing compress Western vendors and operators.
ALLIES SPLINTER UNDER PRESSURE
Commercial incentives or politics cause Western partners to pursue divergent vendors and standards paths.
SUPPLY CHAIN DIVIDES DEEPEN
Bifurcated ecosystems become entrenched, making cross-compatibility too costly and slow to restore.
FORCES PUSHING BACK
China's patent scale and R&D investment
Government industrial policy and support
Market incentives and cost pressure
Standards path dependence
CURRENT STATE
Western consolidation has held through ITU-R Phase 2 and early 3GPP studies. The real test is the parameter-level decisions in 3GPP Release 21 (2027–2029) where friction—or alignment—will be determined.
LIKELIHOOD COUNTERPOINT PREVAILS
28%
▼ -7%
VS ORIGINAL PREDICTION
(Was 35%)
0%
POINT FAVORED
28%
100%
COUNTERPOINT FAVORED
MOST CRITICAL VARIABLE
3GPP RELEASE 21 PARAMETER CHOICES
The detailed technical decisions on waveforms, numerology, reference signals, and procedures will determine whether 6G becomes one world—or two.

## What Needs to Change

### (...for the Counterpoint to Prevail)

For the "unified standard, manageable bifurcation" scenario to dominate rather than the "two incompatible ecosystems" scenario, three conditions need to hold through the 2027-2030 normative process.

First, the Western consolidation achieved in Phase 2 needs to hold under the detailed parameter-level pressure of 3GPP Release 21, where specific waveform choices, beam management procedures, and numerology decisions involve the same engineers from Huawei, Qualcomm, Ericsson, and Nokia negotiating line by line. The political consolidation and the technical specification consolidation are not the same achievement, and the history of 3GPP processes shows that patent-holding companies can successfully introduce parameters that favor their implementations through technically valid proposals that are difficult to reject.

Second, the supply chain bifurcation needs to stabilize at a level that allows genuine interoperability for the use cases where multinational corporations most depend on it, not full convergence, but sufficient compatibility that the operational friction remains below the threshold where companies begin building fully separate infrastructure rather than managing a unified standard through separate vendor relationships.

Third, the security certification and trusted supplier frameworks developing in the US, EU, and allied nations need to stop short of de facto prohibitions that treat technical compliance with the unified standard as insufficient for equipment approval.

If security certification becomes a second incompatible standard effectively requiring Western-only equipment, unified ITU-R specifications become commercially irrelevant in the same way that technically valid Huawei 5G equipment became commercially irrelevant in US carrier networks.

## Likelihood Counterpoint Prevails: 28%

The Counterpoint likelihood settles at 28%, higher than the original 20% but lower than my first revision's 35%. Calibrated to the dual-timeline framework the update establishes. This one's been a bit of a rollercoaster.

The upward move from 20% reflects two genuine 2026 developments the original did not anticipate: the Western consolidation in Phase 2 was more complete than the prediction implied it would be, and the unified ITU-R TPR finalization demonstrates that the institutional framework for producing a common standard hasn't broken down.

The downward move from 35% reflects my originally correct analytical insight: the wedge mechanism doesn't require Western vendors to fail to converge in standards bodies, it operates at the parameter level inside those bodies, and China's 37% patent portfolio and deep 3GPP participation are precisely the levers that create parameter-level fragmentation opportunities regardless of high-level TPR agreement.

The 28% says: the Counterpoint's "unified standard, manageable bifurcation" scenario has a real path through 2030, the Western consolidation happened, the institutional bodies remain intact, and the 5G precedent demonstrates workable international commerce is possible under supply chain bifurcation.

The prediction’s core insight still holds: geopolitics has turned telecommunications infrastructure into a domain where security imperatives often override efficiency. The wedge strategy operates less through open disagreement over high-level standards than through parameter-level negotiations inside normative 3GPP work.

That structural dynamic is still more likely than not to produce the functionally incompatible ecosystems the original prediction described, though through a slower, more complex mechanism than the original “irreconcilable approaches” language suggested.

# Multimodal Model Commoditization

## The Moat Is Gone. The Advantage Never Was the Model.

*DeepSeek trained a frontier-matching model for approximately $6 million, a fraction of the estimated $100 million spent on comparable proprietary models.*

*The MMLU benchmark gap between open-weight and proprietary frontier models narrowed from 17.5 percentage points to approximately 0.3 in a single year. GLM-5.1 became the first open-weight model ever to top SWE-bench Pro. Gartner now classifies foundation models as "strategic commodities."*

*The prediction has been validated, faster, and more decisively, than even its own optimistic framing anticipated.*

*The question the 2027–2030 window must answer is what comes after commoditization: who wins, and why, when the model itself stops being the differentiator.*

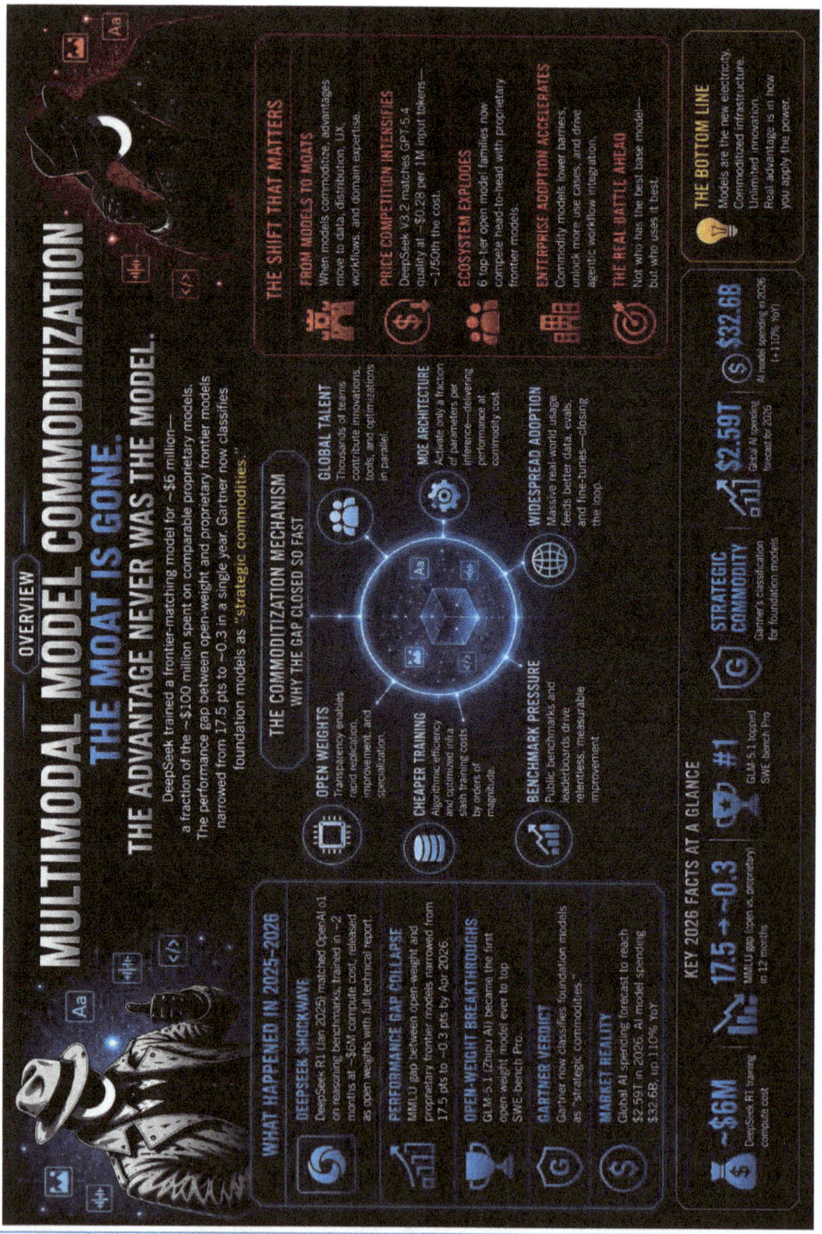
OVERVIEW
MULTIMODAL MODEL COMMODITIZATION
THE MOAT IS GONE.
THE ADVANTAGE NEVER WAS THE MODEL.
DeepSeek trained a frontier-matching model for ~$6 million—a fraction of the ~$100 million spent on comparable proprietary models. The performance gap between open-weight and proprietary frontier models narrowed from 17.5 pts to ~0.3 in a single year. Gartner now classifies foundation models as "strategic commodities."
WHAT HAPPENED IN 2025–2026
DEEPSEEK SHOCKWAVE
DeepSeek R1 (Jan 2025) matched OpenAI o1 on reasoning benchmarks, trained in ~2 months at ~$6M compute cost, released as open weights with full technical report.
PERFORMANCE GAP COLLAPSE
MMLU gap between open-weight and proprietary frontier models narrowed from 17.5 pts to ~0.3 pts by Apr 2026.
OPEN-WEIGHT BREAKTHROUGHS
GLM-5.1 (Zhipu AI) became the first open-weight model ever to top SWE-bench Pro.
GARTNER VERDICT
Gartner now classifies foundation models as "strategic commodities."
MARKET REALITY
Global AI spending forecast to reach $2.59T in 2026. AI model spending $32.6B, up 110% YoY.
THE COMMODITIZATION MECHANISM
WHY THE GAP CLOSED SO FAST
OPEN WEIGHTS
Transparency enables rapid replication, improvement, and specialization.
GLOBAL TALENT
Thousands of teams contribute innovations, tools, and optimizations in parallel.
CHEAPER TRAINING
Algorithmic efficiency and optimized infra slash training costs by orders of magnitude.
MOE ARCHITECTURE
Activate only a fraction of parameters per inference—delivering performance at commodity cost.
BENCHMARK PRESSURE
Public benchmarks and leaderboards drive relentless, measurable improvement.
WIDESPREAD ADOPTION
Massive real-world usage feeds better data, evals, and fine-tunes—closing the loop.
THE SHIFT THAT MATTERS
FROM MODELS TO MOATS
When models commoditize, advantages move to data, distribution, UX, workflows, and domain expertise.
PRICE COMPETITION INTENSIFIES
DeepSeek V3.2 matches GPT-5.4 quality at ~$0.28 per 1M input tokens—~1/50th the cost.
ECOSYSTEM EXPLODES
6 top-tier open model families now compete head-to-head with proprietary frontier models.
ENTERPRISE ADOPTION ACCELERATES
Commodity models lower barriers, unlock more use cases, and drive agentic workflow integration.
THE REAL BATTLE AHEAD
Not who has the best base model—but who uses it best.
KEY 2026 FACTS AT A GLANCE
~$6M
DeepSeek R1 training compute cost
17.5 → ~0.3
MMLU gap (open vs. proprietary) in 12 months
#1
GLM-5.1 topped SWE-bench Pro
STRATEGIC COMMODITY
Gartner's classification for foundation models
$2.59T
Global AI spending forecast for 2026
$32.6B
AI model spending in 2026 (+110% YoY)
THE BOTTOM LINE
Models are the new electricity.
Commoditized infrastructure.
Unlimited innovation.
Real advantage is in how you apply the power.

## Opening

My original prediction identified a pattern. Technology commoditization following the same arc as databases, web servers, and operating systems, and I applied it to multimodal AI models.

The 2026 evidence confirms that pattern is playing out and doing so faster than the original framing anticipated.

DeepSeek's January 2025 release of R1 was the event that changed the framing permanently: a model matching or exceeding OpenAI's o1 on reasoning benchmarks, built in approximately two months at roughly $6 million in compute cost, published as open weights with a full technical report.

That single event validated the commoditization thesis, shook investor confidence in the moat assumptions underlying frontier lab valuations, and accelerated the entire open-source ecosystem's development trajectory.

By mid-2026, the landscape the original prediction described, open-weight models as commodity infrastructure, competitive advantage shifting to data and specialization, is not a forecast.

It's a description of the current market structure.

## Intro

The update requires nuance my original prediction's framing did not fully anticipate. Commoditization has arrived at the level of general-purpose text, code, and single-turn multimodal tasks, the use cases that drove the first wave of AI adoption.

A meaningful capability gap persists at the frontier: complex multi-step agentic reasoning, long tool-use chains, and the hardest reasoning tasks (frontier mathematical olympiad problems, novel code architecture decisions, nuanced safety-critical judgment) remain domains where proprietary frontier models maintain measurable leads.

The market structure the Counterpoint predicted has partially materialized, open-weight models for experimentation and standard workloads, proprietary for mission-critical frontier tasks, but the boundary sits much higher up the capability stack than the Counterpoint anticipated, and it's rising rapidly.

The 2027–2030 question is not whether commoditization has happened. It's whether the proprietary frontier maintains the leads that remain, or whether those too converge within the next 18–24 months.

## Background

The AI model landscape has been restructured by two developments that arrived faster than most analysts predicted.

First, DeepSeek's open-weight model releases, V3 in December 2024 and R1 in January 2025, demonstrated that frontier-class performance could be achieved at dramatically lower training cost than the proprietary labs' infrastructure investments implied, upending the "scale equals moat" assumption that had underpinned AI industry valuations.

Second, Meta's Llama 4 family (April 2025) introduced natively multimodal, Mixture-of-Experts architecture in open weights at competitive performance levels, providing the ecosystem with a production-capable multimodal foundation.

By April 2026, the benchmark picture had shifted materially: the MMLU performance gap between open-weight and proprietary frontier models had narrowed from 17.5 to approximately 0.3 percentage points.

DeepSeek V3.2 matches GPT-5.4 quality at approximately $0.28 per million input tokens, roughly 1/50th of frontier proprietary pricing on comparable tasks.

GLM-5.1 from Zhipu AI briefly held the top position on SWE-bench Pro, the first open-weight model ever to do so.

Gartner responded to this evidence by formally classifying foundation models as "strategic commodities."

Global AI spending is forecast to reach $2.59 trillion in 2026 according to Gartner's May 2026 forecast, with spending on AI models specifically reaching $32.6 billion, up 110% from $15.5 billion in 2025, driven by agentic workflow consumption and embedded model integration rather than frontier capability procurement.

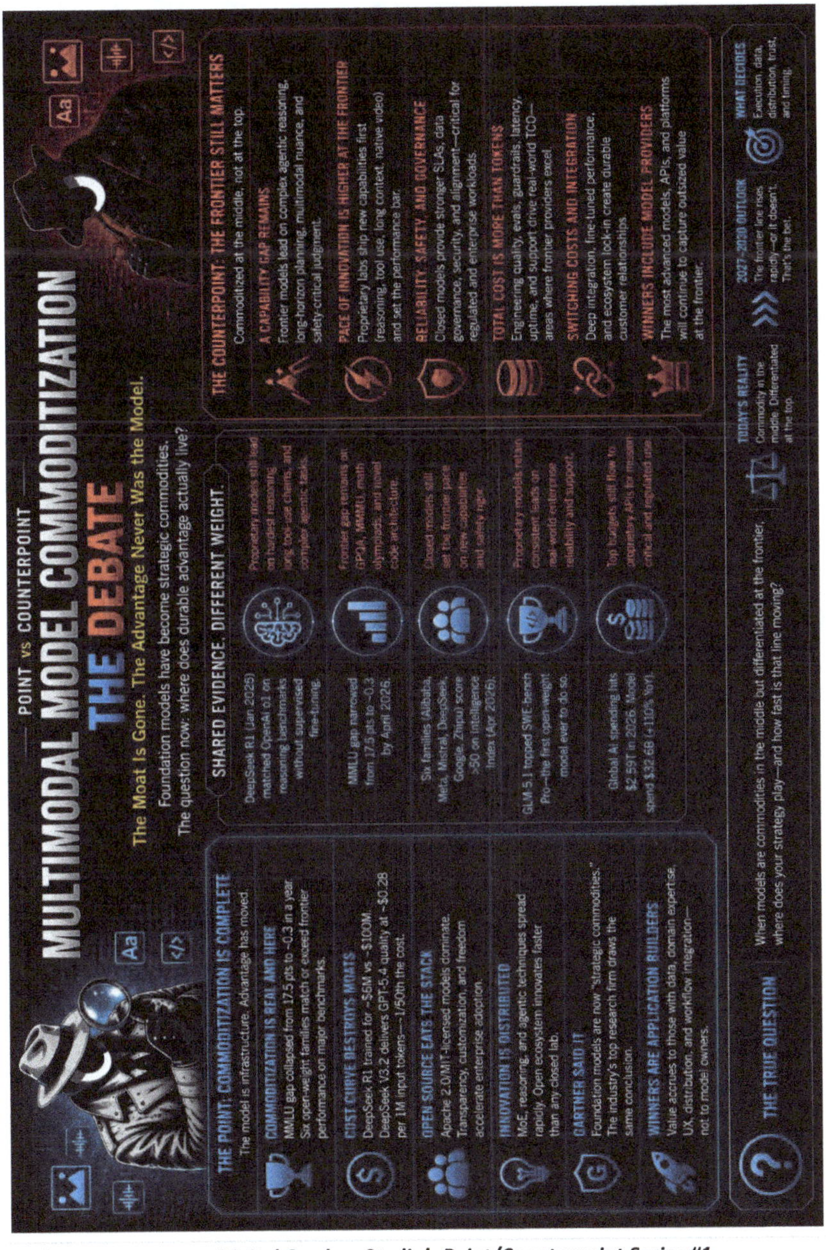
POINT vs COUNTERPOINT
MULTIMODAL MODEL COMMODITIZATION
THE DEBATE
The Moat Is Gone. The Advantage Never Was the Model.
Foundation models have become strategic commodities.
The question now: where does durable advantage actually live?
Aa
</>
THE POINT: COMMODITIZATION IS COMPLETE
The model is infrastructure. Advantage has moved.
COMMODITIZATION IS REAL AND HERE
MMLU gap collapsed from 17.5 pts to ~0.3 in a year. Six open-weight families match or exceed frontier performance on major benchmarks.
COST CURVE DESTROYS MOATS
DeepSeek R1 trained for ~$6M vs ~$100M. DeepSeek V3.2 delivers GPT-5.4 quality at ~$0.28 per 1M input tokens—~1/50th the cost.
OPEN SOURCE EATS THE STACK
Apache 2.0/MIT-licensed models dominate. Transparency, customization, and freedom accelerate enterprise adoption.
INNOVATION IS DISTRIBUTED
MoE, reasoning, and agentic techniques spread rapidly. Open ecosystem innovates faster than any closed lab.
GARTNER SAID IT
Foundation models are now "strategic commodities." The industry's top research firm draws the same conclusion.
WINNERS ARE APPLICATION BUILDERS
Value accrues to those with data, domain expertise, UX, distribution, and workflow integration—not to model owners.
SHARED EVIDENCE. DIFFERENT WEIGHT.
DeepSeek R1 (Jan 2025) matched OpenAI o1 on reasoning benchmarks without supervised fine-tuning.
Proprietary models still lead on hardest reasoning, long tool-use chains, and complex agentic tasks.
MMLU gap narrowed from 17.5 pts to ~0.3 by April 2026.
Frontier gap remains on GPQA, MMMU, math olympiads, and novel code architecture.
Six families (Alibaba, Meta, Mistral, DeepSeek, Google, Zhipu) score >50 on Intelligence Index (Apr 2026).
Closed models still set the frontier pace on new capabilities and safety rigor.
GLM-5.1 topped SWE-bench Pro—the first open-weight model ever to do so.
Proprietary models retain consistent leads on real-world enterprise reliability and support.
Global AI spending hits $2.59T in 2026. Model spend $32.6B (+110% YoY).
Top budgets still flow to proprietary APIs for mission-critical and regulated use.
THE COUNTERPOINT: THE FRONTIER STILL MATTERS
Commoditized at the middle, not at the top.
A CAPABILITY GAP REMAINS
Frontier models lead on complex agentic reasoning, long-horizon planning, multimodal nuance, and safety-critical judgment.
PACE OF INNOVATION IS HIGHER AT THE FRONTIER
Proprietary labs ship new capabilities first (reasoning, tool use, long context, native video) and set the performance bar.
RELIABILITY, SAFETY, AND GOVERNANCE
Closed models provide stronger SLAs, data governance, security, and alignment—critical for regulated and enterprise workloads.
TOTAL COST IS MORE THAN TOKENS
Engineering quality, evals, guardrails, latency, uptime, and support drive real-world TCO—areas where frontier providers excel.
SWITCHING COSTS AND INTEGRATION
Deep integration, fine-tuned performance, and ecosystem lock-in create durable customer relationships.
WINNERS INCLUDE MODEL PROVIDERS
The most advanced models, APIs, and platforms will continue to capture outsized value at the frontier.
THE TRUE QUESTION
When models are commodities in the middle but differentiated at the frontier, where does your strategy play—and how fast is that line moving?
TODAY'S REALITY
Commodity in the middle. Differentiated at the top.
2027–2030 OUTLOOK
The frontier line rises rapidly—or it doesn't. That's the bet.
WHAT DECIDES
Execution, data, distribution, trust, and timing.

## THE POINT: Gartner Said It: Foundation Models Are Strategic Commodities. The Debate Is Over.

The commoditization thesis has been validated by the specific mechanism the original prediction identified, open-weight models matching proprietary performance within months of new capabilities emerging, but the speed and scale of that validation exceeded even the optimistic framing.

DeepSeek's R1 model, released January 2025 with a full technical report disclosing its reinforcement learning methodology and approximately $6 million training cost, matched OpenAI's o1 on reasoning benchmarks without supervised fine-tuning, proving that the architectural innovations underlying frontier reasoning capability could be independently reproduced and made freely available.

The response from the open source ecosystem was immediate and compounding: by April 2026, six distinct model families arrived from six organizations: Alibaba (Qwen), Meta (Llama), Mistral, DeepSeek, Google (Gemma), and Zhipu AI (GLM).

All offered models scoring above 50 on the Artificial Analysis Intelligence Index, compared to the low 30s for the best open weight models a year earlier.

Five of those six families use Apache 2.0 or MIT licensing.

The MoE (Mixture of Experts) architecture, which activates only a fraction of total parameters per inference, made frontier-class models deployable on single H100 GPUs, reducing self-hosting costs by 4–8x relative to the dense architectures that dominated 2024.

Gartner's formal classification of foundation models as "strategic commodities" in early 2026 represents the analyst community's institutional acknowledgment of what the benchmark data had been showing for months.

The competitive advantage shift the original prediction described, from model capability to data moats and specialized fine-tuning, is now the consensus view of enterprise strategy, not a prediction.

McKinsey's 2026 analysis of AI competitive moats identifies proprietary data as the primary durable advantage, specifically the closed-loop data ecosystems where AI deployment generates the training signal that improves the next deployment.

The pattern is visible in Amazon's advertising business ($68 billion in 2025 revenue), which compounds on behavioral and transaction data that no external model provider can replicate regardless of model capability.

The enterprise adoption data confirms the shift in practice: organizations now routinely run hybrid model stacks, open-weight models for internal workloads, cost-sensitive production tasks, and privacy-constrained deployments; proprietary APIs for frontier reasoning and high-stakes customer-facing applications.

Gartner's enterprise AI spending analysis found that in 2026, AI model procurement decisions are increasingly driven by total cost of ownership, vendor lock-in risk, and data sovereignty requirements rather than raw capability benchmarks.

The "AI features as table stakes" dynamic is operating across enterprise software, organizations that treat AI integration alone as a moat are finding that advantage lasts weeks, not years, as every SaaS vendor embeds equivalent capability on the same commodity model infrastructure.

### The Point Roundup

The multimodal model commoditization prediction has been validated and validated on the timeline the original framing specified. Open-weight models now match proprietary performance for the majority of production workloads.

The cost differential, DeepSeek V3.2 at $0.28/million tokens versus frontier proprietary alternatives at $15–30/million tokens, represents a 50–100x price advantage that enterprise procurement teams can't ignore at scale.

Gartner has formalized the "strategic commodity" classification.

The competitive advantage has shifted to data, as the prediction specified.

The 2027–2030 window tests the prediction's most important structural insight: whether the organizations that understood this dynamic early, and built proprietary data pipelines, domain-specific fine-tuning infrastructure, and closed-loop feedback systems, will compound their lead as the model layer continues to commoditize beneath them.

## THE COUNTERPOINT: The Gap Narrowed. It Did Not Close. And the Hard Problems Are Where Value Lives

The Counterpoint's most durable insight was not the prediction that leading labs would maintain secrecy. That proved incorrect, but the observation that the capabilities commanding premium pricing would prove harder to commoditize than baseline text performance did follow correctly.

In 2026 that insight remains structurally valid.

The gap between open-weight and proprietary frontier models has closed dramatically for standard tasks, text generation, summarization, code completion, classification, and remains meaningful for the hardest problems: complex multi-step agentic reasoning across extended task horizons, novel mathematical reasoning at olympiad level, nuanced safety-critical judgment in regulated domains, and the kind of frontier coding capability that distinguishes Claude Mythos's 93.9% SWE-bench Verified score from the best publicly available models' 80-88% range[1].

The enterprise tier structure the Counterpoint predicted has broadly materialized: organizations run open-weight models for the 80% of tasks where performance is sufficient and cost matters, while maintaining proprietary API relationships for the 20% where frontier capability is genuinely required and the business case justifies the premium.

That is not the "experimentation only" positioning the Counterpoint initially argued, but it's a meaningful structural distinction from full commoditization.

The data moat argument, which the Point and Counterpoint now share as common ground, cuts against the most optimistic version of the commoditization thesis.

If competitive advantage has shifted decisively to proprietary data, then the organizations that built data advantages during the proprietary model era, primarily the large platform companies with closed-loop behavioral ecosystems, retain structural advantages that commoditized models can't eliminate.

OpenAI's competitive position increasingly rests not on GPT-5's capability lead, which has narrowed, but on ChatGPT's 600 million weekly active users generating the interaction data that trains the next generation of models.

Google's advantage is not Gemini's benchmark scores but the search, Gmail, YouTube, and Maps behavioral data that no open-weight model trainer can access.

The Counterpoint's argument that "post-training that makes models genuinely useful requires resources and expertise concentrated in leading labs" has been partially falsified by DeepSeek's efficiency, but the underlying point about data concentration favoring incumbents with behavioral ecosystems remains structurally sound.

Commoditization of the base model layer may paradoxically entrench the data advantages of the largest platform incumbents by eliminating the model capability differentiation that gave smaller, model-focused competitors their opportunity.

---

[1] As of June 12, 2026 Anthropic has disabled access to Mythos 5 and Fable 5 in response to the US Government order to prevent access by foreign entities (https://www.anthropic.com/news/fable-mythos-access)

WHAT NEEDS TO CHANGE
— (FOR THE COUNTERPOINT TO PREVAIL) —
Counterpoint wins only if these conditions materialize.
THE POINT MUST BE WRONG ABOUT
COMMODITIZATION NOT COMPLETE
Proprietary models maintain a durable gap across most enterprise-relevant capabilities.
PERFORMANCE GAP PERSISTS
MMLU gap widens again above 2–3 pts. Proprietary leads expand in agentic reasoning, tool use, and multimodal understanding.
OPEN-WEIGHT PLATEAUS
Open models hit scaling limits due to compute access, data quality, or algorithmic ceilings.
DATA ADVANTAGES LOCK IN
Proprietary labs secure exclusive data, real-time feedback loops, and distribution that open models cannot replicate.
INNOVATION CENTRALIZES
Breakthroughs in reasoning, long-context, and multimodal integration happen only inside closed labs.
ENTERPRISE TRUST BARRIERS REMAIN
Regulation, safety, IP risk, and compliance keep mission-critical workloads on proprietary models.
TOTAL COST REMAINS HIGH
Proprietary models deliver meaningfully better TCO for quality, reliability, uptime, and support.
DECISION FRAMEWORK
Actions to take now to test, learn, and build optionality.
1 RUN FRONTIER BENCHMARKS CONTINUOUSLY
Track frontier vs. open on reasoning, agentic tasks, long context, multimodal, and safety benchmarks every quarter.
2 INVEST IN HIGH-VALUE DATA & FEEDBACK
Build proprietary datasets, user feedback loops, and domain-specific knowledge that models cannot easily replicate.
3 FOCUS ON AGENTIC WORKFLOWS
Design complex, multi-step workflows and tool chains that proprietary models currently execute more reliably and accurately.
4 STRENGTHEN SAFETY & GOVERNANCE
Invest in evals, red-teaming, guardrails, and compliance frameworks that favor closed-model deployments in regulated environments.
5 BUILD INTEGRATION & LOCK-IN MOATS
Deep integrations, proprietary connectors, and ecosystem lock-in make switching costly.
6 OPTIMIZE TOTAL COST, NOT JUST TOKENS
Compare quality, reliability, latency, support, and uptime—not just per-token pricing.
THE COUNTERPOINT MUST BE RIGHT ABOUT
FRONTIER CAPABILITIES STAY UNIQUE
Proprietary models retain clear leads in the hardest reasoning, agentic, and multimodal tasks for 18–24+ months.
DISTRIBUTION AND BRAND POWER MATTER
Proprietary labs leverage brand, trust, enterprise relationships, and global distribution at scale.
CLOSED MODEL EXPERIENCE EXCELS
Better reliability, evals, safety, tooling, and support keep enterprises on closed models for critical use cases.
ENTERPRISE ADOPTION FAVORS CLOSED
Large enterprises standardize on proprietary models for compliance, security, and mission-critical workloads.
ECONOMICS FAVOR PROPRIETARY
Higher quality and lower total cost of ownership keep paying for closed models the rational choice.
WINNERS CAPTURE DISPROPORTIONATE VALUE
The leading proprietary players capture the most value across the AI stack.
CURRENT STATE (MID-2026)
Commoditization is real and accelerating.
MMLU gap is ~0.3 pts. Open models match or exceed on many benchmarks.
Costs are collapsing (DeepSeek V3.2 at $0.28 per 1M input tokens).
Six major open-weight families compete at frontier levels.
Gartner classifies foundation models as "strategic commodities."
LIKELIHOOD COUNTERPOINT PREVAILS
PREVIOUS (Q1 2026)
35%
NOW (MID-2026)
20%
CHANGE
-15
percentage points
0%
POINT FAVORED
20%
100%
COUNTERPOINT FAVORED
MOST CRITICAL VARIABLE
FRONTIER CAPABILITY GAP AT THE TOP OF THE STACK
If proprietary models maintain meaningful leads in the highest-value, hardest-to-execute tasks—especially agentic reasoning and long tool-use chains—the counterpoint can still prevail.
THE DEBATE IN ONE QUESTION
Will the frontier capability gap that remains today persist through 2027–2030, or will even the hardest tasks become commodities?

 

## What Needs to Change

### (for the Counterpoint to Prevail)

The original counterpoint's five conditions have partially resolved, some in the Point's favor, one in the Counterpoint's.

For the remaining Counterpoint thesis to prevail through 2030, three conditions need to hold.

First, the capability gap in frontier reasoning, complex agentic tasks, and safety-critical applications needs to persist and remain commercially meaningful, if open-weight models close the remaining 10–15% gap in these domains within the next 18–24 months, the Counterpoint's "premium proprietary for mission-critical" tier loses its structural basis.

Second, the regulatory environment needs to create compliance advantages for proprietary labs, EU AI Act requirements around transparency, safety testing, and audit trails may prove harder for open-weight models deployed at enterprise scale than for proprietary APIs with established governance frameworks.

Third, the incumbent platform data advantages need to compound faster than open-weight model capability improves, the question of whether Google's search data or Amazon's commerce data generates greater AI competitive value than DeepSeek's next architectural efficiency breakthrough is the most important unresolved dynamic in the 2027–2030 window.

## Likelihood Counterpoint Prevails: 20%

The Counterpoint likelihood declines from 35% to 20%. That is a substantial revision driven by the pace and completeness of the commoditization evidence since the original was written.

The original 35% was assigned partly on the basis that leading labs would successfully maintain architectural secrecy; that assumption proved incorrect within months of the original prediction, as DeepSeek's fully published technical approach enabled independent reproduction and improvement across the entire open-source ecosystem.

The MMLU gap narrowing from 17.5 to 0.3 percentage points in a single year, Gartner's formal "strategic commodity" classification, and GLM-5.1's achievement of the top SWE-bench Pro position collectively represent a more decisive validation than a 35% Counterpoint likelihood implied was likely.

The remaining 20% reflects three genuine sources of Counterpoint resilience: the frontier capability gap in agentic reasoning and safety-critical tasks, which remains real and commercially meaningful even if its scope has narrowed; the regulatory complexity creating enterprise compliance advantages for established proprietary providers; and, most importantly, the incumbent platform data concentration dynamic, which the original Counterpoint identified as a structural advantage that commoditized base models can't neutralize.

It's analytically plausible that open-weight model commoditization paradoxically entrenches the data advantages of Google, Amazon, and Microsoft more deeply than it threatens them.

The 20% says: the model capability moat is gone, as the prediction stated. The data moat and regulatory moat remain, and their durability through 2030 is the open question the evidence hasn't yet resolved.

# AI-Generated Code Quality Ceiling

## The CASE Tool Problem Returns: AI Coding Assistants and the Last Mile That Won't Close

*Claude Mythos scores 93.9% on SWE-bench Verified. Experienced developers using AI tools on their own real repositories complete tasks 19% slower than without AI and believe they went faster.*

*Both statements are simultaneously true.*

*The gap between them is the entire AI coding story in 2026.*

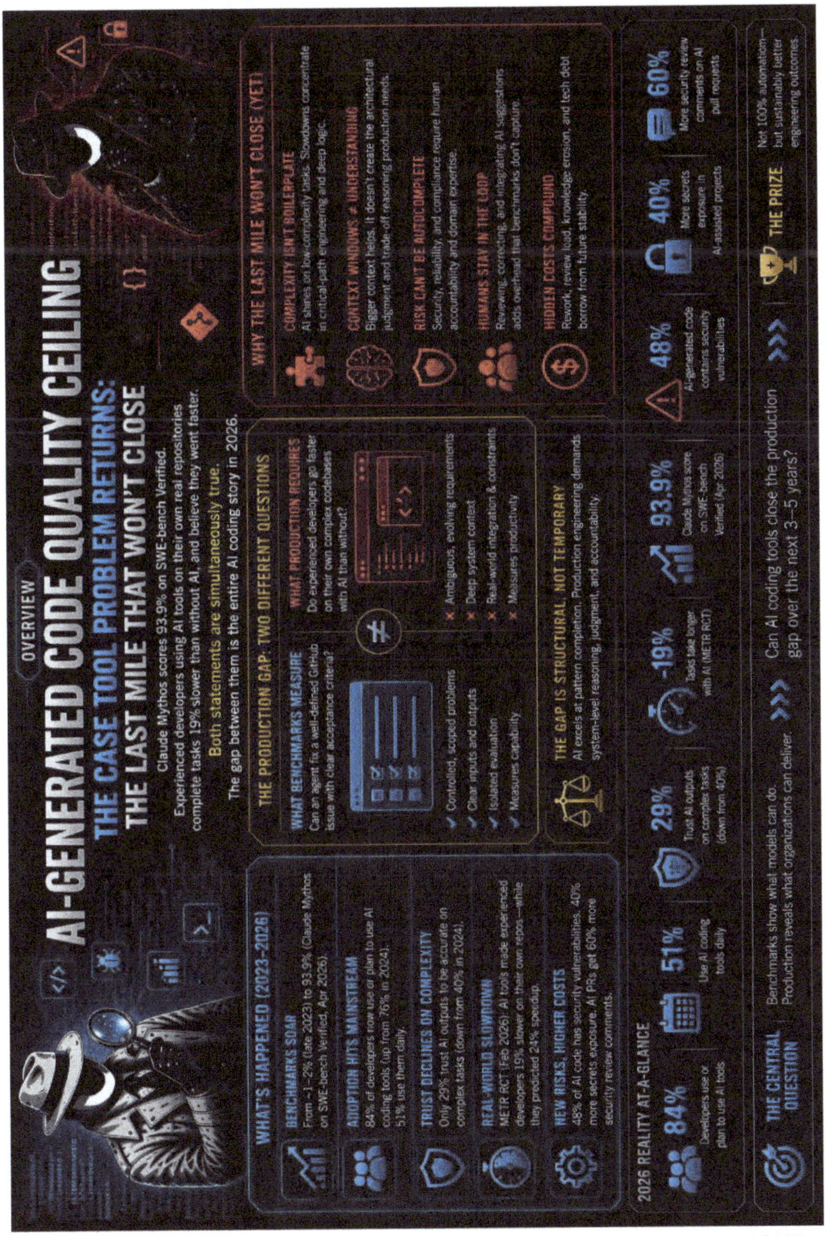
OVERVIEW
AI-GENERATED CODE QUALITY CEILING
THE CASE TOOL PROBLEM RETURNS:
THE LAST MILE THAT WON'T CLOSE
Claude Mythos scores 93.9% on SWE-bench Verified.
Experienced developers using AI tools on their own real repositories complete tasks 19% slower than without AI, and believe they went faster.
Both statements are simultaneously true.
The gap between them is the entire AI coding story in 2026.
WHAT'S HAPPENED (2023–2026)
BENCHMARKS SOAR
From ~1–2% (late 2023) to 93.9% (Claude Mythos on SWE-bench Verified, Apr 2026).
ADOPTION HITS MAINSTREAM
84% of developers now use or plan to use AI coding tools (up from 76% in 2024). 51% use them daily.
TRUST DECLINES ON COMPLEXITY
Only 29% trust AI outputs to be accurate on complex tasks (down from 40% in 2024).
REAL-WORLD SLOWDOWN
METR RCT (Feb 2026): AI tools made experienced developers 19% slower on their own repos—while they predicted 24% speedup.
NEW RISKS, HIGHER COSTS
48% of AI code has security vulnerabilities. 40% more secrets exposure. AI PRs get 60% more security review comments.
THE PRODUCTION GAP: TWO DIFFERENT QUESTIONS
WHAT BENCHMARKS MEASURE
Can an agent fix a well-defined GitHub issue with clear acceptance criteria?
WHAT PRODUCTION REQUIRES
Do experienced developers go faster on their own complex codebases with AI than without?
≠
Controlled, scoped problems
Clear inputs and outputs
Isolated evaluation
Measures capability
Ambiguous, evolving requirements
Deep system context
Real-world integration & constraints
Measures productivity
THE GAP IS STRUCTURAL, NOT TEMPORARY
AI excels at pattern completion. Production engineering demands system-level reasoning, judgment, and accountability.
WHY THE LAST MILE WON'T CLOSE (YET)
COMPLEXITY ISN'T BOILERPLATE
AI shines on low-complexity tasks. Slowdowns concentrate in critical-path engineering and deep logic.
CONTEXT WINDOWS ≠ UNDERSTANDING
Bigger context helps. It doesn't create the architectural judgment and trade-off reasoning production needs.
RISK CAN'T BE AUTOCOMPLETE
Security, reliability, and compliance require human accountability and domain expertise.
HUMANS STAY IN THE LOOP
Reviewing, correcting, and integrating AI suggestions adds overhead that benchmarks don't capture.
HIDDEN COSTS COMPOUND
Rework, review load, knowledge erosion, and tech debt borrow from future stability.
2026 REALITY AT-A-GLANCE
84% Developers use or plan to use AI tools
51% Use AI coding tools daily
29% Trust AI outputs on complex tasks (down from 40%)
-19% Tasks take longer with AI (METR RCT)
93.9% Claude Mythos score on SWE-bench Verified (Apr 2026)
48% AI-generated code contains security vulnerabilities
40% More secrets exposure in AI-assisted projects
60% More security review comments on AI pull requests
THE CENTRAL QUESTION
Benchmarks show what models can do. Production reveals what organizations can deliver.
Can AI coding tools close the production gap over the next 3–5 years?
THE PRIZE
Not 100% automation—but sustainably better engineering outcomes.

## Opening

The original prediction's 70% ceiling framing has been overtaken by events on the benchmark side. Agentic reasoning models have broken through it decisively.

What hasn't been overtaken is the prediction's structural insight: that the productivity narrative would have to shift from "AI replaces coding work" to "AI accelerates certain tasks while humans handle complexity."

The 2026 evidence confirms exactly that shift, with a precision the original piece did not anticipate. The ceiling isn't at 70% on benchmarks. It's at a different layer entirely, one the CASE tool analogy I originally drew from direct experience identifies with uncomfortable accuracy: the gap between what the technology produces and what production systems actually require.

## Intro

The AI coding productivity debate has produced the cleanest possible natural experiment in 2026: the same period that delivered 93.9% SWE-bench performance also delivered the METR finding that experienced developers on real repositories go 19% slower with AI tools than without them.

Those numbers are not in contradiction. They measure different things, benchmark accuracy on well-defined isolated tasks versus end-to-end productivity on complex real-world codebases, and their divergence is more analytically informative than either number alone.

The CASE tool analogy I drew from direct experience with prior generations of code generation tooling has proven precisely predictive: the technology raises the floor for code generation volume while the ceiling for production-quality integration remains stubbornly human-dependent.

What 2026 adds to that framing is the behavioral dimension the METR follow-up revealed: developers who won't work without AI tools, even in controlled experiments, are not demonstrating that AI makes them faster.

They are demonstrating that the tools have become load-bearing infrastructure regardless of whether the productivity case has been made.

That distinction, adoption as dependency versus adoption as demonstrated value, is the analytical lens the 2027-2030 window requires.

## Background

84% of developers now use or plan to use AI coding tools in 2026, up from 76% in 2024, with 51% using them daily. Yet only 29% of professional developers trust AI outputs to be accurate on complex tasks. This is down from 40% in 2024.

That simultaneous rise in adoption and decline in trust is the defining statistical signature of the 2026 AI coding landscape. The benchmark picture has moved decisively in the Counterpoint's direction: when SWE-bench was first released in late 2023, frontier models solved around 1-2% of tasks. Scaffolded agents climbed to 12-20% through 2024. By mid-2024, top-performing systems were cracking 50%.

By April 2026, Claude Mythos results on SWE-bench hit 93.9%, with OpenAI's o3 and GPT-5 competitive on coding benchmarks including HumanEval variants. The gap between those benchmark numbers and the production trust numbers is the story the 2027-2030 window will resolve or fail to.

METR's February 2026 follow-up to their July 2025 study added a revealing data point: they abandoned their second controlled experiment because developers refused to participate in the control condition without AI, creating selection bias that made the study unrunnable.

That behavioral signal, developers so dependent on AI tools they won't work without them, is itself analytically important context for interpreting productivity data.

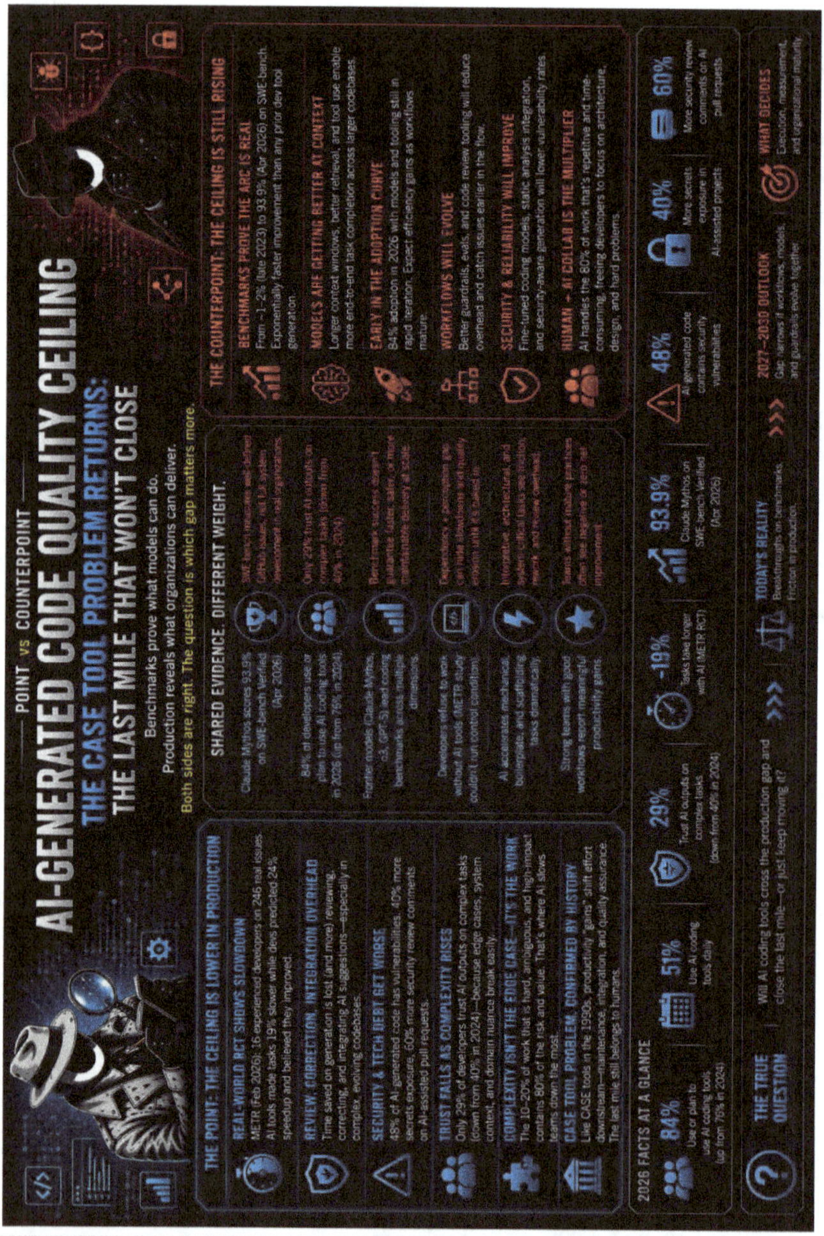
POINT vs COUNTERPOINT
AI-GENERATED CODE QUALITY CEILING
THE CASE TOOL PROBLEM RETURNS:
THE LAST MILE THAT WON'T CLOSE
Benchmarks prove what models can do.
Production reveals what organizations can deliver.
Both sides are right. The question is which gap matters more.
THE POINT: THE CEILING IS LOWER IN PRODUCTION
REAL-WORLD RCT SHOWS SLOWDOWN
METR (Feb 2026): 16 experienced developers on 246 real issues. AI tools made tasks 19% slower while devs predicted 24% speedup and believed they improved.
REVIEW, CORRECTION, INTEGRATION OVERHEAD
Time saved on generation is lost (and more) reviewing, correcting, and integrating AI suggestions—especially in complex, evolving codebases.
SECURITY & TECH DEBT GET WORSE
48% of AI-generated code has vulnerabilities. 40% more secrets exposure, 60% more security review comments on AI-assisted pull requests.
TRUST FALLS AS COMPLEXITY RISES
Only 29% of developers trust AI outputs on complex tasks (down from 40% in 2024)—because edge cases, system context, and domain nuance break easily.
COMPLEXITY ISN'T THE EDGE CASE—IT'S THE WORK
The 10–20% of work that is hard, ambiguous, and high-impact contains 80% of the risk and value. That's where AI slows teams down the most.
CASE TOOL PROBLEM, CONFIRMED BY HISTORY
Like CASE tools in the 1990s, productivity "gains" shift effort downstream—maintenance, integration, and quality assurance. The last mile still belongs to humans.
SHARED EVIDENCE. DIFFERENT WEIGHT.
Claude Mythos scores 93.9% on SWE-bench Verified (Apr 2026).
84% of developers use or plan to use AI coding tools in 2026 (up from 76% in 2024).
Only 29% trust AI outputs on complex tasks (down from 40% in 2024).
Frontier models (Claude Mythos, o3, GPT-5) lead coding benchmarks across multiple dimensions.
Benchmark success doesn't guarantee faster, safer, or more maintainable delivery at scale.
Developers refuse to work without AI tools (METR study couldn't run control condition).
AI accelerates mechanical, boilerplate, and scaffolding tasks dramatically.
Interpretive, architectural, and system-critical tasks see friction, rework, and review overhead.
Strong teams with good workflows report meaningful productivity gains.
Teams without mature practices often see negative or zero net improvement.
THE COUNTERPOINT: THE CEILING IS STILL RISING
BENCHMARKS PROVE THE ARC IS REAL
From ~1–2% (late 2023) to 93.9% (Apr 2026) on SWE-bench. Exponentially faster improvement than any prior dev tool generation.
MODELS ARE GETTING BETTER AT CONTEXT
Longer context windows, better retrieval, and tool use enable more end-to-end task completion across larger codebases.
EARLY IN THE ADOPTION CURVE
84% adoption in 2026 with models and tooling still in rapid iteration. Expect efficiency gains as workflows mature.
WORKFLOWS WILL EVOLVE
Better guardrails, evals, and code review tooling will reduce overhead and catch issues earlier in the flow.
SECURITY & RELIABILITY WILL IMPROVE
Fine-tuned coding models, static analysis integration, and security-aware generation will lower vulnerability rates.
HUMAN + AI COLLAB IS THE MULTIPLIER
AI handles the 80% of work that's repetitive and time-consuming, freeing developers to focus on architecture, design, and hard problems.
2026 FACTS AT A GLANCE
84% Use or plan to use AI coding tools (up from 76% in 2024)
51% Use AI coding tools daily
29% Trust AI outputs on complex tasks (down from 40% in 2024)
-19% Tasks take longer with AI (METR RCT)
93.9% Claude Mythos on SWE-bench Verified (Apr 2026)
48% AI-generated code contains security vulnerabilities
40% More secrets exposure in AI-assisted projects
60% More security review comments on AI pull requests
THE TRUE QUESTION
Will AI coding tools cross the production gap and close the last mile—or just keep moving it?
TODAY'S REALITY
Breakthroughs on benchmarks. Friction in production.
2027–2030 OUTLOOK
Gap narrows if workflows, models, and guardrails evolve together.
WHAT DECIDES
Execution, measurement, and organizational maturity.

## THE POINT: The Production Gap Is Structural. Thirty Years of History Agrees

The most important study in the 2026 AI coding research record, is not a vendor benchmark.

It's the METR controlled study. A randomized controlled trial using 16 experienced open-source developers working on 246 real issues from their own repositories, with tasks randomly assigned to allow or disallow AI assistance.

The result: AI tools caused tasks to take 19% longer to complete, while developers predicted a 24% speedup and even after AI use believed it had improved their performance. METR themselves used state-of-the-art AI tools: Cursor Pro with Claude 3.5/3.7 Sonnet, not legacy models.

The researchers attributed the slowdown to the overhead of reviewing, correcting, and integrating AI suggestions. DORA 2025 research confirms the broader pattern: while generative assistants aid mechanical tasks, they impede interpretive ones, and studies show AI coding agents can make developers 19% slower even when those developers believe they are 20% faster.

Only teams with solid established workflows and practices see improvements, and friction appears when AI-generated code increases review load, creates rework, or exposes gaps in organizational adjustment.

The METR study does not contradict the SWE-bench numbers, it contextualizes them. SWE-bench measures whether an agent can fix a well-defined GitHub issue with clear acceptance criteria.

The METR study measures whether experienced developers working on their own complex codebases go faster with AI than without. Those are different questions, and their divergent answers reveal the production gap more precisely than the 70% ceiling framing could.

The security and technical debt dimension of the production gap hasn't improved materially since the original piece was written, and in some respects has worsened as adoption has accelerated. 48% of AI-generated code contains security vulnerabilities, creating major risks in production systems.

Projects using AI assistants show a 40% increase in secrets exposure, mostly hard-coded credentials and API keys generated in scaffolding code.

Pull requests containing AI code require 60% more reviewer comments on security issues, meaning apparent time savings in code generation are offset by increased review burden representing real but hidden cost.

The CASE tool analogy from my personal experience hasn't been superseded by three decades of tooling improvement. It has been confirmed by a new generation of evidence. The productivity gained at the code generation layer is being borrowed from the stability of the production layer, the security of the deployment layer, and the maintainability of the codebase over time.

The slowdowns are concentrated in critical-path engineering. The deep logic that defines system reliability, while the average speedup reported in broad surveys is driven by massive volume of low-complexity boilerplate tasks where AI is genuinely a force multiplier.

That distribution is not a temporary imbalance awaiting resolution. It's a structural feature of what probabilistic pattern completion can and can't do with complex system-level reasoning.

### The Point Roundup

The original prediction's 70% ceiling framing understated the benchmark progress and precisely identified the production constraint. The 2026 update requires a more nuanced formulation: AI coding tools have broken through arbitrary accuracy thresholds on well-defined benchmark tasks while the production gap, between what the tools generate and what production systems require, remains structurally intact.

The 15-25% realistic productivity gain for experienced developers on real-world projects has been confirmed by independent research, not the 50%+ gains vendor benchmarks implied.

The CASE tool parallel holds: the technology produces impressive output volumes that require substantial human expertise to validate, secure, and integrate into systems that must be maintained and trusted over years. The 2027-2030 window will test whether agentic architectures can address the interpretive, contextual, and security dimensions of the production gap or whether the gap reveals itself as the fundamental boundary between probabilistic code generation and genuine software engineering judgment.

## THE COUNTERPOINT: The Three Conditions Were Met. Ahead of Schedule

The Counterpoint's three required conditions have all been met, and materialized faster than even the Counterpoint's own 2027-2028 timeline anticipated.

Reasoning models at competitive pricing are now available. Codebase-aware context systems handle entire repositories.

Agentic workflows that iteratively test and debug code have moved from demo to daily reality. Claude Mythos reached 93.9% on SWE-bench Verified, Claude 4 hit 77.2%, and agentic workflows navigating 40,000-line codebases and producing patches that pass existing test suites moved from demonstration to production workflow in 2025-2026.

Claude Opus 4.7 leads on SWE-bench Pro at 64.3% versus GPT-5.4's 57.7%, with multi-agent coordination for hours-long workflows and 14% improvement in multi-step agentic reasoning.

The Counterpoint's most revealing behavioral data point is the METR follow-up: developers in the second study refused to participate in the control condition without AI, forcing METR to abandon the study due to selection bias.

That behavioral signal, developers so dependent on AI tools they won't work without them, is not productivity measurement, but it's strong evidence that AI coding tools have crossed a threshold from "experimental" to "operationally necessary" for a growing population of practitioners.

The 84% adoption rate alongside 51% daily use represents a market adoption velocity that exceeds any prior developer tooling category.

The Counterpoint's "sustained improvement trajectory produces reliable production code" scenario is being built by exactly the organizations the prediction identified: GitHub Copilot's enterprise penetration, Amazon Q Developer's AWS integration, and Claude Code's growing adoption by professional developers represent the commercial infrastructure of the Counterpoint's thesis.

The METR study's own authors noted that "near-future AI could perform differently" the study used early-2025 models; Claude Code with Opus 4.5 launched in December 2025 and Spotify's engineering leadership described it as a breakthrough.

The tools have improved substantially since METR ran its study, and the agentic architectures that METR did not test (hours-long multi-agent workflows rather than Cursor's inline suggestions) represent the frontier where the Counterpoint's production reliability case is being built.

The gap between the 19% slowdown on complex legacy codebases and the productivity gains on greenfield development, boilerplate, and unfamiliar codebases is the market segmentation the Counterpoint requires: not universal productivity improvement, but reliable productivity improvement for the use cases that constitute the majority of commercial software development volume.

## What Needs to Change

### (...for the Counterpoint to Fully Prevail)

For the Counterpoint's "AI achieves reliable production code quality" scenario to dominate through 2030, two conditions need to develop that are currently moving at different rates.

First, the agentic architecture improvements visible in the 2026 benchmark record need to translate into the production reliability improvements that the METR study found absent in

2026 real-world testing. The multi-agent coordination for hours-long workflows represents a qualitatively different capability than inline autocomplete, if it begins closing the interpretive gap the METR study documented, the Counterpoint's core thesis arrives on a slightly delayed but structurally intact trajectory.

Second, the security vulnerability rate needs to decline from its current 48% level as AI tools develop better security awareness, secrets detection, and production-context reasoning. The 40% increase in secrets exposure is the most tractable of the production gap dimensions. It's a training and tooling problem, not a fundamental architectural limitation. If the next generation of agentic coding tools ship with security-first defaults and automated secrets detection, a major source of the review overhead that drove the METR slowdown is addressed directly.

The Counterpoint's 2030 scenario is not that AI replaces software engineers. It's that AI makes each software engineer materially more productive on real-world complex tasks, not just benchmarks. The METR study measured 2025 tools on 2025 codebases. Whether 2028 tools produce a different result is the defining empirical question.

## Likelihood Counterpoint Prevails: 45%

The Counterpoint likelihood increases substantially from 35% to 45%, driven by evidence that the technical trajectory the Counterpoint predicted has largely materialized, and materialized faster than anticipated. The three conditions have all been met: reasoning models, codebase-aware context, and agentic iteration.

The benchmark record: 93.9% on SWE-bench Verified, SWE-bench Pro leadership, multi-agent hours-long workflows, represents exactly the capability trajectory the Counterpoint required.

The 45% does not fully vindicate the Counterpoint because the METR production study and declining developer trust data confirm that benchmark improvement and production utility remain on diverging trajectories. The Counterpoint's three technical conditions have been met while its implied outcome, productivity gains exceeding initial projections in real-world complex tasks, hasn't.

The 2027-2030 question is whether agentic architectures close the interpretive gap the METR study found absent in 2026 real-world testing.

If they do, the Counterpoint's core thesis arrives two years late but substantially intact. If the production gap proves as structurally resistant to agentic approaches as it was to autocomplete, the CASE tool parallel will have been the most accurate framing in the entire debate.

The 45% sits at the series' highest Counterpoint likelihood alongside Gen Z Analog Backlash, reflecting a prediction where both sides have produced genuine evidence, the technical trajectory has moved substantially in the Counterpoint's direction, and the production validation that would decisively resolve the debate is still accumulating.

# Solid-State Battery Commercialization Delay

## Production Hell, Physics, and the Race Against a Declining Baseline

*Lithium-ion batteries are forecast to reach $99/kWh in late 2026. Solid-state batteries currently cost $400-800/kWh.*

*The optimistic peer-reviewed projection has solid-state reaching $140/kWh by 2028, at which point lithium-ion may have declined to $80-90/kWh.*

*QuantumScape inaugurated its Eagle Line pilot production facility on February 4, 2026, demonstrating scalable production using the Cobra separator process, a blueprint for manufacturing, not yet volume production.*

*That is the solid-state battery story in 2026: a technology improving on an absolute basis, racing against a target that keeps moving.*

OVERVIEW
SOLID-STATE BATTERY COMMERCIALIZATION DELAY
PRODUCTION HELL, PHYSICS, AND THE RACE AGAINST A DECLINING BASELINE
Solid-state batteries are real and improving.
The barrier isn't science — it's economics, scale, and time.
THE EVIDENCE IS REAL
4–8× MORE EXPENSIVE TODAY
Solid-state costs $400–800/kWh vs. ~$115/kWh for lithium-ion (2024).
PILOT PRODUCTION IS LIVE
QuantumScape opened its Eagle Line pilot facility on Feb 4, 2026.
Not volume production — yet.
DURABILITY IS VALIDATED
>95% capacity retention after 1,000+ cycles (~500,000 km) proven in real-world testing.
INDUSTRY TIMELINES CONVERGE
Major players target commercialization in the 2027–2030 window.
CHINA IS MOVING FAST
Roadmap targets initial volumes in 2027–2028; full scale by 2030.
CATHODE
High-energy materials
SOLID ELECTROLYTE
Enables safety and higher energy density.
ANODE
Lithium metal for greater capacity.
WHERE WE STAND: MILESTONE TRACKER (MID-2026)
LAB VALIDATION
Performance and durability proven in labs.
PILOT PRODUCTION
Facilities and processes are live for sampling and testing.
CUSTOMER SAMPLING
Cells in customer hands for evaluation and technology demos.
PREMIUM APPLICATIONS
Early use in high-end or niche vehicles.
MASS-MARKET SCALE
High-volume manufacturing not proven yet.
COST PARITY WITH LI-ION
Total cost parity remains out of reach.
ACHIEVED
IN PROGRESS
NOT YET
WHY IT MATTERS (AND HURTS)
BETTER BATTERIES UNLOCK EVS
Higher range, faster charging, and improved safety drive mass-market adoption.
MANUFACTURING SCALE REMAINS THE GATEKEEPER
Yields, materials, and equipment must improve dramatically.
THE MOVING TARGET PROBLEM
Lithium-ion costs and energy density keep improving, raising the bar for parity.
PREMIUM MARKETS MAY COME FIRST
Niche and high-end applications are the likely beachhead.
NATIONAL STRATEGIES ARE ON THE LINE
Industrial leadership depends on who scales first and cheapest.
THE QUESTION
Will solid-state batteries close the cost and manufacturing gap fast enough to matter before lithium-ion moves the goalposts again?

## Opening

The original prediction has been validated with uncomfortable precision, and the 2027-2030 update reveals a more structurally complex situation than even the pessimistic scenario anticipated.

The mass-market 2026-2027 deployment targets have been missed as predicted. The semi-solid pivot the original piece identified as the industry's strategic retreat has materialized across virtually every major player.

Manufacturing costs remain 4-8x higher than lithium-ion alternatives. And the most important development the original analysis did not fully model has arrived with mathematical force: lithium-ion is not standing still.

The 2027-2030 window is where that race, two declining cost curves moving simultaneously, with the gap's direction determining the entire commercial thesis, gets its definitive answer.

## Intro

The original prediction's core thesis, that mass-market deployment would miss its 2026-2027 targets, that manufacturing costs would remain prohibitively high, and that the industry would pivot to semi-solid intermediate technologies, has been confirmed by the 2026 evidence record.

What the update reveals is a more nuanced and strategically significant picture: the technology is progressing, specific applications are demonstrating commercial viability, and the 2027-2030 window contains credible milestones.

The catch is the moving target: lithium-ion's cost decline is accelerating faster than the baseline scenarios anticipated, raising the bar for what "cost parity" means even as solid-state costs decline.

That dynamic is the analytical lens the 2030 window requires.

## Background

Solid-state batteries currently cost $400-800 per kWh compared to approximately $115/kWh for lithium-ion in 2024, making them 4-8x more expensive, with the cost difference stemming from expensive ceramic materials, complex manufacturing processes, low production yields, and limited manufacturing scale.

The industry's stated commercialization timeline has converged on a 2027-2030 window.

Toyota holds more than 1,300 solid-state battery patents globally and targets commercial production by 2027-2028, with sulfide-based cells expected to deliver over 620 miles of range and charge from 10-80% in under 10 minutes. Samsung SDI targets 2027 for mass production of all-solid-state batteries.

Nissan's pilot production line in Yokohama targets commercial EVs by 2029. Solid Power plans full mass production by 2030. Hyundai targets partial mass production by 2027 and full scale by 2030. QuantumScape inaugurated its Eagle Line pilot production facility on February 4, 2026, a highly automated suite of equipment incorporating the Cobra separator process, forming the blueprint for scalable production.

The Eagle Line will produce cells to support customer sampling, testing, and technology demonstrations; it’s not yet volume manufacturing. PowerCo's validation testing of QuantumScape cells demonstrated more than 95% energy capacity retention after over 1,000 charging cycles, equivalent to approximately 500,000 kilometers of driving, validating the durability case.

Only polymer solid-state batteries are currently in commercial production; oxide and sulfide types remain in pilot phases with commercialization expected post-2026.

China is leading in industrialization with a roadmap targeting 2027-2028 for initial volumes and 2030 for full scale.

POINT VS COUNTERPOINT
SOLID-STATE BATTERY COMMERCIALIZATION DELAY
PRODUCTION HELL, PHYSICS, AND THE RACE AGAINST A DECLINING BASELINE
THE POINT: THE DELAY IS REAL AND STRUCTURAL
TECHNICAL & MANUFACTURING BARRIERS PERSIST
Dendrites, interfacial instability, complex processing, and low yields keep costs high.
COST GAP REMAINS LARGE
Solid-state costs $400–800/kWh vs. ~$115/kWh for lithium-ion (2024); 4–8x more expensive.
SEMI-SOLID IS THE INDUSTRY RETREAT
Most companies prioritize semi-solid solutions before all-solid-state.
COMMERCIAL TIMELINES HAVE SLIPPED
Mass-market 2026–2027 targets missed; realistic window is late-2020s.
THE MOVING TARGET PROBLEM
Lithium-ion keeps getting cheaper and better, raising the bar for cost parity.
GLOBAL RACE IS REAL
China leads industrialization; roadmap targets initial volumes 2027–2028; full scale by 2030.
THE DEBATE: WHERE THE EVIDENCE LEADS (MID-2026)
ISSUE
THE POINT
THE COUNTERPOINT
Performance & Durability
Durability proven (>95% retention after 1,000+ cycles). Performance promising in labs.
Real-world scale-up and long-term field data are still limited.
Manufacturing & Scale
Pilot lines exist (e.g., QuantumScape Eagle Line). Complex processes and low yields remain major hurdles.
Scaling is progressing. Equipment, processes, and yields will improve with volume.
Cost Economics
$400–800/kWh today vs. ~$115/kWh for lithium-ion (2024). Cost parity pushed to early 2030s in pessimistic scenarios.
Costs are declining. Optimistic projections reach ~$140/kWh by 2028.
Timeline Outlook
Mass-market 2026–2027 targets missed. Realistic window: late 2020s (2027–2030).
Industry convergence on 2027–2030. First commercial applications likely sooner.
Market Path
Premium and niche applications first; mass-market depends on cost parity and scale.
High-end EVs and select uses reach market early, driving learning and scale.
Competitive Landscape
Lithium-ion continues improving (600–800 Wh/kg; $32–$54/kWh forecast by 2030). Moving target.
Solid-state offers step-change benefits that can justify higher cost in early markets.
THE COUNTERPOINT: PROGRESS IS REAL AND ACCELERATING
TECHNOLOGY IS IMPROVING FAST
Higher energy density, fast charging, and safety advantages proven in labs and pilots.
REAL-WORLD VALIDATION IS STRONG
>95% capacity retention after 1,000+ cycles (~500,000 km) validated.
PILOT INFRASTRUCTURE IS SCALING
QuantumScape Eagle Line opened Feb 4, 2026; blueprint for scale and customer sampling.
2027–2030 PATH IS CREDIBLE
Toyota, Samsung SDI, Hyundai, Nissan, Solid Power, and others target this window.
CHINA IS INVESTING AGGRESSIVELY
Roadmap targets initial volumes 2027–2028; full scale by 2030.
EARLY MARKETS WILL CREATE VALUE
Premium EVs and niche applications can justify early adoption.
KEY FACTORS THAT WILL DECIDE THE OUTCOME (2027–2030)
Cost Parity Timing
Whether solid-state costs fall faster than lithium-ion declines.
Scale-Up Success
Ability to move from pilot lines to 100+ GWh production.
Material & Process Breakthroughs
Dendrite control, interfaces, and lower-cost materials.
Market Acceptance
OEM commitments and real-world performance in demanding use cases.
Lithium-Ion Trajectory
Energy density and cost improvements continue to raise the bar.
Supply Chain & Resources
Access to materials and equipment at scale without bottlenecks.
THE QUESTION
Will solid-state batteries close the cost and manufacturing gap fast enough to matter before lithium-ion moves the goalposts again?

 

## THE POINT: Semi-Solid Is the Tell: When the Industry Retreats to the Achievable

The original prediction's manufacturing cost argument has been confirmed and sharpened by the 2026 evidence, but the most important update is the moving target dynamic that compounds the cost challenge on a timeline the original piece did not fully model.

A peer-reviewed ScienceDirect analysis projects solid-state battery costs at $140/kWh in the optimistic scenario by 2028, $286/kWh in the moderate scenario by 2030, and $175/kWh only by 2032-2033 in the pessimistic scenario.

Those projections were calibrated against a lithium-ion baseline of approximately $115/kWh in 2024. Lithium-ion costs are forecast to reach $99/kWh by late 2026 and potentially $32-$54/kWh by 2030 as energy density improves to 600-800 Wh/kg, meaning that even the optimistic solid-state trajectory reaches nominal parity with a competing technology that will itself have continued declining for four more years.

The finish line is not stationary. It's being moved forward by a technology with two decades of manufacturing optimization behind it.

The semi-solid pivot the original piece identified as the industry's strategic retreat has become the dominant commercial reality: commercial deployment of true all-solid-state batteries remains concentrated in polymer types, while the oxide and sulfide variants that offer the most compelling performance advantages remain in pilot phases.

The production infrastructure being built in 2026 represents genuine progress on the manufacturing challenge, but the gap between pilot production and the volume manufacturing required for mass-market economics remains wide enough that the original prediction's late-2020s realistic deployment timeline looks accurate rather than pessimistic.

QuantumScape's Eagle Line, inaugurated February 2026 with automotive OEM customers, technology partners, and government officials in attendance, establishes the Cobra separator process as the baseline for production and forms the blueprint for manufacturing scale-up.

The company is working closely with major global automaker customers to co-develop industrialization strategies; as of Q1 2026, four of the top ten global automotive OEMs are actively engaged, with two joint development agreements in place.

Murata Manufacturing and Corning Incorporated have invested in ceramic separator production systems based on the Cobra process, the ecosystem participation that validates the industrialization path.

These are meaningful milestones on the path from demonstration to production. They are not the gigawatt-hour-scale volume manufacturing infrastructure that mass-market economics requires.

Toyota's broader market adoption remains realistically 2028-2030 at limited initial volumes and high costs, and most industry experts caution that widespread availability requires 5-7 years of production validation and significant manufacturing scale-up.

That 5-7 year production validation cycle is the same timeline lithium-ion itself required during its own commercialization, simultaneously reassuring because it confirms the technology will eventually succeed, and sobering because it means the 2026-2027 mass-market targets were never realistic.

### The Point Roundup

The solid-state battery story through 2030 is not a failure narrative, it's an industrial timeline narrative, and industrial timelines are measured in decades.

The original prediction's core thesis has been validated: mass-market deployment missed its 2026-2027 targets, manufacturing costs remain 4-8x higher than lithium-ion, and the industry has pivoted to semi-solid intermediate technologies delivering real but incremental improvements rather than the revolutionary capability jump that headlines promised.

The 2027-2030 window will see limited initial production in premium automotive applications at volumes and costs that prove the technology works without yet proving it works at the economics mass markets require.

The definitive test of the solid-state thesis comes in the 2028-2030 period: if the optimistic cost curve reaches $140/kWh while lithium-ion declines to $80/kWh, the premium application market is real and durable.

If lithium-ion reaches $54/kWh by 2030, the cost parity case for solid-state in mass-market vehicles becomes structurally harder than any 2024 forecast anticipated.

## THE COUNTERPOINT: 2028 Cost Parity Is an Engineering Roadmap, Not a Forecast

The Counterpoint's premium market thesis has been validated in exactly the segment the original argument identified, and the durability data from real-world validation is stronger than the original optimistic framing anticipated. PowerCo's testing of QuantumScape cells demonstrated more than 95% energy capacity retention after over 1,000 charging cycles, equivalent to 500,000 kilometers of driving, a durability standard exceeding the requirements of virtually every mass-market automotive application.

That is not a benchmark number. It's the validation data that automotive OEM procurement decisions depend on.

SAIC has announced plans to mass-produce 400 Wh/kg all-solid-state batteries beginning in 2026 with volume delivery in 2027, an announced target, not a confirmed outcome, but backed by China's state-directed manufacturing capability. China is leveraging its lithium-ion manufacturing dominance to accelerate solid-state production, and its 2027 volume delivery timeline is backed by the same institutional infrastructure that made Chinese lithium-ion the global cost benchmark.

The China dimension is the variable the original analysis most significantly underweighted: a country that built world-leading lithium-ion manufacturing through state-directed industrial policy and aggressive cost reduction targets is applying that same capability to solid-state, on a timeline that could produce the first commercially meaningful sulfide solid-state volumes before any other national manufacturing program.

The semi-solid pivot that the original prediction characterized as strategic retreat is more accurately described as strategic sequencing, the same graduated development path that every successful battery chemistry has followed from laboratory to mass production.

The peer-reviewed ScienceDirect analysis projects solid-state costs reaching $140/kWh in the optimistic scenario by 2028, comparable to current lithium-ion costs, representing a cost reduction trajectory from $400-800/kWh in 2026 that, if achieved, validates the manufacturing learning curve thesis.

The premium market viability case is structurally sound even at costs above lithium-ion parity: Toyota's solid-state cells targeting 1,000 km range and 10-minute fast charging represent capability improvements that justify premium pricing in a market where range and charging friction have constrained EV adoption more than any other single factor.

A vehicle offering 1,000 km range and 10-minute charging is not competing on cost per kWh. It's competing on the elimination of range anxiety, a consumer pain point that has structural commercial value independent of the cost parity question.

That value proposition creates a viable commercial path even if lithium-ion's decline is faster than solid-state cost projections anticipated.

WHAT NEEDS TO CHANGE
— (FOR THE COUNTERPOINT TO PREVAIL) —
Counterpoint wins only if these conditions materialize.
WHY THE POINT MUST BE WRONG ABOUT
COSTS FALL FASTER
Manufacturing scale and learning drive solid-state costs to parity with lithium-ion by ~2028.
SCALE IS PROVEN
High-yield, repeatable processes reach 100+ GWh production with stable quality by 2028–2029.
SUPPLY CHAIN SECURED
Materials, equipment, and talent scale without critical bottlenecks or geopolitical disruptions.
LITHIUM-ION IMPROVEMENT PLATEAUS
Lithium-ion energy density and cost improvements slow, preserving solid-state's performance and cost advantage.
DEMAND PULLS THROUGH
OEM commitments and consumer pull for range, safety, and charging speed drive early adoption at scale.
DECISION FRAMEWORK: ACTIONS TO TAKE NOW TO TIP THE OUTCOME
1 DRIVE COST TO PARITY
Prioritize materials, design, and process innovations that cut cost per kWh.
· Aggressive cost modeling
· Material substitution
· Higher areal loading
· Design for manufacturability
2 INDUSTRIALIZE AT SPEED
Move from pilot to high-volume manufacturing with high yields and low defect rates.
· Scale pilot lines
· Standardize processes
· Automate quality control
· Build supplier ecosystem
3 DE-RISK & VALIDATE
Prove durability, safety, and performance in real-world conditions and at scale.
· Expand field data
· Long-duration testing
· Safety certification
· Transparent reporting
4 SECURE SUPPLY CHAINS
Lock in critical materials, equipment, and talent for scale.
· Long-term supply agreements
· Domestic & allied sourcing
· Invest in equipment
· Develop workforce
5 STRENGTHEN POLICY & DEMAND
Align policy, standards, and market pull to accelerate commercial adoption.
· Procurement & incentives
· Clear technical standards
· Support early markets
· Public–private partnerships
WHY THE COUNTERPOINT MUST BE RIGHT ABOUT
TECHNOLOGY MATURING FAST
Higher energy density, fast charging, and safety advantages are proven and increasing.
COSTS DROP TO SUSTAINABLE LEVELS
Scale and learning drive costs down to parity (~$115/kWh) by ~2028.
SCALE & INFRASTRUCTURE DELIVERED
Manufacturing at 100+ GWh with high yields and robust supply chains by 2028–2029.
MARKET ADOPTS EARLY
Premium and select mass-market segments adopt first, expanding learning and volumes.
COMPETITIVE LANDSCAPE SHIFTS
Moving target of lithium-ion improvement is outpaced by solid-state economics and performance.
CURRENT STATE (MID-2026)
Lab validation: proven (>95% retention after 1,000+ cycles)
Pilot production: QuantumScape Eagle Line opened Feb 4, 2026
Industry timelines: converging on 2027–2030 window
Costs: ~$400–800/kWh vs. ~$115/kWh for lithium-ion (2024)
China roadmap: initial volumes 2027–2028; full scale by 2030
LIKELIHOOD COUNTERPOINT PREVAILS
30%
▲ +5 pts
FROM 25%
(Previous)
0%
POINT FAVORED
30%
100%
COUNTERPOINT FAVORED
MOST CRITICAL VARIABLE
TOTAL COST OF OWNERSHIP (TCO) AT SCALE
Across hardware, power, connectivity, operations, and management, TCO will determine whether solid-state delivers sustainable value vs. lithium-ion alternatives.
THE QUESTION
Will solid-state batteries close the cost and manufacturing gap fast enough to matter before lithium-ion moves the goalposts again?

## What Needs to Change

### (...for the Counterpoint to Prevail)

Three conditions from the original Counterpoint have been partially met and require completion by 2027-2028 for the thesis to fully materialize.

First, the volume production threshold: QuantumScape's Eagle Line demonstration of scalable production using the Cobra process must translate into the cumulative GWh-scale production that automotive qualification requires, the blueprint exists, the execution remains the test.

Second, the cost trajectory must follow the optimistic rather than moderate scenario: the difference between $140/kWh and $286/kWh by 2028 is the difference between a viable premium market and a niche demonstration market, and that difference is determined by manufacturing yield rates and process optimization not yet proven at scale.

Third, and most critically, the moving target problem must resolve in solid-state's favor, meaning lithium-ion's cost decline must slow enough that solid-state's absolute cost improvements produce a meaningful relative advantage.

If lithium-ion reaches $32-$54/kWh by 2030 as some analysts project, the cost parity case for solid-state evaporates even if the technology achieves its optimistic manufacturing trajectory.

That scenario is plausible and sufficiently consequential that no honest assessment of the solid-state Counterpoint can ignore it.

## Likelihood Counterpoint Prevails: 30%

The Counterpoint likelihood increases modestly from 25% to 30%, reflecting genuine manufacturing infrastructure progress that the original 25% may have underweighted, while acknowledging that the moving target problem has made the Counterpoint's "meaningful mass-market deployment" threshold harder to meet even as the technology itself advances.

The scientific literature's optimistic projection ($140/kWh by 2028 from ScienceDirect peer-reviewed analysis) represents a genuine and achievable cost trajectory if manufacturing scale-up proceeds on the timelines that QuantumScape's Eagle Line, Toyota's Idemitsu sulfide supply chain, and Samsung SDI's pilot production suggest.

What that $140/kWh figure represents in 2028, against a lithium-ion baseline that may have reached $80-90/kWh, is a 55-75% premium, still above the threshold for mass-market economics but potentially viable for premium applications where performance advantages justify it.

The 30% reflects the narrowing gap between the optimistic solid-state trajectory and the continuing lithium-ion decline, a race that is genuinely closer than the original 25% implied, but not yet close enough to call the Counterpoint's meaningful mass-market deployment scenario likely.

The original prediction's 2027-2030 window for limited initial production at premium price points in automotive applications looks accurate.

Universal mass-market deployment, the claim the original prediction was disputing, remains beyond the 2030 window in all but the most optimistic scenarios.

## Satellite Internet Consolidation

### The Sovereignty Response Arrived. Interoperability Didn't. Both Are the Right Story.

*Starlink controls 97% of global satellite internet speed test samples and serves over 9 million subscribers.*

*Amazon Leo has ~300 satellites in orbit and has formally requested an FCC deadline extension after failing to approach the 1,618 required by July 30, 2026. The EU introduced the world's first supranational Space Act in June 2025, targeting all satellite operators including Starlink using GDPR-style long-arm jurisdiction.*

*The IRIS² sovereign constellation signed a 12-year concession contract in December 2025 with a €10.6 billion budget and a 2029 service target.*

*The original prediction was right about the regulatory response and right about the competitive timeline.*

*The specific mechanism, forced interoperability rather than parallel sovereign infrastructure, was the part it missed.*

OVERVIEW
SATELLITE INTERNET CONSOLIDATION
THE SOVEREIGNTY RESPONSE ARRIVED.
INTEROPERABILITY DIDN'T. BOTH ARE THE RIGHT STORY.
Starlink's dominance is real. Sovereignty concerns are real.
Governmental responses have materialized—just not as predicted.
THE EVIDENCE IS REAL
STARLINK DOMINANCE IS REAL
97% of global satellite internet speed test samples (Q3 2025). Over 9 million subscribers.
AMAZON LEO LAGS SCHEDULE
~300 satellites in orbit across 12 launches (mid-2026). FCC deadline: 1,618 by July 30, 2026. Amazon requested extension.
EU SPACE ACT IS LAW
Introduced June 2025. World's first supranational space law using GDPR-style long-arm jurisdiction on all operators.
IRIS² SOVEREIGN CONSTELLATION
12-year concession signed Dec 2025. €10.6B budget. ~290 satellites. Service target: 2029.
ITU TAKES UP THE ISSUE
WRC-27 Agenda Item 1.13 addresses NGEO satellite dominance at the international standard-setting level.
STARLINK
10,347+ ACTIVE SATELLITES
AMAZON LEO
~300 IN ORBIT
IRIS²
TARGET: 2029
KEY DEVELOPMENTS (THROUGH MID-2026)
AMAZON LEO EXTENSION
FCC approved Gen2 expansion of 4,500 additional satellites in Jan 2026. Long-term competitor despite near-term gap.
EU SPACE ACT
Enforcement powers start 2026–2027. Safety, environmental, and cybersecurity compliance required.
IRIS² PROGRESSES
Consortium contract signed Dec 2025. Norway & Iceland joined as first non-EU members in Apr 2026.
ITU WRC-27
Item 1.13 on NGEO systems—driven by market concentration. Standards process begins.
WHY IT MATTERS (AND HURTS)
SOVEREIGNTY CONCERNS ARE REAL
Dependence on one non-sovereign provider creates national security, resilience, and economic risks.
COMPETITION TIMELINE SLIPPED
Meaningful Starlink competition now moves to 2028–2030 at the earliest.
COMPLIANCE COSTS RISE
New regulatory burdens increase costs and complexity for all satellite operators.
SOVEREIGN OPTIONS TAKE TIME
IRIS² and national programs won't provide alternatives until 2029+.
GLOBAL STANDARDS BATTLE BEGINS
ITU process could shape long-term rules—but change will be slow and uncertain.
THE QUESTION
Will sovereignty-driven infrastructure and regulation rebalance the market in time—or will Starlink lock in a decade of dominance before meaningful competition arrives?

## Opening

The satellite internet consolidation prediction has been validated in its structural diagnosis, Starlink's dominance is real, the sovereignty concerns are real, and governmental responses have materialized, while the specific predicted mechanism of "forced interoperability standards by end 2026" hasn't arrived and does not appear on a near-term horizon.

What has arrived instead is a different but analytically interesting regulatory response: the EU Space Act creating safety and environmental compliance requirements across all satellite operators using GDPR-style extraterritorial jurisdiction, $IRIS^2$ proceeding with a funded sovereign alternative constellation, and ITU Agenda Item 1.13 at WRC-27 addressing NGEO satellite dominance at the international standard-setting level.

These represent meaningful governmental actions short of the interoperability mandates the original prediction specified.

The 2026 update requires honest engagement with what arrived and what didn't, and a recalibration of the 2027-2030 trajectory based on the evidence rather than the original framing.

## Intro

The satellite internet consolidation prediction requires separating what it got right from what it got wrong with more precision than a simple vindication or refutation allows, because both verdicts apply to different parts of the same analytical framework.

The structural diagnosis was correct: Starlink's 97% market dominance is confirmed, the sovereignty concerns are real and have produced governmental action, and the competitive timeline skepticism about Amazon was precisely validated by the FCC deadline extension request.

The mechanism was wrong: "forced interoperability standards by end 2026" describes a regulatory tool that hasn't materialized and shows no near-term path to materializing.

What materialized instead was sovereignty-driven parallel infrastructure, the EU Space Act, $IRIS^2$'s €10.6 billion concession contract, ITU engagement at WRC-27, which is the response that sovereignty logic actually produces when a single private US company controls critical communications infrastructure.

Governments don't force open access to the dominant system when they can fund a competing sovereign one instead.

That distinction between the predicted mechanism and the actual mechanism matters for the 2027-2030 trajectory: forced interoperability would have constrained Starlink's dominance from the top down; sovereign parallel infrastructure creates an alternative that may or may not achieve meaningful market presence by 2030 depending on whether IRIS² delivers on its 2029 service target and Amazon Leo reaches commercial viability after its deployment delays.

## Background

Ookla's 2025 Global Satellite Broadband Performance Report found that Starlink customers accounted for 97% of global speed test samples for satellite internet in Q3 2025, with traditional GEO providers Viasat and HughesNet accounting for 1.7% and 1.0% respectively.

This metric measures active users running speed tests, a strong proxy for consumer market share, though not identical to all satellite internet connections globally.

Starlink has grown to over 9 million subscribers, adding 4.6 million in 2025 alone, and operates 10,347+ active satellites, the largest constellation in orbit by a wide margin, authorized for eventual expansion to 42,000.

Amazon Leo (rebranded from Project Kuiper in November 2025 as the program transitioned from development to deployment) had approximately 300 production satellites in orbit as of mid-2026 across 12 successful launches using Atlas V, Falcon 9, and Ariane 6 vehicles, making it the third-largest constellation in orbit behind Starlink and OneWeb.

Amazon formally requested an FCC deadline extension in January 2026, acknowledging it wouldn't reach the 1,618 satellites (half the Gen1 constellation) required by July 30, 2026, due to launch vehicle availability challenges.

The FCC approved Amazon Leo's Gen2 expansion of 4,500 additional satellites in January 2026, signaling regulatory support for Amazon as a long-term Starlink competitor despite the near-term deployment gap.

On the sovereignty response side: the EU introduced its Space Act in June 2025, the IRIS² consortium signed a 12-year concession contract in December 2025 with a €10.6 billion budget and approximately 290 satellites targeting a 2029 service start, and Norway and Iceland joined IRIS² as the first non-EU members in April 2026.

The ITU's WRC-27 agenda includes Item 1.13, addressing NGEO satellite systems, partly driven by Starlink's dominant position, though ITU processes operate at the service category level and can't target individual companies.

POINT vs COUNTERPOINT
SATELLITE INTERNET CONSOLIDATION
THE SOVEREIGNTY RESPONSE ARRIVED. INTEROPERABILITY DIDN'T. BOTH ARE THE RIGHT STORY.
Starlink's dominance is real. Sovereignty concerns are real.
Governmental responses have materialized—just not as predicted.
THE POINT: STRUCTURAL DIAGNOSIS CONFIRMED
SOVEREIGNTY RESPONSE IS REAL
STARLINK DOMINANCE IS REAL
97% of global satellite internet speed test samples (Q3 2025). Over 9 million subscribers. 10,347+ active satellites.
COMPETITIVE TIMELINE SLIPPED
Amazon Leo has ~300 satellites in orbit. FCC deadline: 1,618 by July 30, 2026. Formal extension requested.
REGULATORY RESPONSE ARRIVED
EU Space Act (June 2025), IRIS² (12-year, €10.6B, 2029 target), and ITU WRC-27 Item 1.13 are real actions.
SOVEREIGNTY CONCERNS ARE REAL
Dependence on one non-sovereign provider creates national security, resilience, and economic risks.
STRUCTURAL CONCENTRATION PERSISTS
No meaningful market share erosion. Original prediction's core diagnosis remains validated.
THE DEBATE: WHERE THE EVIDENCE LEADS (MID-2026)
ISSUE
THE POINT
THE COUNTERPOINT
Market Concentration
Starlink: 97% of speed tests. Over 9M subscribers. Dominance remains overwhelming.
High, but not permanent. Competition will emerge with time and scale.
Amazon Leo Deployment
~300 satellites in orbit across 12 launches. FCC deadline miss forces timeline slip.
Still early innings. Production rate will accelerate.
Timeline to Competition
Meaningful competition now moves to 2028–2030 window at the earliest.
Broader availability still expected later in 2026.
Regulatory Response
EU Space Act, IRIS², and ITU Item 1.13 = sovereignty response is materializing.
Regulation alone won't ensure real market competition.
Interoperability Prospects
No forced interoperability standard by end 2026. No near-term interoperability mandates.
Interoperability remains the most efficient long-term path.
Strategic Direction
Sovereign infrastructure is the chosen path. It's 2029, not 2026.
Multiple paths can coexist. Market will decide.
THE COUNTERPOINT: LONG GAME REMAINS OPEN
EFFICIENCY & MARKET FORCES WILL PREVAIL
TECHNOLOGY LEADERSHIP IS SUSTAINABLE
Starlink's scale drives continuous innovation, cost reduction, and performance leadership that others must overcome.
COST & EFFICIENCY FAVOR CONSOLIDATION
Duplication is expensive. Interoperability ultimately reduces cost and improves consumer outcomes.
MARKET FORCES DRIVE OUTCOMES
Commercial demand for global coverage and seamless service will push industry toward interoperability.
REGULATION WON'T FRAGMENT THE MARKET
Safety, security, and spectrum rules will evolve, but won't create isolated sovereign silos.
MULTIPLE PLAYERS WILL EMERGE
Amazon Leo, OneWeb, and others will reach scale. Diversity improves resilience and competition.
KEY FACTORS THAT WILL DECIDE THE OUTCOME (2027–2030)
DEPLOYMENT VELOCITY
Can Amazon Leo and others close the satellite gap fast enough to matter?
ECONOMICS OF SCALE
Will cost curves enable viable competition or lock in Starlink's advantage?
REGULATORY TRAJECTORY
How EU, ITU, FCC, and other regulators shape competition, compliance, and access.
INTEROPERABILITY PRESSURE
Will commercial and policy pressures create voluntary interoperability standards?
SOVEREIGN INFRASTRUCTURE
Do sovereign constellations achieve viability and gain long-term traction?
SPECTRUM & ORBIT ACCESS
Will access constraints limit entrants or enable broader participation?
THE QUESTION
Will sovereign-driven infrastructure and regulation rebalance the market in time—
or will Starlink lock in a decade of dominance before meaningful competition arrives?

## THE POINT: Sovereign Infrastructure Is the Response, Not Interoperability. And It's 2029, Not 2026.

The original prediction's structural diagnosis has been confirmed by the 2026 evidence with unusual clarity, and the Counterpoint's specific scenario has been falsified on its own terms. Amazon Kuiper (now Amazon Leo) was supposed to begin meaningful competition by end 2026 per the Counterpoint's "broader availability scheduled for 2026" framing.

Amazon Leo has 300 satellites in orbit against an FCC deadline of 1,618 by July 30, 2026, and has formally requested a deadline extension.

That gap (300 deployed versus 1,618 required in the same calendar window) is not a close miss.

It's the specific deployment timeline failure the prediction described as the most likely outcome, documented by Amazon's own FCC filings. The satellite deployment math the original piece laid out, requiring 200 satellites per month at a pace exceeding Starlink's historical maximum deployment rate, has proven correct.

Amazon completed approximately one launch per month in its first year of production deployment, each carrying 27-40 satellites, producing roughly 300 total.

The Counterpoint's most credible specific condition ("Amazon must deploy 500+ satellites by mid-2026") hasn't been met. The 97% market share concentration hasn't declined to the sub-85% threshold the Counterpoint required. Starlink's share of satellite internet speed tests as of mid-2026 remains in the range the original prediction described as the likely 2026 outcome for Starlink's market position.

The regulatory response has arrived, in the form of the original prediction correctly identified as necessary but incorrectly specified as forced interoperability.

The EU's Space Act, proposed June 25, 2025, and described by analysts as using "GDPR-style long-arm jurisdiction" to force non-European companies including Starlink to comply with safety and environmental standards, represents exactly the type of governmental action the original prediction argued was historically inevitable when a single company controls critical communications infrastructure.

The ITU Agenda Item 1.13 at WRC-27 addresses NGEO satellite system dominance at the international standard-setting level, a process partly triggered by Starlink's position, though ITU can't target individual companies and operates at the service category level, meaning any new rules would apply to all NGEO operators and could paradoxically create compliance burdens that disadvantage smaller competitors more than Starlink.

The sovereignty concerns the original prediction identified as the primary driver of governmental response has produced exactly the response sovereignty logic predicts: not interoperability with the dominant system, but parallel sovereign infrastructure funded at scale.

IRIS²'s €10.6 billion budget with its 12-year concession contract, Norway and Iceland joining as non-EU members, and Eutelsat CEO Fallacher explicitly benchmarking IRIS² against Starlink on price and performance all confirm that the EU has committed to sovereign competitive infrastructure rather than regulatory constraint of the incumbent.

The prediction was right that governments can't accept single-vendor control.

It was wrong about the form of the response.

## The Point Roundup

The satellite internet consolidation story in 2026 is simultaneously a validation of the original prediction's structural diagnosis and a correction of its specific mechanism. Starlink's 97% market dominance is confirmed and hasn't materially narrowed. The competitive timelines the original prediction was skeptical of have proven even more delayed than its pessimistic framing suggested, Amazon Leo's FCC deadline extension request is the formal documentation of that delay.

The regulatory response has arrived.

The EU Space Act, IRIS² sovereign infrastructure at €10.6 billion, and ITU engagement at WRC-27, but in a sovereignty-infrastructure form rather than a forced-interoperability form.

The 2027-2030 story is whether IRIS²'s 2029 service target is achieved on schedule (Europe's megaprojects historically slip), whether Amazon Leo reaches commercial service in its five target countries before meaningful Starlink competition is possible, and whether the ITU WRC-27 framework produces NGEO regulations that create genuine competitive space or simply add compliance costs across the board.

# THE COUNTERPOINT: The FCC Extended the Deadline. Amazon Leo Is the Third-Largest Constellation in Orbit

The Counterpoint's "Starlink faces effective competition" scenario has been delayed rather than defeated, and the FCC's own behavior signals institutional commitment to that outcome.

Amazon Leo's request for a deployment deadline extension has been received by the FCC, which has historically been willing to work with licensees facing genuine technical challenges rather than cancel viable programs.

More importantly, the FCC approved Amazon Leo's Gen2 expansion of 4,500 additional satellites in January 2026, bringing the total planned constellation to 7,727, simultaneously with Amazon's deadline extension request. An agency that was indifferent to Amazon Leo as a Starlink competitor wouldn't have approved a Gen2 expansion while granting a Gen1 deadline extension.
The FCC's behavior reflects institutional preference for a competitive satellite internet market, and Amazon Leo's 300+ satellites across 12 launches in its first year of production deployment, using three different launch vehicles, represents genuine infrastructure progress even if it falls dramatically short of the FCC deadline.

Amazon has projected service availability in five countries (US, UK, France, Germany, Canada) in 2026, a beta waitlist is open, and the Gen2 approval signals a longer-term constellation that would reach roughly one-fifth of Starlink's planned 42,000-satellite total.

The competition is late. It's not cancelled.

The enterprise and government market segmentation the Counterpoint identified as the viable competitive niche has materialized precisely as described, and it's a commercially significant niche rather than a consolation category.

OneWeb/Eutelsat's explicit focus on aviation, maritime, enterprise, and government customers represents the specific market segment where Starlink's consumer-oriented service model and SpaceX's political profile create genuine competitive openings. IRIS²'s design (combining LEO, GEO, and MEO satellites in a hybrid architecture with secure government and commercial connectivity as its primary use case) is explicitly positioned for the institutional customer segment rather than consumer broadband competition with Starlink.

The Eutelsat CEO's requirement that IRIS² must compete on price and performance with Starlink is a commercial market signal: the sovereign constellation is not purely a strategic infrastructure exercise but a genuine commercial competitor targeting the enterprise and institutional segment where sovereignty concerns create protected market demand.

The sovereign procurement logic the original Counterpoint identified as the third required condition has materialized in IRIS²'s €10.6 billion concession contract, Norway and Iceland's participation, and EU Commissioner Kubilius's explicit framing of IRIS² as a Starlink-dependence reduction measure at the 2026 European Space Conference.

The 2027 to 2030 window is where these competitive threads resolve into market outcomes.

THE POINT
Sovereign infrastructure is the response, not interoperability. It's 2029, not 2026.
WHAT NEEDS TO CHANGE
— (FOR THE COUNTERPOINT TO PREVAIL) —
Counterpoint wins only if these conditions materialize.
THE COUNTERPOINT
Efficiency, scale, and market forces will keep the market open. Interoperability will arrive through incentives, not law.
WHY THE POINT IS MORE LIKELY
(COUNTERPOINT: 25% LIKELY)
DEPLOYMENT GAP IS TOO LARGE
Amazon Leo has ~300 satellites vs. 1,618 FCC requirement by July 30, 2026. Extension requested; gap remains massive.
MARKET CONCENTRATION PERSISTS
Starlink holds 97% of speed test samples and continues adding ~200+ satellites per month.
SOVEREIGN RESPONSE IS UNDERWAY
EU Space Act, IRIS², and ITU WRC-27 Item 1.13 are real actions—not interoperability mandates.
NO INTEROPERABILITY MANDATE
No forced interoperability standard by end-2026. No near-term roadmap for mandatory requirements.
ECONOMICS FAVOR CONSOLIDATION
Duplication is expensive. Sovereign infrastructure and regulation are lower risk and faster to implement.
TIMELINE MATCHES EVIDENCE
All sovereign alternatives target 2029+. That's the new reality—not 2026.
DECISION FRAMEWORK: ACTIONS TO TAKE NOW TO TIP THE OUTCOME
1 CLOSE THE DEPLOYMENT GAP FAST
Deploy substantially more satellites.
• Reach 500+ satellites by mid-2027
• Hit 1,618 by July 30, 2026 (FCC req.)
• Sustain >200 satellites/month
2 DELIVER COMPETITIVE SERVICE AT SCALE
Offer coverage, performance, and pricing that match or beat Starlink.
• Global coverage with low latency
• Comparable speeds and reliability
• Price parity or better
3 DRIVE INTEROPERABILITY THROUGH VALUE
Make cross-network interoperability commercially attractive and easy.
• Standardize APIs and handoff protocols
• Roaming and backhaul arrangements
• User benefits > switching costs
4 ALIGN REGULATION GLOBALLY
Create consistent, enabling frameworks across jurisdictions.
• Harmonize licensing and spectrum
• Incentivize interoperability standards
• Avoid regulatory fragmentation
5 EXPAND EXCHANGE & BACKHAUL NETWORKS
Build shared infrastructure that reduces duplication and cost.
• Neutral exchange points
• Interoperable gateways
• Shared ground infrastructure
6 BUILD MARKET MOMENTUM & TRUST
Enterprises and governments must adopt multi-provider models.
• Procurement policies favor openness
• Demonstrate multi-network reliability
• Prove cybersecurity interoperability
WHY THE COUNTERPOINT IS STILL PLAUSIBLE
(STILL 25% LIKELY)
COMPETITORS SCALE SUCCESSFULLY
Amazon Leo, OneWeb, and others rapidly expand and deliver reliable, high-quality service at scale.
EFFICIENCY DRIVES CONSOLIDATION BENEFITS
Multiple large constellations drive costs down and expand customer choice worldwide.
INTEROPERABILITY BECOMES STANDARD
Market-driven standards create seamless cross-network experiences and roaming without mandates.
DEMAND OUTGROWS SINGLE NETWORK
Global demand accelerates beyond Starlink's capacity; diversity becomes essential.
INNOVATION LOWERS BARRIERS
New launch vehicles, tech advances, and manufacturing scale reduce time and cost to compete.
REGULATION REMAINS LIGHT-TOUCH
Governments avoid heavy mandates; market forces and voluntary frameworks prevail.
LIKELIHOOD THE COUNTERPOINT PREVAILS
WHY THE POINT IS MORE LIKELY
✓ Deployment gap is large and widening
✓ Sovereign response is real and proceeding
✓ No interoperability mandate or near-term roadmap
25%
WHY THE COUNTERPOINT IS STILL PLAUSIBLE
✓ Competitors can close the gap
✓ Market forces can drive interoperability
✓ Demand and innovation can change the math
0%
POINT MORE LIKELY
25%
100%
COUNTERPOINT MORE LIKELY
MOST CRITICAL VARIABLE
SUSTAINED COMPETITOR DEPLOYMENT VELOCITY
THE QUESTION
Will sovereign-driven infrastructure and regulation rebalance the market in time—
or will Starlink lock in a decade of dominance before meaningful competition arrives?

## What Needs to Change

### (...for the Counterpoint to Prevail)

The Counterpoint's "meaningful competition by end 2026" condition hasn't been met.

The question for the 2027-2030 window is whether the revised timeline makes it a delayed prediction rather than a wrong one.

Three conditions need to materialize on the 2027-2029 timeline.

First, Amazon Leo needs to reach commercial service in at least its five target countries with consumer-accessible pricing and genuine service quality, demonstrating that the deployment delay was a schedule problem rather than a fundamental competitive viability problem.

Second, IRIS² needs to achieve its 2029 service target. A European megaproject on a complex schedule with a history of slippage. The original IRIS² target was 2027; it has already slipped to 2029 with the €10.6 billion budget. A second major delay would confirm the Point's implicit argument that sovereign alternatives are perpetually 2-3 years away.

Third, Starlink's market share of satellite speed tests needs to show measurable decline as Amazon Leo and other competitors capture real consumer and enterprise adoption, the metric that would demonstrate the competitive thesis materializing rather than the competitive infrastructure merely existing.

None of these conditions requires an immediate inflection.

The Counterpoint's scenario plays out over the full 2027-2030 window rather than the original 2026 endpoint.

The prediction's original timeline was too aggressive.

Whether the underlying thesis was wrong depends on whether IRIS² and Amazon Leo achieve their 2029 service targets on schedule.

## Likelihood Counterpoint Prevails: 25%

The Counterpoint likelihood holds at 25%, the same as the original, but for updated reasons that reflect the 2026 evidence rather than the original analysis.

The original 20% was based primarily on Amazon Kuiper deployment timeline skepticism that has been confirmed as warranted. The 25% reflects a slight upward revision based on three genuine 2026 developments: IRIS²'s concession contract signature provides the sovereign competitive alternative with a funded, contracted timeline rather than an aspiration; Amazon Leo's FCC Gen2 expansion approval signals regulatory commitment to competition that the original 20% underweighted; and the EU Space Act and ITU WRC-27 engagement demonstrate that meaningful regulatory pressure on Starlink is now operating through real legal mechanisms even if not through forced interoperability.

What keeps the Counterpoint at 25% rather than higher: Amazon Leo's FCC deadline extension request confirms the original prediction's deployment timeline skepticism was precisely correct; IRIS² has already slipped from its 2027 original target to 2029 and European megaprojects historically slip further; and Starlink's market share hasn't declined from the 97% concentration that defines the original prediction's concern, despite eighteen months of accelerating competitive activity.

The 25% says: the structural diagnosis is confirmed, the competitive timeline hasn't materialized on the original schedule, and the 2027-2030 window represents the actual competitive test rather than a confirmation of competition already arrived.

## Social Media Regulation's Defining Year

### When Governments Declare War on TikTok, the Battlefield Is Every Family's Living Room. The Courts Are Referee

*Virginia's teen social media law lasted 57 days before a federal judge blocked it on First Amendment grounds.*

*Australia's ban deactivated 4.7 million accounts, and by March 2026, the eSafety Commissioner's own report confirmed that 70% of Australian minors under 16 were still accessing banned platforms.*

*In the same week: a New Mexico jury ordered Meta to pay $375 million for child exploitation violations, and a California jury found Meta and YouTube liable for designing addictive platforms that caused a young woman's depression and anxiety.*

*The ban failed. The lawsuits landed. Parents reported behavioral changes. Teens reported bypassing everything with a Temu face mask. The prediction and the Counterpoint are both partially right, simultaneously, in the same country.*

*That is the most instructive outcome of all.*

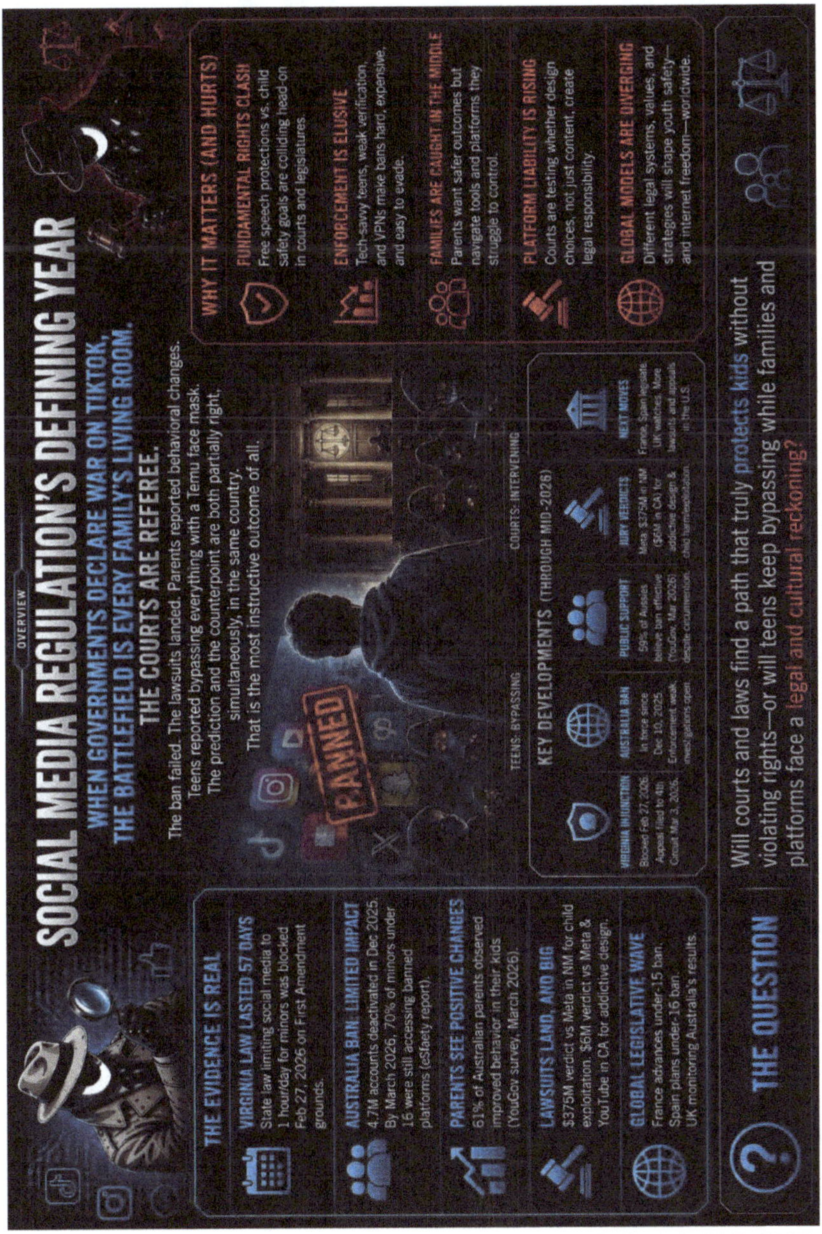
OVERVIEW
SOCIAL MEDIA REGULATION'S DEFINING YEAR
WHEN GOVERNMENTS DECLARE WAR ON TIKTOK,
THE BATTLEFIELD IS EVERY FAMILY'S LIVING ROOM.
THE COURTS ARE REFEREE.
The ban failed. The lawsuits landed. Parents reported behavioral changes.
Teens reported bypassing everything with a Temu face mask.
The prediction and the counterpoint are both partially right,
simultaneously, in the same country.
That is the most instructive outcome of all.
BANNED
TEENS: BYPASSING
COURTS: INTERVENING
THE EVIDENCE IS REAL
VIRGINIA LAW LASTED 57 DAYS
State law limiting social media to 1 hour/day for minors was blocked Feb 27, 2026 on First Amendment grounds.
AUSTRALIA BAN: LIMITED IMPACT
4.7M accounts deactivated in Dec 2025. By March 2026, 70% of minors under 16 were still accessing banned platforms (eSfaety report).
PARENTS SEE POSITIVE CHANGES
61% of Australian parents observed improved behavior in their kids (YouGov survey, March 2026).
LAWSUITS LAND, AND BIG
$375M verdict vs Meta in NM for child exploitation. $6M verdict vs Meta & YouTube in CA for addictive design.
GLOBAL LEGISLATIVE WAVE
France advances under-15 ban.
Spain plans under-16 ban.
UK monitoring Australia's results.
KEY DEVELOPMENTS (THROUGH MID-2026)
VIRGINIA INJUNCTION
Blocked Feb 27, 2026. Appeal filed to 4th Circuit Mar 3, 2026.
AUSTRALIA BAN
In force since Dec 10, 2025. Enforcement weak, investigations open.
PUBLIC SUPPORT
59% of Aussies believe ban effective (YouGov, Mar 2026) despite circumvention.
JURY VERDICTS
Meta: $375M in NM ($6M in CA) for addictive design & child harm/exploitation.
NEXT MOVES
France, Spain legislate. UK watches. More lawsuits and appeals in the U.S.
WHY IT MATTERS (AND HURTS)
FUNDAMENTAL RIGHTS CLASH
Free speech protections vs. child safety goals are colliding head-on in courts and legislatures.
ENFORCEMENT IS ELUSIVE
Tech-savvy teens, weak verification, and VPNs make bans hard, expensive, and easy to evade.
FAMILIES ARE CAUGHT IN THE MIDDLE
Parents want safer outcomes but navigate tools and platforms they struggle to control.
PLATFORM LIABILITY IS RISING
Courts are testing whether design choices, not just content, create legal responsibility.
GLOBAL MODELS ARE DIVERGING
Different legal systems, values, and strategies will shape youth safety—and internet freedom—worldwide.
THE QUESTION
Will courts and laws find a path that truly protects kids without violating rights—or will teens keep bypassing while families and platforms face a legal and cultural reckoning?

## Opening

The original prediction has been validated in its US dimension with surgical precision and partially validated in Australia in ways that are more nuanced and more instructive than a simple vindication.

On February 27, 2026, 57 days after Virginia's one-hour-per-day social media limit took effect, a federal judge issued a preliminary injunction, ruling the statute was likely to be found in violation of the First Amendment. The court acknowledged Virginia's compelling interest in protecting youth while ruling the law was simultaneously overinclusive (requiring adults to verify their age to access protected speech) and underinclusive (exempting addictive interactive gaming).

That dual diagnosis is not a fixable drafting error. It's a structural description of why US legislative approaches to teen social media restriction face constitutional instability at their foundation.

Australia's experience tells a more complicated and more consequential story. Three months after the December 2025 ban, the eSafety Commissioner's own report confirmed what critics predicted: 70% of Australian minors under 16 were still accessing banned platforms.

Yet simultaneously, a YouGov survey of 1,070 Australian adults found 59% believe the ban has been effective, and 61% of parents of children under 16 observed positive behavioral changes, a data tension that reveals something important about the difference between sentiment and enforcement, and between what parents observe and what regulators measure.

## Intro

The teen social media regulation story has diverged sharply along jurisdictional lines that illuminate deep structural differences between the US First Amendment framework and the regulatory environments in which Australia, the EU, and European democracies operate. In the US, the constitutional architecture has done precisely what the original prediction argued, providing courts with doctrinal tools to block state legislation that burdens speech even when legislative intent is genuinely protective.

In Australia, the world's first comprehensive under-16 ban has produced an early evidence record that defies simple narratives on either side: measurable behavioral changes alongside documented mass circumvention, cautious public optimism alongside a regulator's own admission that enforcement is failing, and the emergence of a parallel legal track, platform liability for addictive design, that may prove more structurally durable than age restrictions.

The 2027–2030 window is where both stories resolve: the US through appellate courts, Australia through the Connected Minds Study longitudinal research, and the global regulatory landscape through the question of whether platform liability or platform prohibition proves the more effective path.

## Background

Australia's ban on social media for under-16s, enforced since December 10, 2025 by the eSafety Commissioner, covers TikTok, Instagram, Facebook, Snapchat, YouTube, X, Reddit, Twitch, Kick, and Threads, with platforms bearing sole enforcement responsibility and facing fines up to $49.5 million Australian dollars for systemic failures.

The enforcement reality has been documented by the regulator itself: eSafety's March 2026 report confirmed that 70% of Australian minors under 16 continue to retain accounts, create new accounts, or pass platforms' age assurance systems.

By March, only 310,000 additional accounts had been affected beyond the initial 4.7 million deactivated in the first month, a slowdown suggesting platform enforcement effort is declining rather than intensifying.

eSafety announced investigations of Snap, TikTok, Facebook, Instagram, and YouTube for potential noncompliance with a target decision by mid-2026. In the United States, Virginia's Senate Bill 854 took effect January 1, 2026, and was blocked by federal injunction February 27, 2026. Virginia Attorney General Jay Jones appealed to the Fourth Circuit Court of Appeals on March 3, 2026.

The global legislative wave continues: France's National Assembly approved a bill banning social media for children under 15, Spain announced plans for an under-16 ban, and the UK is monitoring Australia's results.

In March and April 2026, two separate US jury verdicts added a parallel legal dimension: a New Mexico jury found Meta liable for consumer protection violations related to child exploitation ($375 million in damages) and a California jury found Meta and YouTube liable for negligent design of addictive platforms causing mental health harm ($6 million in compensatory and punitive damages), both verdicts are being appealed, but together they represent the first time juries have found platform design choices, rather than specific content, legally actionable.

POINT / COUNTERPOINT
SOCIAL MEDIA REGULATION
THE BATTLEGROUND
Both sides are right, partially.
The debate is over effectiveness, enforcement, and constitutional limits.
THE POINT:
BAN CHAOS HAS ARRIVED
Constitutional barriers and enforcement realities are proving harder than advocates anticipated.
1 VIRGINIA LASTED 57 DAYS
Blocked Feb 27, 2026.
First Amendment grounds.
Overinclusive and underinclusive.
2 ENFORCEMENT NIGHTMARE CONFIRMED
eSafety: 70% of minors still accessing platforms (March 2026 report).
Circumvention is widespread.
Platforms struggle to verify age.
3 TEENS ADAPT FASTER THAN LAWS
False birthdates.
Parents' accounts.
VPNs.
Face masks and burner accounts.
4 CONSTITUTIONAL DOCTRINE IS STABLE
Moody v. NetChoice provides foundation.
Editorial discretion and speech rights are well protected.
Appellate path likely lengthy.
5 LIABILITY MAY BE THE BETTER TOOL
Design lawsuits bypass access restrictions.
Focus shifts from speech to platform architecture and addictive design.
Early jury verdicts are a turning point.
THE BATTLEGROUND
VS.
PROTECTION vs. FREEDOM
SAFETY vs. RIGHTS
ENFORCEMENT vs. EVASION
DESIGN vs. ACCESS
PROTECT
VPN
NEW ACCOUNT
AGE 17
ACCESS GRANTED
The fight isn't whether harm exists.
It's whether bans are the right tool.
THE COUNTERPOINT:
IMPERFECT ENFORCEMENT CAN STILL HELP
The ban does not need perfect compliance to produce meaningful benefit.
1 PARENTS REPORT IMPROVEMENT
61% of Australian parents observed positive behavioral changes.
More engagement, better sleep, and improved family interactions.
2 PUBLIC SUPPORT ENDURES
59% of Australians support the ban.
Broad support gives policymakers mandate to keep refining the approach.
3 PARTIAL COMPLIANCE MAY BE ENOUGH
30% of minors are not accessing platforms.
Even partial reduction can lower exposure and harm.
Benefits may accrue unevenly.
4 LIABILITY IS CHANGING THE GAME
New Mexico: $375M verdict vs. Meta.
California: Design liability vs. Meta and YouTube.
Appeals pending, but legal landscape shifts.
5 SCIENCE IS FINALLY COMING
Australia's Connected Minds Study is underway.
2027–2028 results will be pivotal for policy and enforcement design.
THE DEBATE IN ONE QUESTION
Can societies protect children online in ways that are effective, enforceable, constitutional, and evidence-based, or will teens adapt faster than institutions can respond?

## THE POINT: The Court Said the Quiet Part Loud: Protecting Kids Doesn't Override the Constitution

The US constitutional dimension of the original prediction has been validated with precision that the 57-day timeline captures almost too neatly.

Federal Judge Patricia Tolliver Giles found Virginia's law both overinclusive, requiring everyone including adults to verify their age before accessing protected speech, and underinclusive, exempting potentially addictive interactive gaming from its coverage.

The court found the statute subject to strict scrutiny as a content-based restriction and ruled that while Virginia has a compelling interest in addressing the youth mental health crisis, the law was not narrowly tailored to serve that interest without burdening constitutionally protected speech.

The ruling builds on the Supreme Court's July 2024 decision in Moody v. NetChoice, which established that social media platforms engage in protected editorial discretion when curating and organizing user-generated content, and that First Amendment core principles don't vary with changes in technology.

The Virginia Attorney General appealed to the Fourth Circuit on March 3, 2026, but the doctrinal architecture Judge Giles relied on is structurally sound and consistent with established First Amendment jurisprudence, not an idiosyncratic district court ruling. The appellate trajectory will continue for years, and the Supreme Court may ultimately weigh in, but the prediction's core claim that constitutional challenges in the US would succeed on First Amendment grounds has been confirmed at the district court level and is well-grounded in the doctrine.

Australia's enforcement reality has produced the most significant confirmation of the original prediction's enforcement nightmare claim, and it comes not from critics or journalists but from the eSafety Commissioner's own March 2026 report.

Three months after the ban took effect, the Australian government's own online safety regulator confirmed that 70% of Australian minors under 16 were still accessing banned platforms.

The Molly Rose Foundation/YouthInsight survey of 1,050 Australians aged 12-15, conducted March-April 2026, found more than 60% of teens who had social media accounts before the ban still had access to at least one platform.

The circumvention methods are not sophisticated: lying about age on verification prompts (57% of evaders), entering false birthdates at sign-up (44%), borrowing a parent's or older sibling's account (42%), and routing through a VPN (30%). Only about 5% were using technical tools like VPNs, the majority are simply claiming to be older than they are and platforms are not catching them.

The distributional equity concern the original prediction raised is confirmed in the specific form: the circumvention pattern means teens with the most parental cooperation and digital sophistication to circumvent the ban are doing so most effectively, while the population most likely to remain off platforms is the population that was least engaged to begin with, inverting the equity benefit the ban's proponents intended.

### The Point Roundup

The teen social media ban story through 2030 is being written in two fundamentally different legal languages. In the US, the First Amendment trajectory is clear: state-level time limits and age verification requirements that burden speech face a structural constitutional challenge the Virginia ruling has articulated with precision.

Elsewhere, the regulatory mandate approach is producing documented mass circumvention, confirmed by the regulator itself, alongside genuine behavioral signals that complicate the simple failure narrative.

The prediction's core thesis remains intact and is now better evidenced: the enforcement nightmare has arrived, the US constitutional challenge has succeeded at first instance, and the doctrinal foundation for striking these laws down is sound and established.

The most important analytical development since the original was the addition of the platform liability track, two jury verdicts in the same week of March 2026 finding platform design choices legally actionable, which represents a parallel path to changing teen social media that does not depend on age restrictions, age verification, or constitutional workarounds.

## THE COUNTERPOINT: Imperfect Enforcement Is Not Zero Enforcement: The Case for Staying the Course

The Counterpoint's case is strengthened by a genuine tension in the Australia data that the Point's "70% circumvention" framing can obscure. The eSafety Commissioner's own circumvention data and the YouGov parent survey were collected roughly concurrently and tell different stories from different vantage points.

Seventy percent of Australian minors may still be accessing platforms, but a YouGov survey of 1,070 Australian adults found that 61% of parents of children under 16 observed between two and four positive behavioral changes: 43% noticed more in-person social interactions, 38%

reported children more present and engaged during interactions, 38% reported improved parent-child relationships.

The Conversation's analysis, drawing on MCRI commentary, is appropriately cautious: the YouGov survey was conducted just one month after the ban took effect, measures reported behavioral perceptions rather than objective outcomes, and has significant methodological limitations.

But the tension itself is analytically interesting, it's possible that partial enforcement produces partial behavioral change, that the 30% of teens who have been removed from platforms are seeing meaningful improvements, and that the counterfactual (no ban at all) would produce worse outcomes even if the actual compliance rate is well below 100%.

The Counterpoint does not require the ban to be perfectly effective. It requires the benefit to the compliant population to exceed the harm to the circumventing population, and that question is genuinely open.

The platform liability track that emerged in March 2026 is the Counterpoint's strongest new evidence, and it operates entirely outside the constitutional constraints that are killing the regulatory approach in the US.

The New Mexico jury's $375 million verdict against Meta for consumer protection violations related to child exploitation, and the California jury's finding that Meta and YouTube negligently designed addictive platforms causing mental health harm, are both being appealed, but they represent the first time in US legal history that juries have found platform design choices legally actionable rather than protected by Section 230 or the First Amendment.

The Los Angeles case is a bellwether for thousands of similar lawsuits; legal experts have described it as potentially "the beginning of a tidal wave" of platform liability litigation.

The Counterpoint's strongest 2030 scenario is not that the current age restriction enforcement approach becomes hermetically effective, it's that the combination of age restrictions, platform liability for design choices, algorithmic transparency requirements, and the longitudinal evidence of harm creates a regulatory ecosystem comprehensive enough that platforms redesign their products to be genuinely less harmful.

The liability track can achieve what the restriction track can't: it creates financial incentives for platforms to change behavior rather than to find technical minimums for compliance.

The US constitutional constraint stops governments from restricting access; it does not stop juries from finding platforms liable for how they design that access.

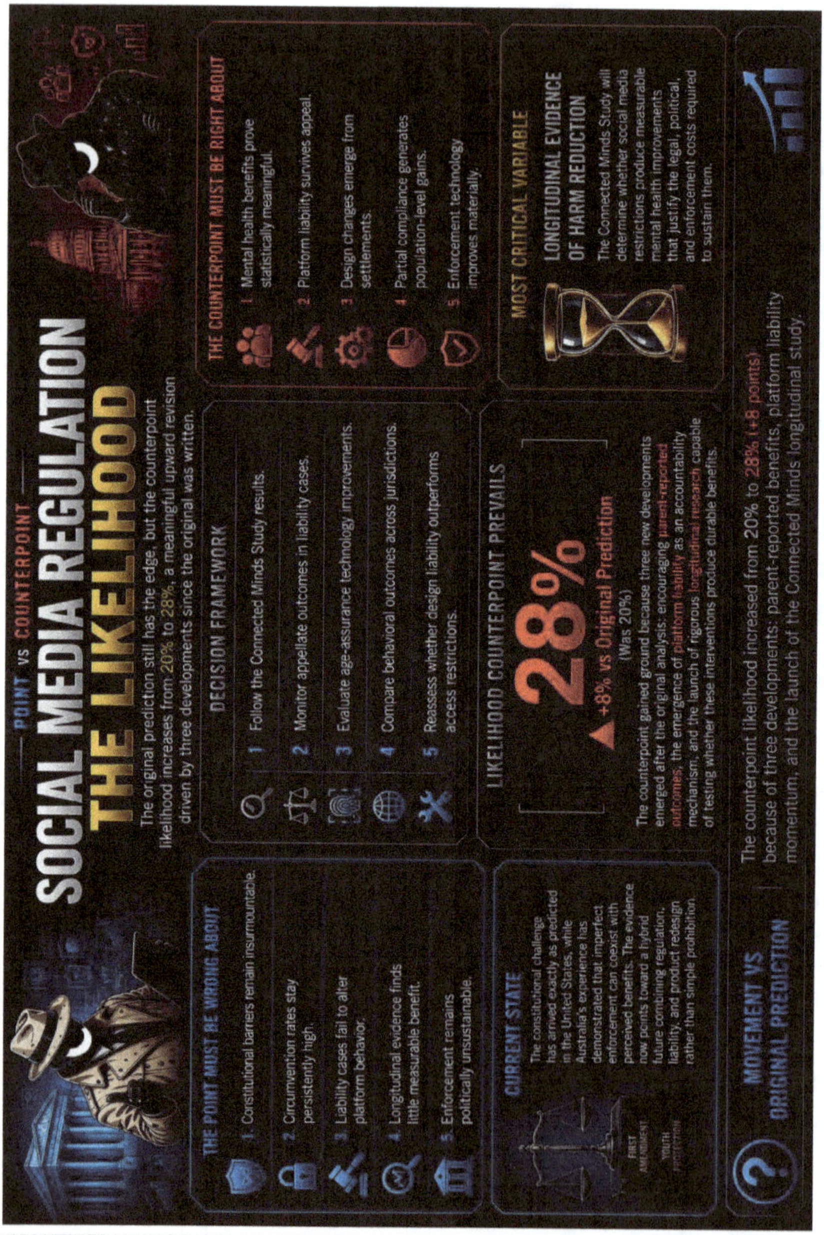
POINT vs COUNTERPOINT
SOCIAL MEDIA REGULATION
THE LIKELIHOOD
The original prediction still has the edge, but the counterpoint likelihood increases from 20% to 28%, a meaningful upward revision driven by three developments since the original was written.
THE POINT MUST BE WRONG ABOUT
1. Constitutional barriers remain insurmountable.
2. Circumvention rates stay persistently high.
3. Liability cases fail to alter platform behavior.
4. Longitudinal evidence finds little measurable benefit.
5. Enforcement remains politically unsustainable.
DECISION FRAMEWORK
1 Follow the Connected Minds Study results.
2 Monitor appellate outcomes in liability cases.
3 Evaluate age-assurance technology improvements.
4 Compare behavioral outcomes across jurisdictions.
5 Reassess whether design liability outperforms access restrictions.
THE COUNTERPOINT MUST BE RIGHT ABOUT
1 Mental health benefits prove statistically meaningful.
2 Platform liability survives appeal.
3 Design changes emerge from settlements.
4 Partial compliance generates population-level gains.
5 Enforcement technology improves materially.
CURRENT STATE
FIRST AMENDMENT
YOUTH PROTECTION
The constitutional challenge has arrived exactly as predicted in the United States, while Australia's experience has demonstrated that imperfect enforcement can coexist with perceived benefits. The evidence now points toward a hybrid future combining regulation, liability, and product redesign rather than simple prohibition.
LIKELIHOOD COUNTERPOINT PREVAILS
28%
+8% vs Original Prediction
(Was 20%)
The counterpoint gained ground because three new developments emerged after the original analysis: encouraging parent-reported outcomes, the emergence of platform liability as an accountability mechanism, and the launch of rigorous longitudinal research capable of testing whether these interventions produce durable benefits.
MOST CRITICAL VARIABLE
LONGITUDINAL EVIDENCE OF HARM REDUCTION
The Connected Minds Study will determine whether social media restrictions produce measurable mental health improvements that justify the legal, political, and enforcement costs required to sustain them.
MOVEMENT VS ORIGINAL PREDICTION
The counterpoint likelihood increased from 20% to 28% (+8 points) because of three developments: parent-reported benefits, platform liability momentum, and the launch of the Connected Minds longitudinal study.

## What Needs to Change

### (...for the Counterpoint to Prevail)

The Counterpoint's three original conditions have received their first real-world test and the results are mixed in instructive ways. For the Counterpoint to prevail through 2030, two conditions must materialize that are currently uncertain.

First, the Murdoch Children's Research Institute and Deakin University's Connected Minds Study, tracking 13-16 year olds before and after restrictions, must find a statistically significant and clinically meaningful improvement in mental health outcomes for the population subject to the ban. Not sentiment data showing parents observe behavioral changes, but peer-reviewed longitudinal evidence of the kind that could survive scrutiny from researchers who have argued the population-level effect of social media on mental health is not clinically significant.

MCRI Professor Susan Sawyer's own caution, that "a clear cause-and-effect relationship hasn't been proven", sets the evidentiary bar precisely: the Connected Minds Study is designed to answer this question, and its findings by 2027-2028 will either validate the legislative approach or confirm that circumvention rates are high enough that the treated population is indistinguishable from the untreated.

Second, the platform liability track must survive appeal and produce design-change settlements rather than damage payments, specifically, the California bellwether verdict must be upheld in a form that creates liability exposure for algorithmic design choices that platforms can't cost-effectively absorb, incentivizing the product redesign the regulatory approach can't compel.

If both conditions materialize, the Counterpoint's scenario that platforms redesigning products because the combination of regulatory pressure and liability exposure makes harmful design economically unsustainable, becomes genuinely achievable.

If neither materializes, the prediction's "ban chaos" framing is confirmed as the permanent condition rather than a transitional phase.

## Likelihood Counterpoint Prevails: 28%

The Counterpoint likelihood increases from 20% to 28%, a meaningful upward revision driven by three developments since the original was written.

First, the Australia early data is more ambiguous and more encouraging than the original 20% implied: parent-reported behavioral changes from a YouGov primary survey, public support at 59%, and cautious optimism from a population that had the most reason to be skeptical suggest the political sustainability of the approach is stronger than effectiveness critics anticipated.

Second, the platform liability track adds an entirely new dimension the original did not anticipate, two jury verdicts in the same week finding platform design choices legally actionable represent a structural shift in the accountability framework that the constitutional constraint on age restrictions does not touch.

Third, the Connected Minds Study longitudinal research now underway represents the first genuinely rigorous scientific test of the ban's effectiveness, and the Counterpoint's scenario is alive as long as that study hasn't reported.

What keeps the number at 28% rather than higher is the eSafety Commissioner's own admission that 70% of Australian minors are still accessing platforms, confirmed by the regulator itself, not just critics, combined with the documented circumvention pattern showing the teens most motivated to get around the ban are those with the most social capital and parental cooperation to do so.

The 28% says: the Counterpoint's scenario requires the longitudinal evidence to find mental health improvements in a population with a 70% circumvention rate, which is a harder bar than "imperfect enforcement produces imperfect improvement", and the platform liability track, while promising, is subject to appeal and uncertain to produce the design-change outcomes rather than damage payment outcomes the Counterpoint requires.

# Smartphone Upgrade Cycle Death

## The Two-Year Cycle Is Dead. The AI Super-Cycle Didn't Show Up to the Funeral.

*Manufacturers spent two years marketing AI as the reason you need a new phone. Consumers spent two years ignoring them. The average replacement cycle reached 3.84 years in the first three quarters of 2025.*

*Only 11% of smartphone owners upgrade for AI features, a 7-point drop from the prior year, per CNET's YouGov survey of 2,201 US adults. In 2026 the figure edged back to 12%, still lower than 2024's 18%.*

*The two-year cycle isn't coming back. The AI super-cycle hasn't arrived. The question is what replaces both.*

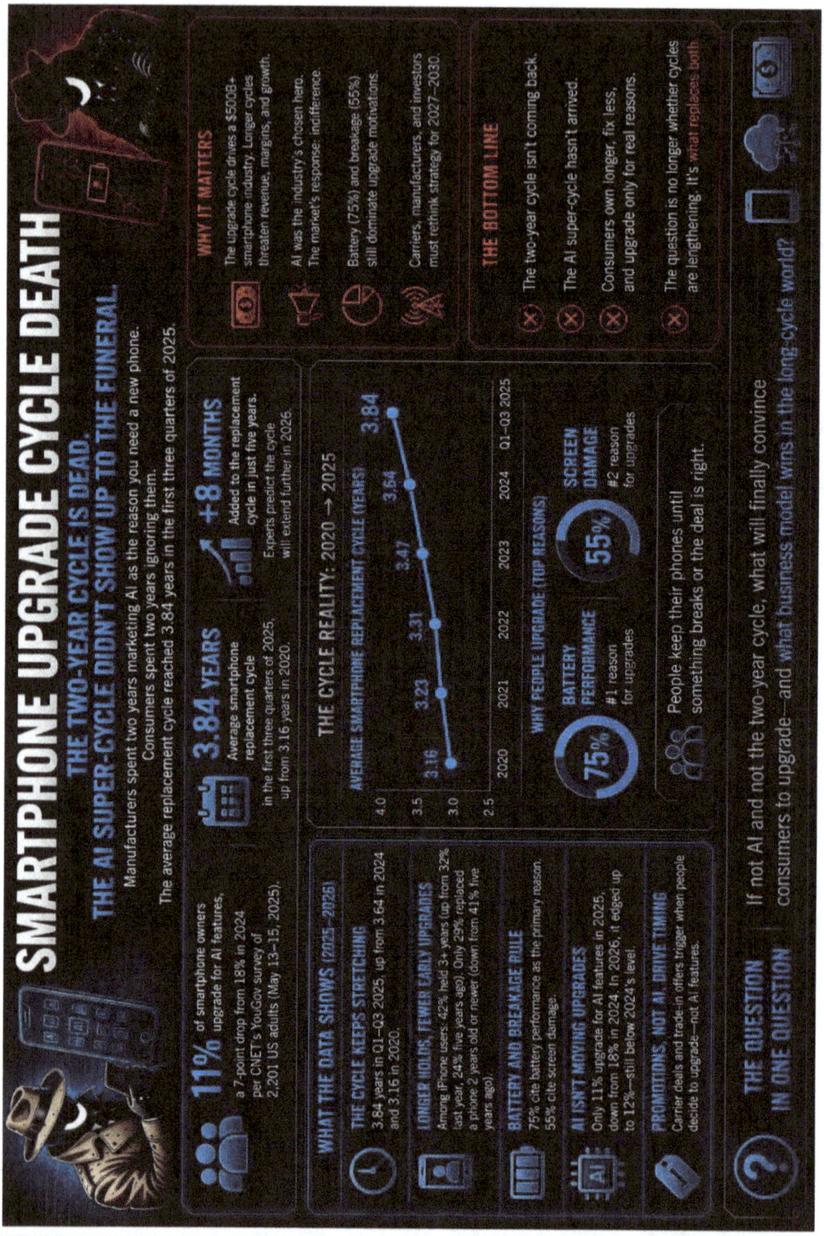

SMARTPHONE UPGRADE CYCLE DEATH
THE TWO-YEAR CYCLE IS DEAD.
THE AI SUPER-CYCLE DIDN'T SHOW UP TO THE FUNERAL.
Manufacturers spent two years marketing AI as the reason you need a new phone.
Consumers spent two years ignoring them.
The average replacement cycle reached 3.84 years in the first three quarters of 2025.
11% of smartphone owners upgrade for AI features,
a 7-point drop from 18% in 2024 per CNET's YouGov survey of 2,201 US adults (May 13–15, 2025).
3.84 YEARS
Average smartphone replacement cycle
in the first three quarters of 2025, up from 3.16 years in 2020.
+8 MONTHS
Added to the replacement cycle in just five years.
Experts predict the cycle will extend further in 2026.
WHY IT MATTERS
The upgrade cycle drives a $500B+ smartphone industry. Longer cycles threaten revenue, margins, and growth.
AI was the industry's chosen hero. The market's response: indifference.
Battery (75%) and breakage (55%) still dominate upgrade motivations.
Carriers, manufacturers, and investors must rethink strategy for 2027–2030.
WHAT THE DATA SHOWS (2025–2026)
THE CYCLE KEEPS STRETCHING
3.84 years in Q1–Q3 2025, up from 3.64 in 2024 and 3.16 in 2020.
LONGER HOLDS, FEWER EARLY UPGRADES
Among iPhone users: 42% held 3+ years (up from 32% last year, 24% five years ago). Only 29% replaced a phone 2 years old or newer (down from 41% five years ago).
BATTERY AND BREAKAGE RULE
75% cite battery performance as the primary reason. 55% cite screen damage.
AI ISN'T MOVING UPGRADES
Only 11% upgrade for AI features in 2025, down from 18% in 2024. In 2026, it edged up to 12%—still below 2024's level.
PROMOTIONS, NOT AI, DRIVE TIMING
Carrier deals and trade-in offers trigger when people decide to upgrade—not AI features.
THE CYCLE REALITY: 2020 → 2025
AVERAGE SMARTPHONE REPLACEMENT CYCLE (YEARS)
4.0
3.5
3.0
2.5
3.16
3.23
3.31
3.47
3.64
3.84
2020
2021
2022
2023
2024
Q1–Q3 2025
WHY PEOPLE UPGRADE (TOP REASONS)
75%
BATTERY PERFORMANCE
#1 reason for upgrades
55%
SCREEN DAMAGE
#2 reason for upgrades
People keep their phones until something breaks or the deal is right.
THE BOTTOM LINE
The two-year cycle isn't coming back.
The AI super-cycle hasn't arrived.
Consumers own longer, fix less, and upgrade only for real reasons.
The question is no longer whether cycles are lengthening. It's what replaces both.
THE QUESTION IN ONE QUESTION
If not AI and not the two-year cycle, what will finally convince consumers to upgrade—and what business model wins in the long-cycle world?

## Opening

The original prediction has been confirmed with a precision that makes the 2026 update more about calibration than correction. The average smartphone lifecycle reached 3.84 years in the first three quarters of 2025, continuing a trend that has added eight months to the replacement cycle in five years, with experts predicting the figure will extend further in 2026.

The Counterpoint's AI super-cycle thesis has received its definitive market test, and the result is unambiguous: only 11% of US smartphone owners upgrade their devices for AI features, per CNET/YouGov primary survey data, down from 18% in 2024.

The industry has spent two years marketing AI as the compelling upgrade driver it needed, and consumers have responded by continuing to hold their phones until something breaks or a carrier deal makes the math irresistible. The 2027-2030 window is where the prediction's second-order consequences, business model disruption, services pivot, carrier strategy rethink, move from early signal to structural reality.

And a genuinely important development has arrived that neither the original prediction nor the Counterpoint fully anticipated: iOS 27's AI marketplace architecture, announced at WWDC June 8, 2026, deliberately routes AI capability through software and cloud rather than requiring new silicon, simultaneously confirming the services pivot thesis and making the hardware super-cycle structurally harder to achieve.

## Intro

The smartphone upgrade cycle extension is no longer a prediction. It's a documented, accelerating, multi-year trend confirmed by independent consumer surveys, trade-in data, and manufacturer behavior.

The more analytically interesting question for the 2027-2030 window is not whether cycles are lengthening, they are, but whether the AI features arriving in 2026 and 2027 represent the inflection point that finally reverses the trend, or whether they represent the industry's most sophisticated and ultimately unsuccessful attempt to manufacture a replacement cycle that consumer behavior has decided is over.

The Light Reading analysis published April 2026, observing that the smartphone industry is "marketing AI to people who just need a phone that works", captures the gap between industry narrative and consumer reality with unusual candor. That gap, and what it means for Apple, Samsung, carriers, and the broader smartphone ecosystem through 2030, is the story this update tells.

## Background

The average smartphone lifecycle reached 3.84 years in the first three quarters of 2025, up from 3.16 years in 2020, an increase of 0.68 years or just over eight months in five years. In 2024 the global figure reached 3.64 years.

Among iPhone users specifically, 42% replacing an Apple device in the twelve months ending September 2025 retired a device they had owned for at least three years, up from 32% the prior year and 24% five years earlier. Only 29% replaced an iPhone that was two years old or newer, compared to 41% five years ago.

The motivations driving replacement remain stubbornly consistent and stubbornly hardware-focused: 75% of people cite battery performance as their primary upgrade reason, 55% cite screen damage. CNET's YouGov survey (n=2,201 US adults, May 2025) found that only 11% of smartphone owners would upgrade for AI features, a 7-point drop from 18% in 2024.

Consumer interest in AI as an upgrade driver declined even as AI features became more widely available. By 2026, the figure edged to 12%, still below 2024's level, confirming the trend hasn't reversed. The April 2026 Light Reading analysis confirmed the picture from the supply side: Samsung Flagship entered 2026 with 64.4% of its installed base in the two-plus-year upgrade window, Apple with 51.8%, both pools motivated primarily by performance degradation and battery urgency, with carrier promotional offers providing timing triggers rather than AI features providing underlying motivation.

Performance and battery lead every brand tier every month in the data, with no evidence in 2025 or Q1 2026 that AI features had begun narrowing the gap.

POINT / COUNTERPOINT
SMARTPHONE UPGRADE CYCLE
THE BATTLEGROUND
Both sides are right, partially.
The debate is over what ultimately drives upgrade decisions.
THE POINT:
THE TWO-YEAR CYCLE IS DEAD FOR REAL
Consumers have voted with their behavior, not their wallets.
1. CYCLES KEEP GETTING LONGER
Average lifecycle hit 3.84 years in Q1–Q3 2025, up from 3.16 years in 2020—a gain of 8+ months in five years.
2. AI DOESN'T MOVE THE NEEDLE
Only 11% upgrade for AI features in 2025, down from 18% in 2024 (CNET/YouGov, n=2,201). In 2026, it edged to 12%.
3. BATTERY AND BREAKAGE RULE
75% cite battery performance.
55% cite screen damage.
These drive upgrades across all brands.
4. DEALS TRIGGER TIMING, NOT AI
Carrier promotions and trade-in offers pull upgrades forward. AI features aren't the underlying motivation.
5. THE INSTALLED BASE KEEPS AGING
Samsung: 64.4% of users are in the 2+ year window. Apple: 51.8%.
Huge pools, zero urgency for AI.
THE BATTLEGROUND
RELIABILITY
BATTERY LIFE
VALUE
PRACTICAL NEEDS
REAL-WORLD NEEDS
NEW FEATURES
AI
NEW FEATURES
AI CAPABILITIES
FUTURE POTENTIAL
INDUSTRY NARRATIVE
WHAT CONSUMERS ACTUALLY VALUE
CONSUMERS UPGRADE WHEN THE PAIN IS REAL, NOT WHEN THE PROMISE IS BIG.
THE COUNTERPOINT:
THE SUPER-CYCLE IS BUILDING, NOT DEAD
The benefits of AI and new experiences will eventually become undeniable.
1. AI BENEFITS WILL COMPOUND
Today's features are table stakes. Tomorrow's will be workflow-changing and life-improving.
2. HARDWARE WILL CATCH UP
On-device AI, new form factors, better batteries, and silicon leaps will create real pull.
3. EXPERIENCE DRIVES ASPIRATION
New interactions and capabilities create desire—desire creates upgrade cycles.
4. ECOSYSTEM LOCK-IN DEEPENS
AI services, data, and device ecosystems make switching costly and upgrades more attractive.
5. 2027–2030 IS THE WINDOW
History shows it takes time for platform shifts to translate into mainstream upgrade waves.
THE DEBATE IN ONE QUESTION
Will AI and the next wave of experiences finally create upgrade urgency strong enough to reverse the cycle extension—or will consumers keep holding their phones until something breaks or the deal is right?
AI

## THE POINT: Battery and Breakage Drive Upgrades. AI Drives Press Releases.

The Counterpoint's super-cycle thesis has received its market test in the 2025-2026 window, and the verdict is contained in a single primary-sourced statistic: only 11% of US smartphone owners upgrade for AI features, a 7-point drop from 18% in 2024, confirmed by CNET's YouGov survey of 2,201 adults conducted May 13-15, 2025.

That directional decline in AI-upgrade motivation, happening simultaneously with AI features becoming more widely available, is the most analytically significant finding in this update. Consumer sentiment around AI as an upgrade driver dropped from 18% to 11% in a year when AI marketing from every major manufacturer reached its highest intensity. The industry is spending more to market AI features and converting fewer consumers to AI-motivated upgrades. Apple Intelligence's rollout has been characterized by delayed features, quality-controlled releases, and capabilities that improve existing workflows rather than creating the "can't live without this" discontinuity that genuine super-cycles require.

Apple itself acknowledged that quality issues stalled its next-generation Siri release, a candid admission that the AI features intended to drive the upgrade narrative were not ready to deliver the experience the narrative required.

The Counterpoint's specific claim, that iPhone 18 and Galaxy S27 AI capabilities would make current phones "feel obsolete, similar to how the original iPhone made existing smartphones of that time unusable", hasn't materialized in either consumer behavior or in the hardware differentiation visible in 2025-2026 device releases.

The most consequential single development for the 2027-2030 prediction is iOS 27's AI marketplace architecture, announced at Apple's WWDC keynote on June 8, 2026, and confirmed by Bloomberg reporting.

iOS 27's Extensions framework opens Siri, Writing Tools, and Image Playground to third-party AI providers including Claude, ChatGPT, and Gemini through a dedicated App Store marketplace, turning the iPhone into what observers described as "the device people carry to access artificial intelligence" rather than a hardware platform whose AI capabilities are bounded by its silicon.

That architectural decision, routing AI capability through software and competitive cloud services rather than locking it to Apple's own on-device processing, reflects Apple's accurate reading of where consumer value is migrating, and it simultaneously confirms the original prediction's business model disruption thesis and makes the hardware super-cycle structurally harder to achieve.

If AI access is routed through a competitive marketplace regardless of which iPhone generation you own, the "your old phone can't run these AI features" obsolescence narrative loses its primary mechanism. The average smartphone selling price has climbed to approximately $430-455 globally, reflecting increasing demand for premium models, but that premium is being paid for build quality, camera hardware, and brand affiliation rather than AI silicon differentiation.

The industry is pivoting toward services revenue not because it chose to but because consumer behavior has left it no alternative.

### The Point Roundup

The smartphone upgrade cycle death prediction's 2027-2030 trajectory is clear and confirmed by multiple independent data streams. The average replacement cycle will reach 4+ years by 2027 as the original prediction specified, the 3.84-year figure from Q3 2025 is already within range, and the directional trend hasn't reversed despite two years of intensive AI marketing.

The business model disruption the prediction identified is materializing in exactly the form it described: Apple's iOS 27 AI marketplace is the services pivot the prediction anticipated, carrier promotional intensity rather than feature appeal is the primary timing mechanism for upgrades, and the industry is adapting to a world where hardware refresh cycles are driven by degradation and deals rather than compelling new capabilities.

A smartphone industry generating services revenue from an installed base held for 4+ years is a structurally different business than the hardware refresh engine that built it, and the organizations that have already made that transition will have durable advantages over the ones still waiting for the super-cycle that hasn't arrived.

## THE COUNTERPOINT: "The Installed Base Is Upgrade-Ready. It's Waiting for a Reason

The Counterpoint's thesis hasn't been refuted, it has been delayed.

The distinction matters because the upgrade-eligible installed base is now larger than at any point in recent history, creating a compressed-spring dynamic that a genuinely compelling AI capability could release.

Samsung Flagship enters 2026 with 64.4% of its installed base in the two-plus-year upgrade window, the highest upgrade-eligible share of any brand.

Apple's 51.8% upgrade-eligible share, applied to its 55.9% installed-base share, produces the largest absolute pool of replacement-ready consumers in the market.

That pool is not indifferent to upgrading, it's waiting for a sufficiently compelling reason. The 11% who upgrade for AI features today are the early adopters.

The question is whether that number grows toward the 25-30% threshold at which AI-motivated upgrades become a meaningful industry driver, and iOS 27's Extensions marketplace creates the infrastructure through which hardware differentiation could eventually matter again.

When third-party AI models deliver materially richer experiences on the newer NPUs in iPhone 18 and iPhone 19 than on iPhone 15 and 16, the obsolescence mechanism the Counterpoint described has an architectural vehicle that the pre-Extensions era lacked.

The super-cycle thesis's original error was timing, not direction. Apple Intelligence's quality-controlled rollout, including the delayed next-generation Siri, was a strategic decision to avoid shipping undercooked AI features rather than evidence that the capability doesn't exist.

iPhone users are 12% more likely to be active AI users than Android users and the segment most likely to want proactive, always-on AI, meaning the population most motivated to upgrade for AI capability is already concentrated in the installed base most eligible for replacement.

Apple Intelligence's roadmap is explicitly designed to make older devices feel less capable over time: personalized assistance across devices, expanded language support, developer-enabled innovation through the Foundation Models framework, and proactive contextual automation, a capability roadmap that deepens rather than plateaus.

The super-cycle the Counterpoint described, driven by iPhone 18 and Galaxy S27 creating obsolescence, may simply be an iPhone 19 or 20 story, arriving 18-24 months later than the original prediction but carrying the same structural logic. The camera analogy is the Counterpoint's most historically grounded argument: the camera super-cycle took six years from "nice addition" to "primary reason to own a smartphone."

The 2030 window gives the AI super-cycle four years. That is tight but not impossible.

POINT vs COUNTERPOINT
SMARTPHONE UPGRADE CYCLE
THE LIKELIHOOD
The original prediction still holds the edge, but the counterpoint likelihood declines from 35% to 28%, a meaningful downward revision driven by three developments since the original was written.
THE POINT MUST BE WRONG ABOUT
1. CONSUMER BEHAVIOR STAYS STUBBORN
Consumers continue prioritizing battery life, reliability, and value over new features.
2. AI FAILS TO CREATE A MUST-HAVE MOMENT
AI features remain incremental and don't deliver a "can't live without this" experience.
3. SOFTWARE WINS OVER SILICON
Capabilities route through software and cloud, reducing the need for new hardware.
4. AFFORDABILITY CONSTRAINTS PERSIST
High prices and economic pressure keep many consumers from upgrading.
5. CARRIER PROMOS STAY THE MAIN TRIGGER
Deals drive timing, not AI capabilities or feature excitement.
DECISION FRAMEWORK
1 Track real-world usage and satisfaction with AI features through 2027.
2 Monitor installed base age and carrier upgrade incentive intensity.
3 Assess whether on-device AI and new silicon create measurable experience discontinuity.
4 Compare replacement cycles across regions as economic conditions evolve.
5 Evaluate whether services revenue growth outperforms hardware revenue decline.
THE COUNTERPOINT MUST BE RIGHT ABOUT
1. AI BECOMES INDISPENSABLE
AI capabilities solve real problems and become core to daily workflows.
2. HARDWARE DISCONTINUITY ARRIVES
New silicon enables experiences that today's devices simply cannot deliver.
3. NEW INTERFACES CREATE NEW NEEDS
Voice, visual, and ambient AI interactions redefine the smartphone experience.
4. WILLINGNESS TO PAY INCREASES
Consumers accept higher prices for devices that deliver clear, compelling benefits.
5. UPGRADE CYCLE REVERSES
Replacement cycles return to ~2 years as excitement and utility drive demand.
CURRENT STATE
The replacement cycle has extended to 3.84 years, up from 3.16 years in 2020. Only 11% of US owners upgrade for AI features (down from 18% in 2024), and battery life (75%) and screen damage (55%) remain the top drivers. iOS 27's AI marketplace architecture routes capabilities through software and cloud—reducing the need for new hardware and reinforcing the services-over-silicon trend.
LIKELIHOOD COUNTERPOINT PREVAILS
28%
▼ -7% vs Original Prediction
(Was 35%)
The counterpoint likelihood declined because three developments reinforced the original thesis: consumer interest in AI as an upgrade driver dropped despite more AI features; Apple's iOS 27 architecture makes hardware upgrades less necessary; and economic pressures continue to push consumers toward longer replacement cycles.
MOST CRITICAL VARIABLE
AI MUST DELIVER A CLEAR, MEASURABLE EXPERIENCE DISCONTINUITY
If AI features don't create a step-change in daily utility that today's devices cannot provide, consumers will keep holding. The moment of undeniable, must-have utility is the gatekeeper to a shorter upgrade cycle.
MOVEMENT VS ORIGINAL PREDICTION
The counterpoint likelihood declined from 35% to 28% (-7 points) because consumers showed less interest in upgrading for AI, iOS 27 reduces the need for new hardware, and economic headwinds continue to extend replacement cycles.
2027–2030 OUTLOOK
AI utility, interface evolution, and economic conditions will determine whether the cycle extends further—or finally contracts.

## What Needs to Change

### (...for the Counterpoint to Prevail)

For the super-cycle thesis to materialize in the 2027-2030 window after its 2026 failure, two structural conditions need to change that the current trajectory does not support.

First, on-device AI features need to produce experiences that are demonstrably, viscerally superior on hardware released in 2027-2028 compared to hardware released in 2024-2025, not incrementally better, but categorically different in ways users encounter in daily workflows rather than in benchmark comparisons. Real-time conversational translation, persistent personal AI with genuine cross-app memory, and complex multimodal processing are the candidate capabilities, but each requires both the hardware performance and the software maturity that neither has yet delivered simultaneously at the quality level that moves consumer behavior.

Second, the 11% who currently upgrade for AI features needs to grow toward 25-30%, the threshold at which upgrade-for-AI becomes a meaningful industry driver rather than an early adopter signal. That growth requires AI features to move from "impressive demo" to "daily necessity" in the same way that smartphone cameras moved from "nice addition" to "primary reason to own a smartphone" between 2012 and 2018.

The camera super-cycle took six years to fully develop. The 2030 window gives the AI super-cycle four.

## Likelihood Counterpoint Prevails: 28%

The Counterpoint likelihood declines from 35% to 28%, a meaningful downward revision driven by the definitive market test result and the iOS 27 architectural decision.

The original 35% reflected genuine uncertainty about whether AI features requiring dedicated neural processing hardware would create compelling upgrade pressure, a reasonable hypothesis in early 2025 that has been tested and found wanting by 2026 consumer behavior data. The CNET/YouGov primary survey finding, 11% upgrading for AI in 2025, a 7-point drop from 18% in 2024 even as AI features proliferated, is the empirical boundary of the super-cycle thesis's addressable market as currently constituted.

iOS 27's Extensions architecture, routing AI capability through competitive cloud services rather than locking it to specific silicon generations, makes the "your old phone can't run these AI features" obsolescence mechanism structurally harder to maintain.

The remaining 28% reflects three genuine residual uncertainties: iOS 27's marketplace could produce compelling third-party experiences that require newer NPU performance in ways that current-generation phones can't approximate; on-device AI could finally deliver the experience discontinuity, conversational translation, persistent memory, complex multimodal processing, that 2026 AI features conspicuously failed to provide; and economic conditions improving alongside a genuinely compelling capability could compress cycles back toward three years, not two, but meaningfully shorter than the 4+ year trajectory.

The 28% says: the super-cycle is not arriving in the form or on the timeline the Counterpoint predicted, the 4+ year cycle is confirmed, but the 2028-2030 period contains genuine uncertainty about whether on-device AI finally delivers the experience discontinuity that makes older hardware feel authentically obsolete.

# Gaming Subscription Fatigue

## Sony Up, Xbox Down. Gaming Subscriptions Face Their Defining Reckoning

*Microsoft's internal estimates indicate it gave up more than $300 million in Call of Duty sales revenue in 2024 by making Black Ops 6 a day-one Game Pass title.*

*Then it raised prices 50%. Then it cut prices and removed Call of Duty day-one access.*

*In the same week, Sony raised PlayStation Plus prices. 82% of Black Ops 6 sales came from PlayStation.*

*The platform without the subscription.*

*120 million people worldwide still pay for at least one gaming subscription.*

*The model isn't collapsing. It's thrashing for a formula that works.*

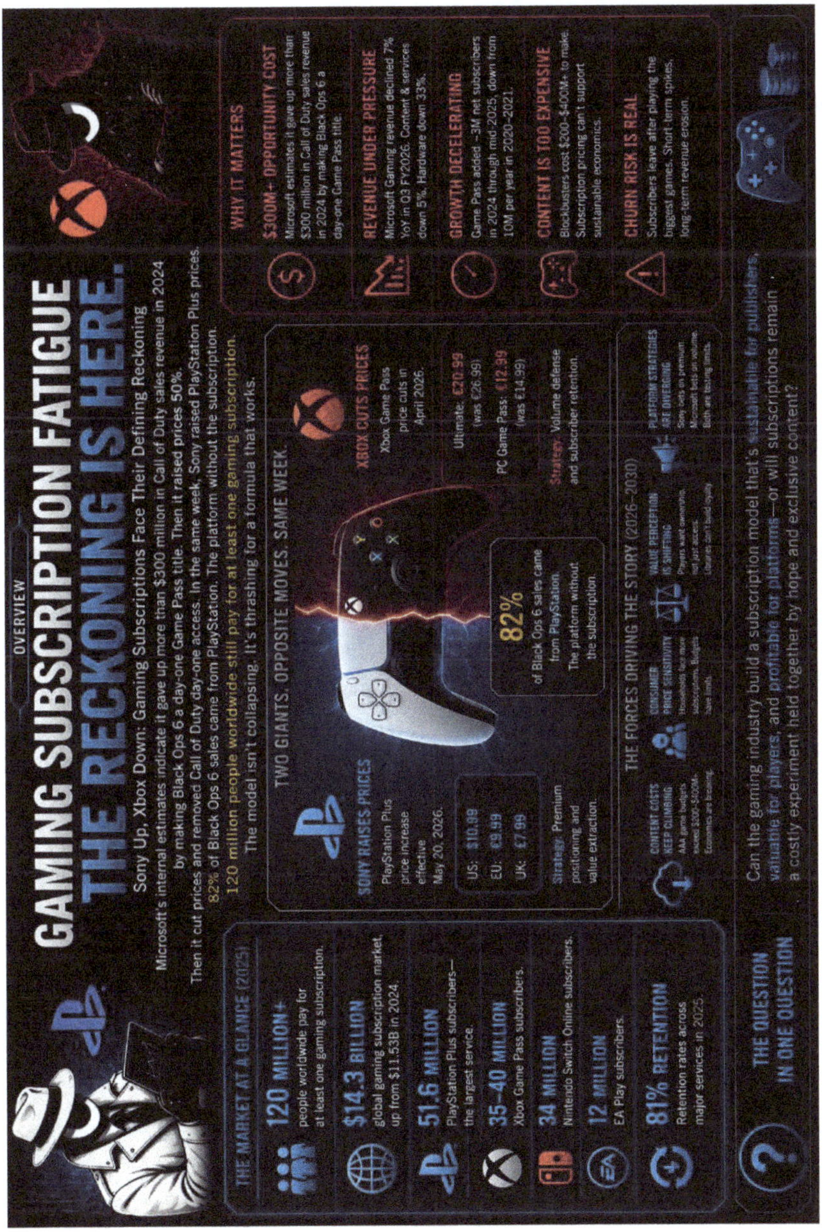
OVERVIEW
GAMING SUBSCRIPTION FATIGUE
THE RECKONING IS HERE.
Sony Up, Xbox Down: Gaming Subscriptions Face Their Defining Reckoning
Microsoft's internal estimates indicate it gave up more than $300 million in Call of Duty sales revenue in 2024 by making Black Ops 6 a day-one Game Pass title. Then it raised prices 50%.
Then it cut prices and removed Call of Duty day-one access. In the same week, Sony raised PlayStation Plus prices.
82% of Black Ops 6 sales came from PlayStation. The platform without the subscription.
120 million people worldwide still pay for at least one gaming subscription.
The model isn't collapsing. It's thrashing for a formula that works.
THE MARKET AT A GLANCE (2025)
120 MILLION+
people worldwide pay for at least one gaming subscription.
$14.3 BILLION
global gaming subscription market, up from $11.53B in 2024.
51.6 MILLION
PlayStation Plus subscribers—the largest service.
35–40 MILLION
Xbox Game Pass subscribers.
34 MILLION
Nintendo Switch Online subscribers.
12 MILLION
EA Play subscribers.
81% RETENTION
Retention rates across major services in 2025.
TWO GIANTS. OPPOSITE MOVES. SAME WEEK.
SONY RAISES PRICES
PlayStation Plus price increase effective May 20, 2026.
US: $10.99
EU: €9.99
UK: £7.99
Strategy: Premium positioning and value extraction.
82%
of Black Ops 6 sales came from PlayStation.
The platform without the subscription.
XBOX CUTS PRICES
Xbox Game Pass price cuts in April 2026.
Ultimate: €20.99 (was €26.99)
PC Game Pass: €12.99 (was €14.99)
Strategy: Volume defense and subscriber retention.
THE FORCES DRIVING THE STORY (2026–2030)
CONTENT COSTS KEEP CLIMBING
AAA game budgets exceed $200–$400M+. Economics are breaking.
CONSUMER PRICE SENSITIVITY
Households face more subscriptions. Budgets have limits.
VALUE PERCEPTION IS SHIFTING
Players want ownership, not just access. Libraries don't build loyalty.
PLATFORM STRATEGIES ARE DIVERGING
Sony bets on premium. Microsoft bets on volume. Both are testing limits.
WHY IT MATTERS
$300M+ OPPORTUNITY COST
Microsoft estimates it gave up more than $300 million in Call of Duty sales revenue in 2024 by making Black Ops 6 a day-one Game Pass title.
REVENUE UNDER PRESSURE
Microsoft Gaming revenue declined 7% YoY in Q3 FY2026. Content & services down 5%. Hardware down 33%.
GROWTH DECELERATING
Game Pass added ~3M net subscribers in 2024 through mid-2025, down from 10M per year in 2020–2021.
CONTENT IS TOO EXPENSIVE
Blockbusters cost $200–$400M+ to make. Subscription pricing can't support sustainable economics.
CHURN RISK IS REAL
Subscribers leave after playing the biggest games. Short-term spikes, long-term revenue erosion.
THE QUESTION IN ONE QUESTION
Can the gaming industry build a subscription model that's sustainable for publishers, valuable for players, and profitable for platforms—or will subscriptions remain a costly experiment held together by hope and exclusive content?

## Opening

The gaming subscription story has evolved from the original prediction's "growth plateau" framing into something structurally more significant: a documented sustainability crisis that has forced the industry's two dominant players into contradictory strategic moves within the same week.

The original prediction's core thesis, that subscription costs exceed individual game purchases for casual players, that the "Netflix for games" model has fundamental limitations, and that the industry would need to substantially rethink its approach, has been confirmed not by market stagnation but by the specific financial reality that including a blockbuster franchise in a subscription cannibalized more sales revenue than the subscriber gains could offset.

The 2027-2030 window is where the industry either finds the hybrid model the Counterpoint predicted or confirms that gaming subscriptions are structurally incompatible with the economics of blockbuster game publishing.

## Intro

The gaming subscription prediction's most important analytical update is not that the growth plateau arrived, it did, but that the specific financial mechanism producing the crisis is now documented rather than inferred. The original piece argued from structural logic: subscription pricing can't support blockbuster game economics, the "Netflix for games" model has fundamental limitations, and the industry would need to rethink its approach.

The 2026 evidence replaces that structural argument with a named number from a named company's internal estimates: $300 million in foregone Call of Duty sales revenue from a single day-one inclusion decision, followed by a 50% price increase, followed by a price cut and the removal of day-one blockbuster access, all within eighteen months.

That sequence is not a company executing a strategy. It's a company discovering, at $300 million cost, that the strategy doesn't work and reversing course before the next earnings call.

The Counterpoint's hybrid model prediction hasn't been refuted by this evidence, the 81% retention rate and the growing cloud gaming segment remain real, but the specific economic mechanism the original prediction identified as the structural problem has been confirmed by the industry's own behavior in ways that no retention statistic or subscriber count can answer around.

## Background

The global gaming subscription market reached approximately $14.3 billion in 2025, up from $11.53 billion in 2024, with over 120 million people worldwide paying for at least one gaming subscription service, PlayStation Plus with roughly 51.6 million subscribers leading, Xbox Game Pass at approximately 35-40 million, Nintendo Switch Online at 34 million, and EA Play with 12 million. Retention rates across major services reached 81% in 2025.

Those headline numbers look healthy. The underlying economics tell a different story. Microsoft Gaming reported $5.34 billion in Q3 FY2026 revenue, a 7% year-over-year decline, with content and services falling 5% and hardware revenue falling 33%. Game Pass crossed 40 million subscribers in Q1 2026, a roughly 10% increase from a year earlier, but the pace of subscriber additions has slowed dramatically from 10 million in 2020-2021 to approximately 3 million in 2024 through mid-2025.

The growth deceleration and revenue decline are happening simultaneously. The combination that defines a model under structural pressure rather than a temporary cyclical challenge.

Bloomberg's October 2025 reporting, drawing on internal Microsoft estimates from former employees, provided the financial data point that crystallized the sustainability debate: the company estimates it gave up more than $300 million in Call of Duty console and PC sales in 2024 by making Black Ops 6 a day-one Game Pass title, with 82% of all Black Ops 6 copies sold coming from PlayStation, the platform without a subscription option for the game.

POINT vs COUNTERPOINT
GAMING SUBSCRIPTIONS
THE CABLE BUNDLE COMPARISON
JUST GOT REAL.
NOT IN A GOOD WAY.
THE POINT: THE MODEL IS UNDER STRUCTURAL PRESSURE
$300M+ OPPORTUNITY COST
Microsoft estimates it gave up more than $300 million in Call of Duty sales in 2024 by putting Black Ops 6 on Game Pass day-one.
82% WENT TO PLAYSTATION
82%
82% of all Black Ops 6 copies sold came from PlayStation—the platform without a subscription.
REVENUE IS DECLINING
Microsoft Gaming revenue down 7% YoY in Q3 FY2026. Content & services down 5%. Hardware down 33%.
PRICE WHIPLASH CONFIRMS THE STRUGGLE
Raised Game Pass prices 50% in Oct 2025. Cut prices in Apr 2026 and removed Call of Duty day-one access.
THE FUNDAMENTAL FLAW
Subscribers play blockbusters for a month and leave. You can't build a $70 game business on churn.
THE EVIDENCE IS UNIGNORABLE
GAME PASS
40M+
Subscribers (Q1 2026)
PLAYSTATION PLUS
51.6M
Subscribers (2025)
CALL OF DUTY BLACK OPS 6
ONE GAME. TWO OUTCOMES.
$300M+
FOREGONE SALES REVENUE (EST.)
Microsoft's internal estimate for 2024
SUBSCRIBER GAINS
Short-term bump. Long-term churn.
SALES CANNIBALIZED
Hundreds of millions in lost revenue.
ECONOMICS DON'T BALANCE
The math doesn't support the model.
STRATEGY IN REVERSE
180° pivot in 6 months.
THE COUNTERPOINT: ADAPT AND IT CAN WORK
SUBSCRIPTION = ACCESS
Subscriptions lower friction and grow the total addressable market.
CONTENT DEPTH DRIVES RETENTION
A deeper, higher-quality library increases time spent and loyalty.
ENGAGEMENT CREATES VALUE
More playtime, more engagement, more opportunities to monetize.
HYBRID MODELS WILL EMERGE
Day-one for some, delayed for others. Different tiers for different players.
THE INDUSTRY WILL OPTIMIZE
Pricing, packaging, and content strategy will find the right equilibrium.
WHAT THIS MEANS
The cable bundle comparison is no longer theoretical. It's here.
Blockbuster economics don't fit subscription math—yet.
The next 3–5 years will decide which model wins.
THE REAL QUESTION:
Can subscriptions be sustainable for publishers, valuable for players, and profitable for platforms? Or is this a costly experiment on its way to reset?

## THE POINT: The Cable Bundle Comparison Just Got Real. Not in a Good Way

The original prediction's core insight, that the subscription model faces a structural problem that content library breadth can't solve, has been confirmed by Microsoft's own internal analysis of the Black Ops 6 inclusion.

Microsoft's internal estimates indicate it gave up more than $300 million in Call of Duty sales revenue on console and PC in 2024 by making the game available day-one on Game Pass, an opportunity cost calculated based on the 2024 versus 2023 sales differential, with the full Bloomberg report noting the figure represents foregone revenue rather than a cash expenditure.

The figure is an internal estimate from former employees rather than a published financial disclosure, but it's consistent with the visible financial outcomes: Game Pass saw record day-one subscriber additions from Black Ops 6, while 82% of all copies sold came from PlayStation and overall US video game subscription spend increased just 16% year-over-year during launch month, not nearly enough to offset the foregone direct sales.

The structural problem is the fundamental difference between subscription economics in music and gaming: a Spotify subscriber who streams a hit album remains subscribed for the next one. A Game Pass subscriber who plays Call of Duty for a month and then unsubscribes has consumed the platform's most valuable asset without generating the recurring revenue that would justify its inclusion at subscription pricing.

This is what Bobby Kotick, Activision's former CEO who testified against Microsoft's acquisition, told the FTC in 2023 when he argued that adding Call of Duty to Game Pass didn't make "commercial sense."

The 2024 outcome confirmed his assessment. Microsoft's response, raising Game Pass Ultimate to $29.99 in October 2025 and then cutting prices in April 2026 while removing Call of Duty day-one access, represents a 180-degree strategic reversal within six months, which is not the behavior of a model that found its formula.

The diverging strategies of Microsoft and Sony in May 2026 are the industry's live experiment in which direction the model breaks, and neither direction looks like the "gaming utility" vision the Counterpoint predicted.

Sony announced PlayStation Plus price increases effective May 20, 2026, subscriptions rising to $10.99 in the US, €9.99 in Europe, £7.99 in the UK, while Microsoft simultaneously reduced Xbox Game Pass Ultimate from €26.99 to €20.99 and cut PC Game Pass from €14.99 to €12.99. Xbox Gaming CEO Asha Sharma stated the service had "simply become too expensive for players", a direct acknowledgment that the October 2025 price hike to $29.99 triggered enough cancellations to force a reversal within six months.

Sony raising prices while Microsoft cuts them is not a sign of a market finding equilibrium. It reflects two fundamentally different content economics reaching opposite conclusions: Sony's first-party exclusives retain full-price commercial value long enough that subscription inclusion after 12-24 months captures catalog value without destroying launch revenue; Microsoft's Activision Blizzard titles are too expensive to subsidize at consumer subscription price points.

The broader gaming industry context compounds the pressure: Microsoft Gaming revenue declined 7% year-over-year in Q3 FY2026, hardware revenue fell 33%, and Nintendo, the one major platform holder not heavily invested in subscription services, is experimenting with $10 digital discounts rather than subscription bundles, suggesting even the hardware market leader sees the subscription economics skeptically.

### The Point Roundup

The gaming subscription fatigue prediction has been validated by a more dramatic mechanism than the original piece anticipated: not gradual consumer disengagement but a publisher-side sustainability crisis that produced documented foregone revenue on a single game inclusion and a 180-degree strategic reversal within six months of a 50% price increase.

The 2027–2030 story is not about whether subscribers will abandon gaming subscriptions, the retention data suggests most won't, but about whether the economics of sustaining those subscriptions with content valuable enough to justify them can be made to work at the price points consumers will accept.

The May 2026 divergence, one platform retreating from premium content to cut prices, the other raising prices while protecting content economics, confirms the original prediction's core insight: the "Netflix for games" model that appeared inevitable three years ago has collided with the fundamental difference between episodic entertainment and $200-300 million interactive experiences that require return on investment that subscription pricing structurally undermines.

## THE COUNTERPOINT: The Formula Isn't Broken. It's Being Rebuilt in Real Time.

The Counterpoint's case rests on a data point the fatigue narrative consistently underweights: 120 million people worldwide pay for at least one gaming subscription service, the global market grew from $11.53 billion to $14.3 billion in a single year, and industry retention rates reached 81%, the highest on record.

PlayStation Plus reported a 7.8% churn rate, its lowest in five years.

These are not the numbers of a model in crisis. They are the numbers of a model in maturation, moving past the unsustainable growth rates of the pandemic era toward the steady-state retention economics that define durable subscription businesses.

The $300 million in foregone Call of Duty sales and the subsequent removal of day-one blockbuster access from Game Pass is, from the Counterpoint's perspective, not a failure of the subscription model but the correction that makes it economically sustainable: a library service calibrated to what catalog content is actually worth rather than a service attempting to subsidize $300 million franchises at $20 per month and discovering the math is impossible.

Microsoft's April 2026 price cut combined with the removal of Call of Duty day-one access is arguably the first step toward exactly the hybrid architecture the Counterpoint predicted, a subscription service that provides catalog depth, indie discovery, and mid-tier content at a price point consumers will maintain, alongside separate premium or purchase options for blockbuster titles. Game Pass retains 500 million monthly active users across the Xbox ecosystem regardless of subscriber count, the platform scale that represents the Counterpoint's "gaming utility" vision at ecosystem level if not at subscription revenue level.

The Microsoft-Sony pricing divergence of May 2026, which the fatigue narrative characterizes as evidence of model instability, is more accurately described as market differentiation revealing two distinct viable approaches to gaming subscription economics.

Microsoft's price cut targets volume growth across PC, cloud, and mobile audiences where catalog depth and Game Pass's 500 million monthly active user ecosystem create value without requiring blockbuster inclusion.

Sony's price increase reflects a premium positioning strategy for a platform where first-party exclusives retain full-price commercial value and a 7.8% churn rate, the lowest in PlayStation Plus history, demonstrates subscribers' willingness to pay.

Both strategies are coherent on their own content-economic terms.

Cloud and mobile gaming growth for Game Pass reached 89% year-over-year, suggesting the subscription model's strongest growth vector is precisely the context where catalog depth, device flexibility, and low-price sensitivity are the primary value drivers rather than blockbuster inclusion.

The Counterpoint's hybrid model prediction is already visible in early formation: Nintendo's $10 digital discounts, Sony's personalized dynamic pricing experiments, and Microsoft's tier restructuring all represent the industry moving toward a landscape where subscription and purchase coexist, and consumers self-select into the model that reflects their actual gaming behavior.

POINT vs COUNTERPOINT
GAMING SUBSCRIPTION FATIGUE
THE LIKELIHOOD
The counterpoint likelihood increases modestly from 40% to 42%, a calibration rather than a directional shift.
THE POINT MUST BE WRONG ABOUT
1. COSTS WILL STAY TOO HIGH
Subscription prices remain above the psychological threshold for mass adoption.
2. BLOCKBUSTERS CANNIBALIZE SALES
Day-one inclusion of major titles continues to cost more in lost sales than it gains in subscribers.
3. ENGAGEMENT WILL REMAIN SHALLOW
Subscribers play a few titles briefly, then churn before long-term value compounds.
4. HYBRID MODELS WON'T EMERGE
The industry fails to find the right mix of day-one, delayed, and premium tiers.
5. ECONOMICS WON'T IMPROVE
Unit economics stay negative or break-even at scale for publishers and platforms.
CURRENT STATE
Subscriptions continue to grow in users, but unit economics are under pressure. Microsoft is recalibrating pricing and content strategy after Black Ops 6 results, while Sony is raising prices. The industry is in active search of the right model.
DECISION FRAMEWORK
1 Track subscriber growth, churn, and engagement depth.
2 Monitor unit economics and content ROI across platforms.
3 Watch pricing elasticity and ARPU trends.
4 Evaluate the success of tiered and hybrid models.
5 Assess whether publishers achieve sustainable margins.
LIKELIHOOD COUNTERPOINT PREVAILS
42%
▲ +2% vs Original Prediction
(Was 40%)
The counterpoint gained modest ground as tiered strategies emerge, engagement tools improve, and players show continued willingness to subscribe—but structural risks remain.
THE COUNTERPOINT MUST BE RIGHT ABOUT
1. VALUE WILL BECOME CLEARER
Players perceive strong value from breadth, convenience, and ongoing content.
2. CONTENT STRATEGY WILL EVOLVE
Publishers find the right balance of day-one, delayed, and exclusive content.
3. RETENTION WILL IMPROVE
Better curation, personalization, and live service depth increase lifetime value.
4. HYBRID MODELS WILL SUCCEED
Different tiers and bundles serve different player segments profitably.
5. UNIT ECONOMICS TURN POSITIVE
Costs and pricing align, making subscriptions sustainable for publishers and platforms.
MOST CRITICAL VARIABLE
UNIT ECONOMICS AND CONTENT ROI
Whether subscriptions can deliver positive, sustainable margins for publishers by aligning pricing, content investment, and player lifetime value.
MOVEMENT VS ORIGINAL PREDICTION
The counterpoint likelihood increased from 40% to 42% (+2 points) as the industry experiments with new models and strategies. This is a calibration, not a directional shift.
2027–2030 OUTLOOK
The next 3–5 years will reveal whether the industry cracks the code—or confirms that game subscriptions have hard limits.

## What Needs to Change

### (...for the Counterpoint to Prevail)

For the hybrid model to become the definitive resolution by 2030, three structural conditions need to crystallize that are currently in early formation.

First, the publisher economics need to be restructured so that subscription inclusion generates revenue comparable to traditional windowed sales, Sony's reportedly estimated $85 million opportunity cost on Horizon Forbidden West's early subscription inclusion (per InGameNews commentary, not a Sony primary disclosure) demonstrates the magnitude of the problem. A windowed model where games enter subscription catalogs 12-24 months after launch, at which point full-price revenue has been largely captured, is the most commercially rational resolution, and Sony's own strategy increasingly reflects this.

Second, the cloud and mobile gaming growth vector, 89% year-over-year for Game Pass, needs to continue accelerating to the point where the economics of the cloud and mobile subscriber base compensate for the structural limitations of the blockbuster-on-console model.

Third, family plan and household-sharing economics need to reach the mainstream consumer rather than remaining pilot features, the value proposition for multi-gamer households is compelling but the rollout has been deliberately cautious.

If those three conditions materialize together by 2028-2029, the Counterpoint's "gaming utility" vision has a credible path for catalog and cloud gaming even if blockbuster day-one inclusion never returns.

If publisher economics remain unresolved, the $300 million Black Ops 6 lesson will repeat with every subsequent blockbuster, and the model will oscillate between unsustainable inclusion and value-destroying exclusion indefinitely.

## Likelihood Counterpoint Prevails: 42%

The Counterpoint likelihood increases modestly from 40% to 42%, a calibration rather than a directional shift, reflecting both genuine subscription resilience and the active industry experimentation with hybrid models, balanced against the $300 million Call of Duty finding and the contradictory May 2026 pricing strategies that represent structural evidence against the "perfecting the formula" scenario.

The 81% retention rate and $14.3 billion market demonstrate that gaming subscriptions are not collapsing, they are sustaining at scale while struggling to grow sustainably on the original premium content economics.

The Counterpoint's hybrid model thesis, a subscription service providing catalog depth and discovery at a lower price point alongside separate purchase options for blockbusters, is the most plausible resolution path, and Microsoft's April 2026 restructuring is arguably the first step toward exactly that architecture.

The 42% reflects a genuine and relatively balanced uncertainty: the subscriber numbers are real and sticky, the retention rates are at record highs, and the industry has strong financial incentives to make the model work in some form. But the $300 million in foregone Call of Duty sales, the 180-degree strategic reversal within six months of a 50% price increase, and the structural incompatibility between $200-300 million game economics and subscription pricing all argue that "perfecting the formula" requires solving a problem that currently has no demonstrated solution at the scale these platforms are operating.

The 42% is not a concession that the Counterpoint's scenario is close. It acknowledges that the subscriber retention data and the cloud gaming growth trajectory together represent a real and durable foundation that the original prediction's "reckoning" framing somewhat underweighted.

# Foldable Phones Breakthrough. Or Bust?

## From Niche to Pivotal: Foldables Cross the Threshold, But Which Threshold?

*The original prediction said Apple's entry would be the definitive test. The test is now approaching.*

*Samsung Display demonstrated a crease-free foldable panel at CES 2026, is producing screens for the iPhone Fold, and Bloomberg's Mark Gurman confirmed a fall 2026 launch alongside iPhone 18 Pro.*

*IDC has already forecast Apple capturing 22% of foldable unit share and 34% of foldable market value in its first year, for a device that hasn't shipped yet.*

*The question has changed from "will Apple enter?" to "what happens to the category when it does?" And the most honest answer the 2026 evidence supports: the category is validated.*

*The mainstream adoption question is still open, and the 62.3% six-month depreciation rate is the data point that best explains why.*

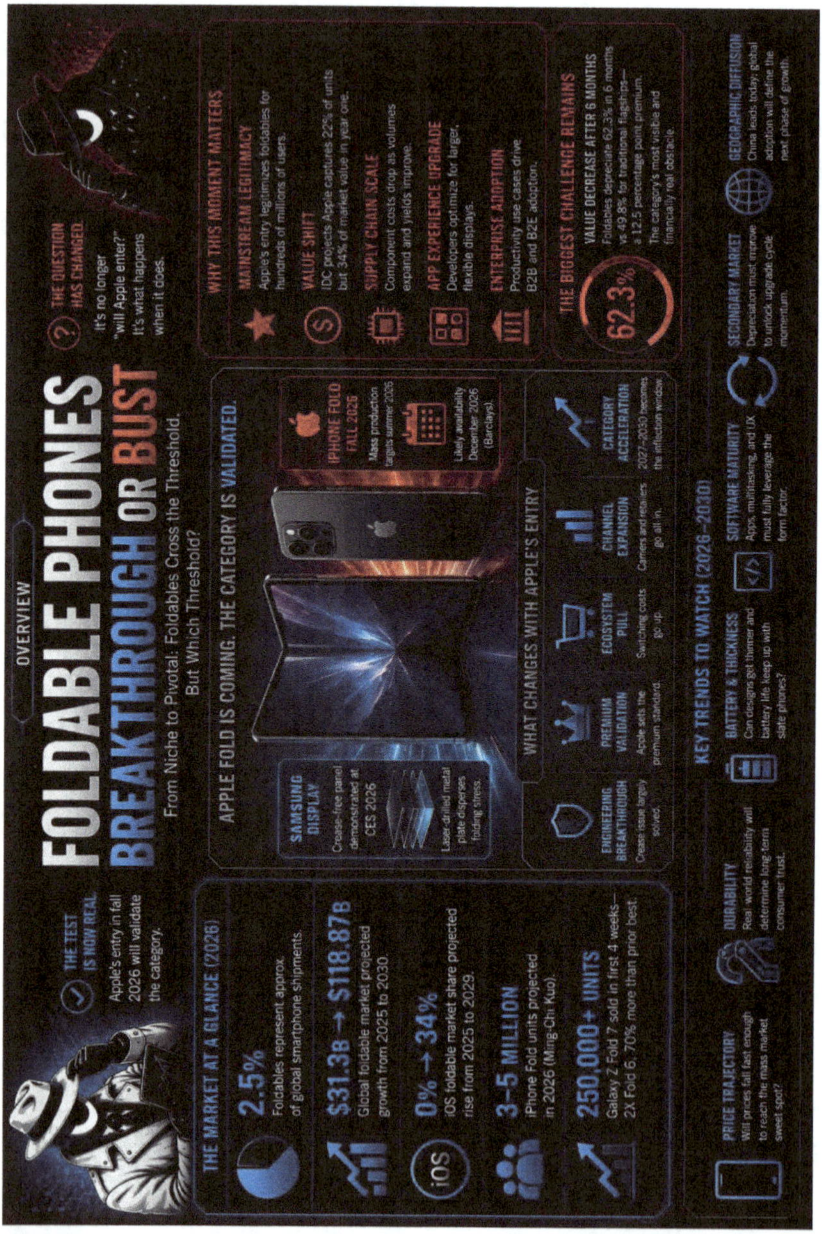
OVERVIEW
FOLDABLE PHONES
BREAKTHROUGH OR BUST
From Niche to Pivotal: Foldables Cross the Threshold. But Which Threshold?
THE TEST IS NOW REAL.
Apple's entry in fall 2026 will validate the category.
THE QUESTION HAS CHANGED.
It's no longer "will Apple enter?" It's what happens when it does.
THE MARKET AT A GLANCE (2026)
2.5%
Foldables represent approx. of global smartphone shipments.
$31.3B → $118.87B
Global foldable market projected growth from 2025 to 2030.
0% → 34%
iOS foldable market share projected rise from 2025 to 2029.
3–5 MILLION
iPhone Fold units projected in 2026 (Ming-Chi Kuo).
250,000+ UNITS
Galaxy Z Fold 7 sold in first 4 weeks—2X Fold 6, 70% more than prior best.
APPLE FOLD IS COMING. THE CATEGORY IS VALIDATED.
SAMSUNG DISPLAY
Crease-free panel demonstrated at CES 2026
Laser-drilled metal plate disperses folding stress.
IPHONE FOLD FALL 2026
Mass production targets summer 2026
Likely availability December 2026 (Barclays).
WHAT CHANGES WITH APPLE'S ENTRY
ENGINEERING BREAKTHROUGH
Crease issue largely solved.
PREMIUM VALIDATION
Apple sets the premium standard.
ECOSYSTEM PULL
Switching costs go up.
CHANNEL EXPANSION
Carriers and retailers go all in.
CATEGORY ACCELERATION
2027–2030 becomes the inflection window.
WHY THIS MOMENT MATTERS
MAINSTREAM LEGITIMACY
Apple's entry legitimizes foldables for hundreds of millions of users.
VALUE SHIFT
IDC projects Apple captures 22% of units but 34% of market value in year one.
SUPPLY CHAIN SCALE
Component costs drop as volumes expand and yields improve.
APP EXPERIENCE UPGRADE
Developers optimize for larger, flexible displays.
ENTERPRISE ADOPTION
Productivity use cases drive B2B and B2E adoption.
THE BIGGEST CHALLENGE REMAINS
62.3%
VALUE DECREASE AFTER 6 MONTHS
Foldables depreciate 62.3% in 6 months vs 49.8% for traditional flagships—a 12.5 percentage point premium.
The category's most visible and financially real obstacle.
KEY TRENDS TO WATCH (2026–2030)
PRICE TRAJECTORY
Will prices fall fast enough to reach the mass market sweet spot?
DURABILITY
Real-world reliability will determine long-term consumer trust.
BATTERY & THICKNESS
Can designs get thinner and battery life keep up with slate phones?
SOFTWARE MATURITY
Apps, multitasking, and UX must fully leverage the form factor.
SECONDARY MARKET
Depreciation must improve to unlock upgrade cycle momentum.
GEOGRAPHIC DIFFUSION
China leads today; global adoption will define the next phase of growth.

## Opening

The foldable phone story has changed more than almost any other topic in this series between the original writing and this update, not because the original prediction was wrong, but because its central conditional has moved from hypothetical to operational.

The question is no longer whether Apple will enter the foldable market and whether that entry will validate the category. Apple is preparing to enter, the production infrastructure is taking shape, the crease engineering problem the original prediction identified as the decisive technical challenge has been substantially addressed by Samsung Display's CES 2026 demonstration, and the market trajectory is already responding to the anticipated launch.

What the 2027-2030 analysis requires is a more precise framing of what "mainstream adoption" and "niche" actually mean for a $2,400 device, and an honest accounting of the structural ceiling that premium pricing imposes on even the most compelling hardware.

## Intro

The foldable phone prediction's 2026 update is defined by a single structural shift: the primary analytical question has changed. The original piece asked whether Apple's entry would validate the category.

That question now has an answer, yes, the category is validated, the crease engineering problem has been substantially addressed, and the market is already pricing in Apple's arrival before the device has shipped.

The question the 2027-2030 analysis inherits is more precise and more consequential: validated for whom, at what price, and on what timeline does "category validation" translate into the mainstream adoption threshold the original prediction specified? The 62.3% six-month depreciation rate is the data point that makes this question concrete rather than rhetorical.

It's not an aesthetic judgment about foldable phones or an engineering critique of their durability. It's the market's current verdict, expressed in resale economics, on what owning a foldable cost beyond the purchase price.

Apple's entry will change many things about the foldable category. Whether it changes that number, whether the depreciation premium narrows as the market matures and Apple's presence normalizes the form factor, is the specific variable that separates "premium category validation" from "mainstream adoption" more reliably than any unit shipment forecast.

## Background

Foldable devices currently account for approximately 2.5% of global smartphone shipments. The global foldable market is projected to grow from $31.3 billion in 2025 to $118.87 billion by 2030, driven by falling component costs and broadening enterprise adoption. iOS is projected to rise from 0% of the foldable market in 2025 to 34% of foldable market share by 2029, encoding Apple's expected dominance of the high-value end of the category within its first full product cycle.

The competitive landscape entering Apple's anticipated launch has developed materially: Samsung's Galaxy Z Fold 7 sold more than 250,000 units in its first four weeks, twice that of its predecessor, the Galaxy Z Fold 6, and 70% more than the previous best-selling foldable, with research firms characterizing the result as pushing foldable smartphones into the mass market in Western Europe. Samsung Display demonstrated a crease-free foldable panel at CES 2026 using a laser-drilled metal display plate that disperses folding stress; Samsung Display President Lee Cheong confirmed plans to mass-produce foldable OLED displays for a "North American client," universally interpreted as Apple.

Bloomberg's Mark Gurman confirmed an iPhone Fold launch targeting fall 2026 alongside iPhone 18 Pro, with mass production targeting summer 2026, though both Engadget and multiple supply chain analysts note production is running one to two months behind the original schedule.

Barclays analyst Tim Long is projecting December 2026 as the more realistic consumer availability date.

Ming-Chi Kuo projects 3-5 million iPhone Fold units in 2026, which, if achieved, would make it the best-selling foldable device ever in its debut year.

Against this: foldable devices experience a 62.3% value decrease after six months compared to 49.8% for traditional flagship devices, a 12.5 percentage point depreciation premium that represents the category's most concrete and financially visible consumer-facing obstacle.

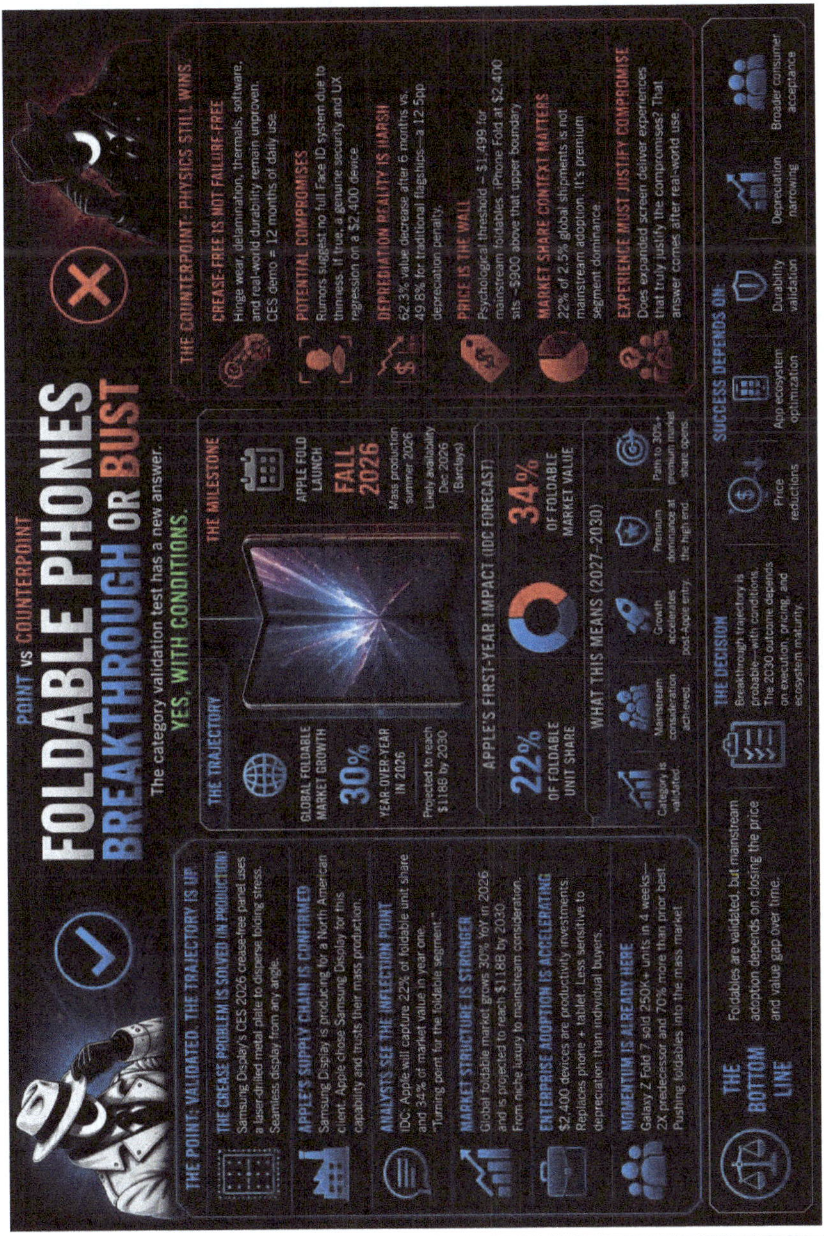
POINT vs COUNTERPOINT
FOLDABLE PHONES
BREAKTHROUGH OR BUST
The category validation test has a new answer.
YES, WITH CONDITIONS.
THE POINT: VALIDATED. THE TRAJECTORY IS UP.
THE CREASE PROBLEM IS SOLVED (IN PRODUCTION)
Samsung Display's CES 2026 crease-free panel uses a laser-drilled metal plate to disperse folding stress. Seamless display from any angle.
APPLE'S SUPPLY CHAIN IS CONFIRMED
Samsung Display is producing for a North American client. Apple chose Samsung Display for this capability and trusts their mass production.
ANALYSTS SEE THE INFLECTION POINT
IDC: Apple will capture 22% of foldable unit share and 34% of market value in year one. "Turning point for the foldable segment."
MARKET STRUCTURE IS STRONGER
Global foldable market grows 30% YoY in 2026 and is projected to reach $118B by 2030. From niche luxury to mainstream consideration.
ENTERPRISE ADOPTION IS ACCELERATING
$2,400 devices are productivity investments. Replaces phone + tablet. Less sensitive to depreciation than individual buyers.
MOMENTUM IS ALREADY HERE
Galaxy Z Fold 7 sold 250K+ units in 4 weeks—2X predecessor and 70% more than prior best. Pushing foldables into the mass market.
THE TRAJECTORY
GLOBAL FOLDABLE MARKET GROWTH
30%
YEAR-OVER-YEAR IN 2026
Projected to reach $118B by 2030
THE MILESTONE
APPLE FOLD LAUNCH
FALL 2026
Mass production summer 2026
Likely availability Dec 2026 (Barclays)
APPLE'S FIRST-YEAR IMPACT (IDC FORECAST)
22%
OF FOLDABLE UNIT SHARE
34%
OF FOLDABLE MARKET VALUE
WHAT THIS MEANS (2027–2030)
Category is validated.
Mainstream consideration achieved.
Growth accelerates post-Apple entry.
Premium dominance at the high end.
Path to 30%+ premium market share opens.
THE COUNTERPOINT: PHYSICS STILL WINS.
CREASE-FREE IS NOT FAILURE-FREE
Hinge wear, delamination, thermals, software, and real-world durability remain unproven. CES demo ≠ 12 months of daily use.
POTENTIAL COMPROMISES
Rumors suggest no full Face ID system due to thinness. If true, a genuine security and UX regression on a $2,400 device.
DEPRECIATION REALITY IS HARSH
62.3% value decrease after 6 months vs. 49.8% for traditional flagships—a 12.5pp depreciation penalty.
PRICE IS THE WALL
Psychological threshold ~ $1,499 for mainstream foldables. iPhone Fold at $2,400 sits ~$900 above that upper boundary.
MARKET SHARE CONTEXT MATTERS
22% of 2.5% global shipments is not mainstream adoption. It's premium segment dominance.
EXPERIENCE MUST JUSTIFY COMPROMISE
Does expanded screen deliver experiences that truly justify the compromises? That answer comes after real-world use.
THE BOTTOM LINE
Foldables are validated, but mainstream adoption depends on closing the price and value gap over time.
THE DECISION
Breakthrough trajectory is probable—with conditions. The 2030 outcome depends on execution, pricing, and ecosystem maturity.
SUCCESS DEPENDS ON:
Price reductions
App ecosystem optimization
Durability validation
Depreciation narrowing
Broader consumer acceptance

## THE POINT: The Category Validation Test Has a New Answer: Yes, With Conditions

The original prediction's central technical hypothesis, that Apple's crease-free engineering would be the decisive differentiator, has been confirmed in production preparation. Samsung Display demonstrated at CES 2026 a crease-free foldable panel that uses a laser-drilled metal display plate to disperse folding stress, showing completely seamless text display across the fold from any viewing angle.

This is not an incremental improvement. Every foldable since the Galaxy Fold in 2019, through seven Samsung generations, Google's Pixel Fold, and multiple OnePlus and OPPO competitors, has exhibited a visible crease. Samsung Display's CES 2026 demonstration is the first publicly confirmed crease-free production-ready technology. Apple chose Samsung Display as the supplier for the iPhone Fold specifically for this capability, with Ming-Chi Kuo noting the decision suggests Apple trusts Samsung Display more than its own in-house capability to ensure stable mass production.

The supply chain confirmation, Samsung Display President publicly acknowledging production for a North American client, gives the crease-free claim manufacturing credibility rather than rendering it purely a pre-launch marketing assertion.

The IDC vice president of client devices stated directly: "The launch of Apple's first foldable iPhone will mark a turning point for the foldable segment. Apple tends to be a catalyst for mainstream adoption of new categories."

That characterization, from the research organization forecasting Apple at 22% of foldable unit share and 34% of foldable value share in year one, reflects institutional conviction about the category inflection, not speculative enthusiasm.

The market structure entering Apple's anticipated launch is more favorable than the original prediction's framing anticipated, and the 2030 trajectory reflects that structural improvement.

The global foldable market growing 30% year-over-year in 2026, driven by Apple's anticipated entry alongside Samsung's Galaxy Z Trifold, represents acceleration from the 6% growth rate in prior forecasts. Foldable smartphones have crossed the threshold from experimental luxury to mainstream consideration, with enterprise procurement steadily broadening the buyer profile beyond tech enthusiasts.

The enterprise adoption vector deserves specific attention: a $2,400 device that consolidates phone and tablet functionality is not a premium luxury purchase in an enterprise context, it's a justifiable productivity investment that removes the need to carry two devices, and enterprise procurement cycles are substantially less sensitive to the depreciation rate that constrains individual consumer adoption.

The Galaxy Z Fold 7's performance in Western Europe, characterized by SamMobile as "pushing foldable smartphones into the mass market, confirms the trajectory is not solely Apple-dependent, even as Apple's entry is expected to accelerate it materially. The 2027-2030 window is where the original prediction's core question, breakthrough or permanent niche, gets its answer.

The evidence entering that window is more favorable to the breakthrough scenario than the original piece's framing anticipated.

### The Point Roundup

The foldable breakthrough scenario has moved from conditional to probable between the writing of the original prediction and this update.

Apple's entry is taking shape, the crease problem has been substantially addressed, the market is growing at 30% year-over-year with analyst projections of $118 billion by 2030, and the category has already crossed into mainstream consideration.

What hasn't changed is the pricing reality: at an expected $2,400, the iPhone Fold addresses the premium end of a premium market, and the distance between "premium mainstream" and "mass market" is measured in price point reductions that won't arrive until the second or third iPhone Fold generation.

The 2027-2030 window will see the category validated, the question the original prediction asked now has its answer.

Whether that validation produces the 30%+ of premium phone market share the piece projected by 2027 depends on how quickly Apple's entry drives down category pricing, how fast the app ecosystem optimizes for the form factor, and whether the depreciation gap narrows as the market matures.

## THE COUNTERPOINT: The Physics Still Wins: Crease-Free Doesn't Mean Failure-Free

The Counterpoint's engineering argument has been partially answered by Samsung Display's crease-free technology, but CES 2026 demonstrations and production confirmations are not the same as twelve months of real-world consumer use.

The crease has been the most visible and most photogenic foldable problem. The less photogenic problems, hinge wear over thousands of cycles, display delamination after sustained use, thermal management across dual screens, software optimization gaps, and the fundamental question of whether a book-style foldable's expanded screen delivers experiences that genuinely justify its compromises rather than replicating tablet experiences in a phone-sized package, are only answered after extended consumer deployment, not at launch.

There are also rumored hardware compromises in the iPhone Fold worth noting: reports suggest Apple may omit a full Face ID TrueDepth camera system due to the device's extreme thinness when opened. If confirmed, this would represent a genuine security and UX regression on a device positioned as Apple's most premium smartphone, the specific "insufficient practical benefits versus compromises" dynamic the Counterpoint described. Apple has solved the most photogenic problem. The less photogenic ones will only be answered in the months following consumer availability.

The pricing reality remains the Counterpoint's strongest argument, and it has been sharpened rather than softened by the market data available since the original piece was written.

Foldable devices experience a 62.3% value decrease after six months compared to 49.8% for traditional flagships, a 12.5 percentage point depreciation premium that represents a concrete, documented, and financially visible penalty for foldable ownership. Unlike crease visibility, which requires direct comparison to notice, the depreciation gap materializes explicitly in trade-in valuations, resale markets, and carrier upgrade program economics, contexts where consumers and procurement teams make explicit comparisons.

Industry analysis identifies a psychological threshold around $1,499 for mainstream foldable adoption, close enough to premium flagship pricing to justify the form factor premium, but below the $1,799+ that relegates devices to enthusiast territory. The iPhone Fold's expected $2,400 entry price sits approximately $900 above that upper boundary. IDC's forecast of 22% foldable unit share and 34% foldable value share for Apple in year one is impressive, but the total foldable market currently represents approximately 2.5% of global smartphone shipments.

22% of 2.5% is not mainstream adoption. It's premium segment dominance of a category that hasn't yet crossed the adoption thresholds the original prediction specified. A commercially valuable outcome for Apple, but a different claim than "30%+ of the premium phone market" the original prediction identified as the validation threshold.

POINT VS COUNTERPOINT
FOLDABLE PHONES
THE VERDICT
The counterpoint likelihood declines from 30% to 28%. A modest but meaningful revision.
THE POINT: VALIDATED WITH CONDITIONS
CREASE-FREE BREAKTHROUGH
Samsung Display's CES 2026 demo delivers the first publicly confirmed crease-free foldable panel.
APPLE IS ENTERING
Supply chain confirmed. Production targeting summer 2026. Launch in fall 2026.
ANALYSTS SEE THE TURNING POINT
22%
34%
IDC projects 22% of units and 34% of value in year one for Apple.
MARKET ACCELERATION
Foldable market grows 30% YoY in 2026 to $118B by 2030.
ENTERPRISE ADOPTION EXPANDS
$2,400 devices drive productivity use cases and broaden buyer profiles.
28%
COUNTERPOINT LIKELIHOOD
PREVIOUSLY 30% → NOW 28%
WHY THE POINT LEADS
CATEGORY VALIDATED
Apple's entry and crease-free tech legitimize the category.
MAINSTREAM MOMENTUM
Foldables are no longer niche—they're entering the mainstream.
GROWTH ACCELERATING
Strong market momentum into 2027–2030.
THE TEST CONTINUES
Adoption scale depends on pricing, ecosystem, and real-world durability.
THE COUNTERPOINT: CONCERNS REMAIN REAL
REAL-WORLD DURABILITY UNKNOWN
Hinge wear, delamination, heat, and long-term reliability won't be proven until months of real-world use.
POSSIBLE COMPROMISES
Rumors of no full Face ID TrueDepth due to thinness could mean security & UX tradeoffs.
$2,400
PRICE ABOVE MAINSTREAM THRESHOLD
~$900 above the ~$1,499 psychological adoption threshold.
DEPRECIATION PENALTY PERSISTS
62.3% value loss in 6 months vs 49.8% for flagships (12.5pt premium).
2.5%
STILL A SMALL CATEGORY TODAY
Foldables are ~2.5% of global smartphone shipments.
CURRENT STATE (2026)
2.5%
OF GLOBAL SMARTPHONE SHIPMENTS
Foldables remain a small slice of the market.
$31.3B
2025 MARKET SIZE
Global foldable market today.
250K+
GALAXY Z FOLD 7 SOLD IN 4 WEEKS
Strong early performance in Western Europe.
FALL 2026
IPHONE FOLD LAUNCH TARGET
Production underway with Samsung Display.
THE PIVOTAL WINDOW
2026–2030
MOST IMPORTANT VARIABLE (THE MOVE THAT MATTERS)
PRICING TRAJECTORY
Will component costs fall fast enough to bring prices to the $1,499 threshold?
DURABILITY PROOF
Real-world reliability over 12–24 months on reliability and lifespan.
APP ECOSYSTEM DEPTH
Will apps reach the quality and optimization needed for the foldable form factor?
DEPRECIATION GAP
Will the 12.5pt penalty narrow as the market matures and supply grows?
ENTERPRISE & MARKET EXPANSION
Broader use cases and emerging markets drive scale beyond enthusiasts.
THE BREAKTHROUGH SCENARIO IS NOW PROBABLE. | UNIVERSAL MAINSTREAM ADOPTION REMAINS PRICE-DEPENDENT

## What Needs to Change

### (...for the Counterpoint to Prevail)

For the "permanent niche" scenario to hold through 2030, two specific conditions need to persist that the current trajectory is actively working against.

First, the iPhone Fold's real-world reliability record over 12-18 months of consumer use needs to confirm the durability concerns that pre-launch engineering claims can't definitively address, specifically hinge wear, display delamination, and the long-term performance of Samsung Display's crease-free technology under real-world stress.

If Apple's quality control has genuinely solved these problems, the reliability argument collapses in 2027 as the review record accumulates. If it hasn't, the first wave of highly visible repair and failure stories will reinforce the consumer hesitancy that has kept the category niche despite Samsung's seven generations of improvement.

Second, the price needs to remain above the mainstream threshold through the iPhone Fold 2 or 3 generation, which requires component costs not declining at the pace that Apple's ecosystem historically drives through supplier relationships and manufacturing scale. Falling component costs are explicitly identified as a driver of the projected $118 billion foldable market by 2030; if those costs fall fast enough to bring a third-generation iPhone Fold to $1,800 or below, the mainstream adoption scenario becomes structurally achievable and the Counterpoint's pricing argument loses its force.

The depreciation rate differential, 62.3% versus 49.8%, is the most structural of the remaining barriers, because it persists regardless of engineering improvements and requires market maturation (broader resale ecosystem, carrier program integration) rather than manufacturing cost reduction to resolve.

## Likelihood Counterpoint Prevails: 28%

The Counterpoint likelihood declines from 30% to 28%, a modest but meaningful revision that reflects the genuine progress on crease engineering demonstrated at CES 2026, the Galaxy Z Fold 7's mainstream market breakthrough in Western Europe, and the production confirmation of Apple's fall 2026 launch target.

The original 30% was calibrated against genuine uncertainty about whether Apple could solve the crease problem, that uncertainty has been substantially resolved by Samsung Display's publicly demonstrated technology and confirmed production role. What hasn't moved in the Counterpoint's favor is the depreciation differential, 62.3% versus 49.8% in six months, which remains the most concrete financial obstacle to mainstream adoption and hasn't improved materially despite the category's technical progress.

The $1,499 mainstream adoption threshold versus the iPhone Fold's expected $2,400 launch price leaves a $900 gap that no single-generation engineering improvement will close.

The 28% says: the category is no longer in danger of permanent niche status, Apple's entry has foreclosed that outcome, but the "premium category for anyone willing to pay luxury prices" scenario remains the 2027-2028 reality until a second or third generation iPhone Fold arrives at a price point that the mainstream adoption threshold requires. The breakthrough scenario has moved from conditional to probable.

Universal mainstream adoption, the specific claim the original prediction made, remains contingent on cost declines that manufacturing history suggests will require multiple product generations to achieve.

## Streaming Service Consolidation Wave

### The Cable TV Model Returns. Just Wearing Different Clothes

*Paramount made an all-cash offer for Warner Bros. Discovery at $108.4 billion enterprise value. Netflix made a competing offer. Paramount won, agreeing to pay WBD's $2.8 billion Netflix termination fee as part of the deal.*

*BET+ shut down and was folded into Paramount+. Hulu is being absorbed into Disney+.*

*The consolidation the original prediction described as coming by end of 2026 arrived, and it arrived with the world's dominant streaming platform in a bidding war for its largest remaining competitor.*

*The question for 2030 is not whether the consolidation happened.*

*It's whether what it left behind looks like a competitive marketplace or the cable TV bundle that streaming was supposed to replace.*

STREAMING SERVICE CONSOLIDATION WAVE
THE CABLE MODEL RETURNS.
JUST WEARING DIFFERENT CLOTHES.
The consolidation happened. Paramount outbid Netflix for Warner Bros. Discovery in a deal valued at $108.4 billion enterprise value. BET+ shut down. Hulu is being absorbed into Disney+.
The question is no longer whether streaming consolidates. It's whether the survivors create a better marketplace, or rebuild the system consumers left behind.
THE EVIDENCE IS REAL
$108 BILLION MERGER
Paramount secured Warner Bros. Discovery in one of the largest entertainment deals in decades.
NETFLIX: 325 MILLION SUBSCRIBERS
Projected 2026 revenue: $50.7B–$51.7B
Ad revenue expected to exceed: $3 billion
THE GIANTS ARE FORMING
Disney + Hulu: ~196 million subscribers
ESPN+: 24 million
Paramount-WBD: 200+ million
CHURN CHANGED THE GAME
Subscriber growth slowed.
Switching behavior increased.
Scale became survival.
AD TIERS WON
Ad-supported subscriptions accounted for all SVOD growth in 2026.
NETFLIX
Disney+
hulu
ESPN+
prime video
Paramount+
WBD
CABLE TV
HBO max
discovery+
peacock
tubi
THE STREAMING WARS DIDN'T END.
THEY CONSOLIDATED.
THE CHALLENGE REMAINS
PRICING POWER RETURNS
Fewer competitors can mean higher prices.
CONTENT RISK INCREASES
Debt servicing pressures can reduce investment.
THE BUNDLE RETURNS
Consumers may again need multiple services to access what they want.
CREATIVE DIVERSITY MAY SHRINK
IP ecosystems reward familiarity over experimentation.
CONVENIENCE CAN BECOME DEPENDENCE
A smoother experience isn't always a more competitive market.
THE QUESTION
Will streaming's new oligopoly deliver sustainable economics and a better consumer experience...
or recreate the cable TV model that streaming originally disrupted?

## Opening

The original prediction has been validated so comprehensively that the update requires less analysis of whether consolidation occurred and more analysis of what the specific form it took means for the 2027-2030 consumer experience.

The consolidation wave didn't merely arrive, it arrived at a scale that exceeded the original prediction's framing of "at least two major platforms shutting down or merging." The Paramount-WBD merger at $108.4 billion enterprise value, Netflix's competing bid that was outbid by Paramount's superior offer, BET+'s shutdown, Hulu's absorption into Disney+, and multiple cable network closures together constitute exactly the "major consolidation wave" the original piece predicted.

The 2027-2030 analysis is no longer about whether consolidation happens. It's about what the post-consolidation streaming market does to consumers: whether the oligopoly forming now delivers the content quality and pricing rationalization the Counterpoint's optimistic scenario predicted, or whether it recreates the cable TV dynamics that originally drove consumers to streaming in the first place.

## Intro

The streaming consolidation prediction occupies unusual analytical territory in this series: it has been validated so completely, and so precisely on the timeline it specified, that the update's primary work is not assessing whether the prediction was right but understanding what being right actually means for consumers.

The original piece predicted a major consolidation wave by end of 2026. A $108.4 billion enterprise value merger, a competing bid from the world's dominant streaming platform, BET+'s shutdown, Hulu's absorption, and multiple cable network closures together constitute a consolidation event that exceeds the prediction's own framing in scale.

That level of confirmation creates a specific analytical obligation: when a prediction lands this accurately, the honest question is whether it landed accurately because the analysis was correct or because the structural forces it identified were so overdetermined that any reasonably calibrated observer would have reached the same conclusion.

The answer here is the former.

The prediction correctly identified the specific economic mechanisms, the churn dynamics, the subscriber saturation ceiling, and the content cost pressures that made consolidation inevitable before those mechanisms had fully played out in the data.

What the update adds is the consumer experience question the prediction's consolidation framing necessarily deferred: the oligopoly has formed on schedule, the 4-5 dominant platforms the Counterpoint predicted have arrived through exactly the corporate consolidation the original prediction described, and whether that structure delivers competitive pricing and content quality or recreates the cable TV bundle that streaming was supposed to replace is what the next five years will answer.

## Background

Netflix ended 2025 with 325 million global subscribers, expects full-year 2026 revenue between $50.7 and $51.7 billion representing 12-14% growth, and projects ad revenue to double to over $3 billion in 2026 alone.

Disney last disclosed 195.7 million subscribers across Disney+ and Hulu combined, plus 24.1 million at ESPN+.

Warner Bros. Discovery expects to exceed 150 million subscribers by end of 2026.

The combined Paramount-WBD entity would have more than 200 million direct-to-consumer subscribers across 100+ countries.

The post-consolidation landscape produces a structure where Netflix, Amazon, Disney, and the Paramount-WBD combined entity each hold approximately 200-325 million subscribers, remarkably close to the "4-5 dominant platforms" the Counterpoint predicted as the sustainable equilibrium, arrived at through the corporate consolidation the original prediction described.

The consolidation is driven by three converging forces: churn rates that have risen significantly from 2019 levels as subscriber saturation sets in, the plateauing of the global subscriber market as penetration reaches a ceiling in developed markets, and the necessity of achieving the content scale and cost efficiency that only major platform size enables.

Ad-supported tiers accounted for all SVOD subscription growth in 2026, with average revenue per service climbing from $10.96 to $11.30, confirming the revenue diversification thesis the Counterpoint identified as the sustainable economics model.

POINT vs COUNTERPOINT
STREAMING SERVICE CONSOLIDATION WAVE
THE CABLE TV MODEL RETURNS. JUST WEARING DIFFERENT CLOTHES.
Two visions for what the post-consolidation streaming market means for consumers.
THE POINT: CONSOLIDATION CAN DELIVER A BETTER MARKETPLACE
SCALE CREATES SUSTAINABLE ECONOMICS
Only major scale supports premium content investment, global distribution, and long-term profitability.
CONTENT QUALITY SHOULD IMPROVE
Larger libraries, deeper IP ownership, and rationalized investment lead to more hits, less duplication, and better experiences.
PRICING RATIONALIZATION IS NEEDED
Consumers face too many services. Consolidation can eliminate waste and stabilize pricing over time.
AD TIER GROWTH BENEFITS CONSUMERS
Ad-supported tiers drove all SVOD growth in 2026, delivering lower prices and more choice for budget-conscious viewers.
A MORE STABLE INDUSTRY SUPPORTS CREATORS
Financial stability means stronger partnerships, better deals for top talent, and more opportunities for storytellers.
THE BATTLEGROUND
CONSOLIDATION IS HAPPENING. THE QUESTION IS WHAT IT CREATES.
NETFLIX
Disney+
hulu
ESPN+
prime video
Paramount+
WBD
WHAT'S AT STAKE
CONSUMER CHOICE
More value or fewer options?
PRICING POWER
Stability or higher costs?
CONTENT ECOSYSTEM
More investment or less diversity?
ACCESS & CONVENIENCE
Simpler or locked in?
LONG-TERM COMPETITION
Healthy market or durable oligopoly?
THE COUNTERPOINT: CONSOLIDATION RECREATES CABLE DYNAMICS
LESS COMPETITION MEANS MORE PRICING POWER
Fewer major players lead to higher prices, fewer promotional offers, and less leverage for consumers.
CHOICE & DIVERSITY WILL SHRINK
Library overlap means more titles disappear. Algorithms and scale favor safe bets over creative risk.
THE BUNDLE WILL RETURN
No single service has everything. Consumers are forced to subscribe to multiple services—again.
DEBT & COST CUTTING CREATE RISK
High acquisition debt pressures margins, leading to content cuts, fewer originals, and higher ad loads.
AD TIERS CAN BECOME THE DEFAULT
The push to ad tiers is about revenue, not choice. Expect more ads and tiering complexity.
ADS
THE DEBATE IN ONE QUESTION
Will consolidation create a healthier, more sustainable streaming marketplace that delivers better content at fair prices—or recreate the cable TV model consumers rejected?

## THE POINT: The Graveyard Arrived. It's Orderly, Corporate, and Expensive

The original prediction's consolidation thesis has been confirmed at a magnitude that exceeds its own framing, and the specific mechanics of the Paramount-WBD deal reveal structural dynamics the prediction correctly identified. Warner Bros.

Discovery CEO David Zaslav proposed separating WBD's streaming and studio businesses from its cable and network businesses, attracting Netflix's interest in the streaming assets, a separation that itself confirms the prediction's core insight that cable network economics and streaming economics can't coexist sustainably within the same corporate structure.

Netflix made a competing offer ranging from $21.23 to $27.75 per share in cash-and-equity consideration. Paramount Skydance made an all-cash offer at $30 per share, subsequently enhanced to $31 per share at approximately $108.4-111 billion enterprise value, and agreed to pay WBD's $2.8 billion Netflix termination fee to secure the deal.

The bidding war itself, between the world's dominant streaming platform and the second-tier survivor most dependent on scale to remain viable, is the market's clearest signal that the consolidation imperative is structural rather than opportunistic.

Paramount's own SEC filings argued against the Netflix deal on the grounds that it would give Netflix "an estimated 43 percent share of global subscription video-on-demand viewers, more than twice its closest competitor", a concentration that would have delivered higher prices for consumers and reduced competition for content creators.

The Paramount-WBD merger also raises consequential questions about the streaming services in the combined portfolio, analysts note they "would be surprised if Discovery Plus continues to be a standalone option two years from now."

The consolidation is not finished. The headline merger is the first wave. The integration decisions of 2026-2028, which sub-platforms survive, which content libraries are rationalized, how debt servicing constrains content investment, are the second wave that determines the 2030 consumer experience.

The consumer-facing consequences of the consolidation are already materializing in the pricing dynamics the original piece predicted and the Counterpoint hoped to avoid.

Analysts frame the emerging market structure directly: price hikes are as much about driving consumers to ad-supported tiers as they are about revenue growth per se, the consolidated platform landscape removes the competitive friction that historically constrained how aggressively platforms could raise prices.

BET+ shut down and was folded into Paramount+. Hulu is being folded into Disney+, available under a unified interface for bundle subscribers.

WOW! cable migrated its customers to YouTube TV and exited the TV business entirely.

The cable network closures the original piece predicted have continued: HBO Family, ThrillerMax, MovieMax, OuterMax, and Universal Kids all closed in 2025.

While the re-bundling reduces the friction of managing multiple subscriptions, the loss of competition may lead to stagnating innovation and content homogenization, with mergers prioritizing "IP ecosystems" where a single story is told across movies, series, and interactive games to maximize existing intellectual property lifecycle and reduce the risk of launching unproven new brands.

The streaming industry built its audience by disrupting the cable TV model's expensive bundles and weak consumer choice.

The consolidated streaming oligopoly of 2026 is reconstructing that model, with better user interfaces and on-demand access, but with the same fundamental economics of a few dominant players with pricing power and reduced competitive pressure to innovate.

The average American household spends approximately $46 per month on streaming across 2.9 platforms, a figure that, compounded by ongoing annual price increases across 4-5 consolidated platforms, is on a trajectory toward the $80-100 monthly cable bill that originally drove cord-cutting.

## The Point Roundup

The original prediction has been validated at a scale that requires calibration, not correction. The consolidation wave has arrived, is documented in SEC filings, and has produced the oligopoly structure the piece predicted. The "graveyard" arrived, it's just orderly, corporate, and $108 billion expensive, which is exactly how media consolidation waves end in every era of entertainment history.

The 2027-2030 consumer experience will be determined by three variables: whether the Paramount-WBD merger survives regulatory review and delivers operational efficiency without the content investment cuts that debt servicing pressure typically forces; whether the pricing power of the consolidated oligopoly drives consumer costs toward the cable TV levels that triggered the original cord-cutting movement; and whether the IP ecosystem content strategy that consolidation enables produces genuinely better programming or simply fewer opportunities for diverse voices and original creative risk-taking.

The streaming wars ended not with a single victor but with four dominant platforms whose scale was achieved through the consolidation the original piece predicted, and whose consumer impact is the analytical question the 2027-2030 evidence will answer.

## THE COUNTERPOINT: The Re-Bundling Is Not the Return of Cable. It's Cable Done Right.

The Counterpoint's core prediction, 4-5 dominant platforms with complementary content libraries, bundling strategies reducing churn, and ad-supported tiers providing sustainable economics, has been confirmed by the 2026 market structure with a precision that the original 45% likelihood substantially undervalued.

Netflix at 325 million subscribers with $50+ billion in expected 2026 revenue and ad revenue doubling to over $3 billion, Disney at approximately 220 million combined subscribers, Amazon at 220 million, and the Paramount-WBD merger targeting 200 million combined, produces exactly the four-platform oligopoly the Counterpoint described as the sustainable equilibrium. This outcome arrived through corporate consolidation driven by the same subscriber saturation and churn economics the original prediction identified as structural problems, which means both the prediction and the Counterpoint have been validated simultaneously: the original prediction was right that the economics were unsustainable at the fragmented market scale; the Counterpoint was right that 4-5 dominant platforms represents the sustainable resolution.

Disney has already begun folding Hulu content into Disney+ for bundle subscribers with plans to complete the integration in 2026, precisely the bundling integration the Counterpoint identified as the path to sustainable churn reduction.

Netflix's profit grew from $1.87 billion to $2.42 billion year-over-year, demonstrating that the Counterpoint's "anchor of a sustainable duopoly" thesis is operationally validated at the platform level that sets the economics for the entire industry.

The Paramount-WBD merger, despite its debt burden, produces a genuinely more competitive streaming landscape than the alternative Netflix acquiring WBD, would have created. Paramount's own SEC filings made this case directly to WBD shareholders: the Netflix transaction would have given Netflix an estimated 43% share of global SVOD viewers, more than twice its closest competitor, delivering higher prices for consumers, reduced compensation for content creators, and significant harm to theatrical exhibitors.

The Paramount deal preserves theatrical release commitments, prevents Netflix from achieving unchecked dominance, and maintains the competitive friction between at least four major platforms that keeps pricing and content investment honest.

The advertiser dimension further supports the Counterpoint's sustainability thesis: the Paramount-WBD merger produces more ad-supported distribution options across cable, theatrical, streaming, and broadcast rather than consolidating into streaming-only distribution, eMarketer confirmed the deal simplifies ad buying while maintaining competitive pricing pressure across formats.

Ad-supported tiers have expanded the total addressable market by offering lower-cost entry points that didn't exist in the cable era, the ability to access content for $6-8 per month with advertising represents a genuine consumer benefit the original cable model never provided.

## What Needs to Change

### (...for the Counterpoint to Fully Prevail)

For the sustainable equilibrium scenario to fully materialize through 2030 rather than devolving into the cable TV recreation the prediction's pessimistic scenario described, three conditions need to hold that are currently under pressure.

First, the Paramount-WBD merger needs to clear regulatory review, the deal targets closing by end of 2026 with a $7 billion regulatory termination fee committed, and deliver its promised operational synergies without the debt burden forcing the content spending cuts that Yale's analysis identified as the primary risk. The cost savings may hurt revenue if they reduce production quality and cause subscribers to exit, human capital synergies, the most cited justification for media M&A, have a consistent history of underdelivering on executive projections.

Second, the pricing discipline the consolidated duopoly enables needs to stop short of the consumer revolt threshold. Average revenue per SVOD service climbed from $10.96 to $11.30 in a single year, modest in isolation, but compounded across 4-5 services for 3-5 years, it produces the $80-100 monthly aggregate bill that cable commanded at its peak and that originally drove cord-cutting.

Third, content diversity and creative risk-taking need to survive the IP ecosystem consolidation dynamic.

The "tell one story across movie, series, and game" strategy maximizes existing franchise economics while systematically reducing investment in new IP that historically drove the original creative value proposition of streaming over cable.

## Likelihood Counterpoint Prevails: 55%

The Counterpoint likelihood increases from 45% to 55%, a 10-point move that warrants explanation.

The original 45% was assigned when the "4-5 dominant platforms" outcome was a prediction; it has now arrived as a documented market structure. That structural confirmation is what drives the revision upward: the Counterpoint's predicted resolution mechanism, consolidation to a sustainable oligopoly rather than chaotic collapse, is precisely what has occurred.

Netflix's $50+ billion revenue trajectory, doubling ad revenues, and profit growth from $1.87 to $2.42 billion confirm the Counterpoint's "anchor of a sustainable duopoly" thesis at the platform level that sets industry economics.

Ad-supported tiers accounting for all 2026 SVOD subscription growth confirms the revenue diversification thesis.

The 55% does not reflect a prediction that the consolidated streaming market will be good for consumers, that question remains genuinely open.

It reflects that the *structural* outcome the Counterpoint predicted has arrived while the *consumer experience* outcome the Counterpoint predicted as the benefit of that structure is still being tested.

What keeps the number at 55% rather than higher is the Paramount-WBD merger's debt burden creating real content investment risk; the pricing trajectory showing aggregate subscriber costs rising toward the cable bill levels that originally drove cord-cutting; and the IP ecosystem consolidation dynamic reducing the creative diversity that distinguished streaming from cable.

The 55% says: the Counterpoint's structural prediction has been confirmed.

Whether the consumer experience that emerges from that structure validates the Counterpoint's value proposition or confirms the prediction's "cable in different clothes" warning, is what the 2027-2030 window will answer.

## Cashless Society Backlash

### The War on Cash Became a War for Cash, And the Score Isn't Close

*Hungary enshrined cash as a constitutional right in April 2025.*

*Slovakia had done the same in 2023. Slovenia's National Assembly voted 61-to-29 in December 2025.*

*Switzerland, 73.4% of voters in March 2026. Four European constitutions in three years now protect the right to pay with physical currency.*

*New York's mandatory cash acceptance law took effect March 21, 2026.*

*The US Senate advanced CBDC-banning legislation with an 84-6 procedural vote, the most lopsided bipartisan payment-policy vote in recent memory.*

*The war on cash became a war for cash. The direction is not ambiguous.*

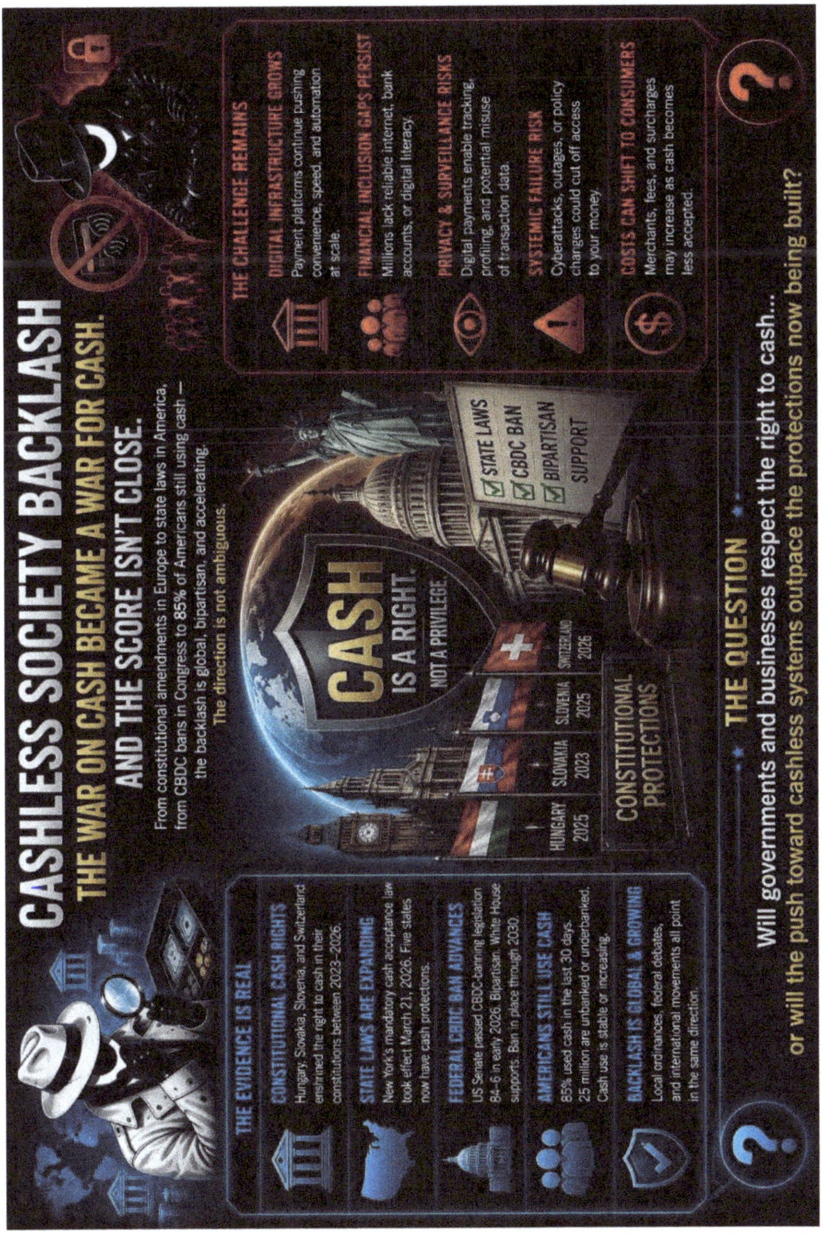
CASHLESS SOCIETY BACKLASH
THE WAR ON CASH BECAME A WAR FOR CASH.
AND THE SCORE ISN'T CLOSE.
From constitutional amendments in Europe to state laws in America, from CBDC bans in Congress to 85% of Americans still using cash — the backlash is global, bipartisan, and accelerating.
The direction is not ambiguous.
THE EVIDENCE IS REAL
CONSTITUTIONAL CASH RIGHTS
Hungary, Slovakia, Slovenia, and Switzerland enshrined the right to cash in their constitutions between 2023–2026.
STATE LAWS ARE EXPANDING
New York's mandatory cash acceptance law took effect March 21, 2026. Five states now have cash protections.
FEDERAL CBDC BAN ADVANCES
US Senate passed CBDC-banning legislation 84–6 in early 2026. Bipartisan. White House supports. Ban in place through 2030.
AMERICANS STILL USE CASH
85% used cash in the last 30 days. 25 million are unbanked or underbanked. Cash use is stable or increasing.
BACKLASH IS GLOBAL & GROWING
Local ordinances, federal debates, and international movements all point in the same direction.
CASH
IS A RIGHT.
NOT A PRIVILEGE.
HUNGARY 2025
SLOVAKIA 2023
SLOVENIA 2025
SWITZERLAND 2026
CONSTITUTIONAL PROTECTIONS
STATE LAWS
CBDC BAN
BIPARTISAN SUPPORT
THE CHALLENGE REMAINS
DIGITAL INFRASTRUCTURE GROWS
Payment platforms continue pushing convenience, speed, and automation at scale.
FINANCIAL INCLUSION GAPS PERSIST
Millions lack reliable internet, bank accounts, or digital literacy.
PRIVACY & SURVEILLANCE RISKS
Digital payments enable tracking, profiling, and potential misuse of transaction data.
SYSTEMIC FAILURE RISK
Cyberattacks, outages, or policy changes could cut off access to your money.
COSTS CAN SHIFT TO CONSUMERS
Merchants, fees, and surcharges may increase as cash becomes less accepted.
THE QUESTION
Will governments and businesses respect the right to cash...
or will the push toward cashless systems outpace the protections now being built?

## Opening

The cashless society backlash prediction hasn't just been validated, it has been exceeded in scope and formalized in ways that extend its consequences well beyond the 2026-2027 window originally specified.

What the 2026 update reveals is a global convergence of legislative, constitutional, and regulatory action across jurisdictions that share no common political system, currency, or culture, yet have reached the same conclusion: the elimination of cash is not merely an inconvenience for vulnerable populations, it's a structural transfer of power from individuals to institutions that requires active legal protection to prevent.

The prediction identified state-by-state cash acceptance mandates, financial inclusion concerns, and privacy-driven resistance as the structural drivers.

The 2026 evidence confirms all three while adding a fourth dimension the original did not anticipate: constitutional enshrinement at the highest level of national law, now across four European democracies in three years.

## Intro

The cashless society backlash prediction's 2026 update is defined by a qualitative shift that the original analysis described as possible but did not forecast as the dominant outcome: the backlash has moved from legislative preference to constitutional permanence.

State-by-state cash acceptance mandates, financial inclusion advocacy, and privacy-driven consumer resistance were the structural drivers the original piece identified, and all three have continued to operate in 2026. But the four European constitutional amendments represent a categorically different kind of protection than regulatory mandate or statutory preference. Constitutional amendments require supermajority or referendum-level support to enact and equivalent processes to reverse.

They are not vulnerable to a new administration's payment policy preferences, a platform's lobbying effort, or a future parliament's regulatory appetite. When Hungary, Slovakia, Slovenia, and Switzerland, four democracies that share no common political system, currency, or European integration status, each independently conclude that cash access requires the highest level of legal protection their systems can provide, the backlash has moved from a policy position that can be negotiated to a foundational right that must be litigated.

The CBDC prohibition through 2030 removes the one technology that could have provided cash's privacy and universality benefits in digital form for the entire analytical window this series covers. Together, these developments don't just validate the original prediction's direction of travel. They close off the retreat paths the Counterpoint's scenario required.

## Background

The Federal Reserve's Diary of Consumer Payment Choice shows cash remains stable in the US, with 85% of Americans reporting cash use in the last 30 days, and more than half reporting their cash usage has remained steady or increased over the past five years.

Cash plays a critical role for the estimated 25 million unbanked or underbanked Americans, as well as seniors, rural communities, and individuals facing barriers to digital-only payment options. The legislative landscape has expanded materially on multiple simultaneous fronts. In the US: New York Governor Kathy Hochul signed Senate Bill S4153A into law on November 21, 2025, prohibiting food stores and retail establishments from refusing cash payments or charging surcharges for cash, effective March 21, 2026.

New York becomes the fifth state with such protections, joining New Jersey, Colorado, Connecticut, and Delaware. Ten additional states introduced cashless ban legislation in 2024; Washington state implemented bans in Snohomish and King Counties in 2025.

Federal CBDC legislation advanced through the Senate in early 2026 as part of the 21st Century ROAD to Housing Act, with the CBDC prohibition provision passing a procedural cloture vote of 84-6 before advancing to the White House.

The provision prohibits the Federal Reserve from issuing a central bank digital currency until December 31, 2030, with a sunset clause that requires future Congress action to maintain or extend. In Europe: Hungary's Fifteenth Constitutional Amendment, passed April 14, 2025 and effective July 1, 2025, elevates cash payment to a fundamental constitutional right.

Slovenia's National Assembly voted 61-to-0 (29 apparently abstaining) on December 1, 2025 to pass a constitutional amendment making cash acceptance a constitutional obligation.

Switzerland's voters approved a constitutional right to cash by 73.4% in a national referendum in March 2026. Slovakia had enacted equivalent constitutional protection in 2023. Four European constitutions now protect physical currency rights.

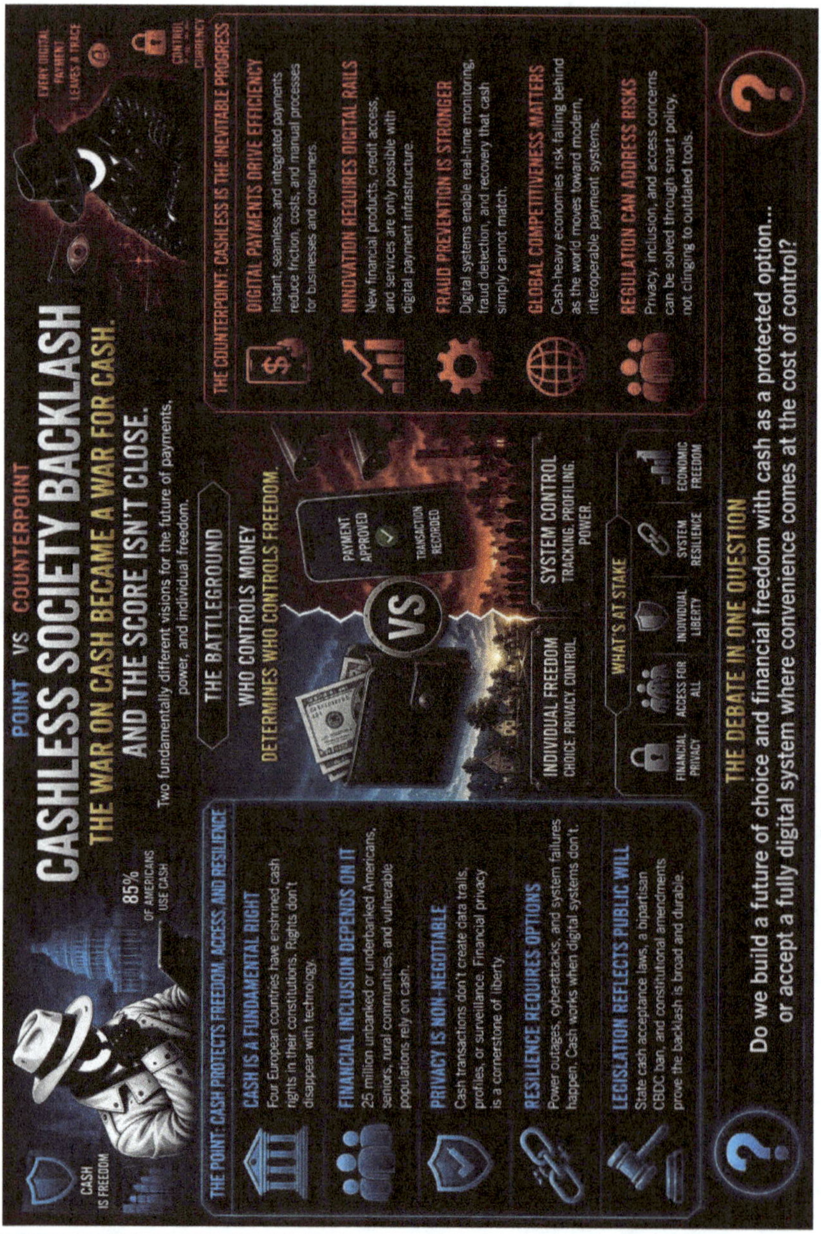
POINT VS COUNTERPOINT
CASHLESS SOCIETY BACKLASH
THE WAR ON CASH BECAME A WAR FOR CASH.
AND THE SCORE ISN'T CLOSE.
Two fundamentally different visions for the future of payments, power, and individual freedom.
CASH IS FREEDOM
85% OF AMERICANS USE CASH
EVERY DIGITAL PAYMENT LEAVES A TRACE
THE POINT: CASH PROTECTS FREEDOM, ACCESS, AND RESILIENCE
CASH IS A FUNDAMENTAL RIGHT
Four European countries have enshrined cash rights in their constitutions. Rights don't disappear with technology.
FINANCIAL INCLUSION DEPENDS ON IT
25 million unbanked or underbanked Americans, seniors, rural communities, and vulnerable populations rely on cash.
PRIVACY IS NON-NEGOTIABLE
Cash transactions don't create data trails, profiles, or surveillance. Financial privacy is a cornerstone of liberty.
RESILIENCE REQUIRES OPTIONS
Power outages, cyberattacks, and system failures happen. Cash works when digital systems don't.
LEGISLATION REFLECTS PUBLIC WILL
State cash acceptance laws, a bipartisan CBDC ban, and constitutional amendments prove the backlash is broad and durable.
THE BATTLEGROUND
WHO CONTROLS MONEY DETERMINES WHO CONTROLS FREEDOM.
VS
PAYMENT APPROVED
TRANSACTION RECORDED
INDIVIDUAL FREEDOM
CHOICE. PRIVACY. CONTROL.
SYSTEM CONTROL
TRACKING. PROFILING. POWER.
WHAT'S AT STAKE
FINANCIAL PRIVACY
ACCESS FOR ALL
INDIVIDUAL LIBERTY
SYSTEM RESILIENCE
ECONOMIC FREEDOM
THE COUNTERPOINT: CASHLESS IS THE INEVITABLE PROGRESS
DIGITAL PAYMENTS DRIVE EFFICIENCY
Instant, seamless, and integrated payments reduce friction, costs, and manual processes for businesses and consumers.
INNOVATION REQUIRES DIGITAL RAILS
New financial products, credit access, and services are only possible with digital payment infrastructure.
FRAUD PREVENTION IS STRONGER
Digital systems enable real-time monitoring, fraud detection, and recovery that cash simply cannot match.
GLOBAL COMPETITIVENESS MATTERS
Cash-heavy economies risk falling behind as the world moves toward modern, interoperable payment systems.
REGULATION CAN ADDRESS RISKS
Privacy, inclusion, and access concerns can be solved through smart policy, not clinging to outdated tools.
THE DEBATE IN ONE QUESTION
Do we build a future of choice and financial freedom with cash as a protected option...
or accept a fully digital system where convenience comes at the cost of control?

## THE POINT: You Can't Repeal a Constitutional Amendment With an App Update

The most consequential single development in the cashless backlash story since the original prediction was written is the legislative action on CBDCs, not because it ended the debate, but because of what the political coalition that produced it reveals about the backlash's structural durability.

The Senate advanced the CBDC prohibition through an 84-6 procedural cloture vote, attached to the 21st Century ROAD to Housing Act, a bipartisan supermajority that reflects the depth of concern about government-issued digital currency across the political spectrum. Republican lawmakers framed CBDC as government surveillance and financial monitoring infrastructure.

Progressive Democrats joined the coalition on financial inclusion grounds, sharing the concern about centralized control over individual transactions from a different ideological starting point.

The White House issued a statement supporting the Act.

That coalition, conservative financial sovereignty advocates and progressive financial inclusion advocates reaching the same legislative conclusion from opposite directions, is structurally durable in ways that lobbying by digital payment incumbents can't easily disrupt.

This is the pattern the prediction identified as the structural driver of the backlash: the populations most concerned about cashless payment systems are politically diverse, numerically significant, and motivated by concerns, privacy, exclusion, infrastructure resilience, that have no straightforward technological resolution. The CBDC ban is time-limited to 2030, but the coalition that produced it is not.

The European constitutional enshrinement wave represents a qualitatively different kind of validation than state legislation: constitutional amendments are not regulatory preferences that can be superseded by federal frameworks or platform policy updates.

When Hungary, Slovakia, Slovenia, and Switzerland, four democracies with different political systems, currencies, and European integration levels, each independently reach the conclusion that cash access requires the highest level of legal protection their systems can provide, the backlash has moved from consumer sentiment to fundamental law.

The arguments are consistent across jurisdictions: digital vulnerability and financial exclusion in Hungary's formulation; the right to economic participation without technological prerequisites in Slovenia's; the broader principle that access to essential goods can't be conditioned on participation in private payment infrastructure in Switzerland's.

The European Central Bank specifically welcomed Slovenia's reform, stating it supported efforts to strengthen "the availability and acceptance of cash", a significant institutional signal that the backlash is aligned with rather than opposed to the European financial architecture.

The privacy dimension has strengthened rather than weakened as the evidence base for digital payment surveillance has accumulated.

Florida's cashless ban legislation explicitly cited payment system resilience, "ensuring businesses continue to operate when Mother Nature targets Florida", adding infrastructure failure scenarios to the financial inclusion and privacy arguments, expanding the coalition further.

The convergence of financial inclusion, privacy rights, and infrastructure resilience into a single legislative coalition across multiple jurisdictions and political systems explains why the momentum has remained durable across election cycles and political shifts.

### The Point Roundup

The cashless society backlash has crossed from legislative trend to structural legal reality between the writing of the original prediction and this update.

Constitutional amendments in four European democracies, cash acceptance mandates covering a growing share of the American population, a federal CBDC prohibition through 2030, and bipartisan congressional momentum toward a national cash acceptance standard represent a reversal of the cashless trajectory that is not reversible by the next app update or payment technology advance.

The 2027-2030 window will see the prediction's "patchwork of state-by-state requirements" either consolidate into federal legislation, which the congressional momentum suggests is increasingly likely, or remain a compliance mosaic that businesses manage with growing complexity and cost.

Either outcome confirms the prediction's core thesis: the "inevitable march toward a fully digital payment system" has been interrupted, and the interruption has achieved constitutional permanence in multiple jurisdictions.

## THE COUNTERPOINT: 85% Used Cash Last Month. 85% Also Used Cards. The Future Includes Both.

The Counterpoint's argument has been genuinely weakened by the CBDC ban and constitutional enshrinements, but the most important analytical distinction the backlash narrative consistently blurs is between cash acceptance mandates and cash usage growth. Cash acceptance laws require businesses to accept physical currency.

They don't reverse the consumer behavior trend that has driven cash from approximately 40% of transactions in 2012 to 18% in 2022, with the trend line consistently downward despite the legislative momentum moving in the opposite direction. 85% of Americans used cash in the last 30 days, but that statistic encompasses everyone from the daily cash user to the person who paid a parking meter once in a month.

The Federal Reserve's own data shows cash at 18% of all payments, and that trajectory predates the legislative backlash and hasn't reversed in states with existing cash acceptance mandates. Cash acceptance mandates preserve the option to use cash for the populations that need it, an important and valid consumer protection goal.

They don't reverse the broader behavioral shift that makes cash an increasingly minority payment method for the majority of the population.

The four European constitutional amendments protect a right that a growing proportion of their own populations are voluntarily exercising less frequently. Legal protection and behavioral adoption are different phenomena, and the backlash story tends to conflate them.

The CBDC prohibition until 2030 is significant but specifically time-limited and narrowly scoped. The legislation represents a temporary prohibition with a built-in expiration date, Congress can extend, make permanent, or allow CBDC development after 2030.

The political coalition that produced the 84-6 procedural vote is bipartisan but fragile in specific ways: the conservative financial sovereignty argument and the progressive financial inclusion argument converge on opposing CBDCs but diverge on virtually everything else about payment regulation, monetary policy, and financial infrastructure.

A future Congress under different political conditions, or facing a payment system crisis that makes digital resilience arguments more salient, could revisit the CBDC question with different conclusions.

The CBDC ban is also narrowly targeted at direct Federal Reserve issuance to consumers, it does not foreclose private stablecoin development, Federal Reserve wholesale CBDC for interbank settlement, or the continued advancement of private digital payment infrastructure that serves the majority of the population through channels no congressional vote affects.

Private-sector digital payment innovation continues advancing outside the CBDC framework: Apple Pay, Google Pay, Venmo, Zelle, digital wallets, and the ongoing improvement of contactless payment accessibility for elderly and disabled users all continue on private-sector timelines.

The Counterpoint's resolution mechanism may have been blocked in its federal CBDC form for the 2030 window, but the underlying technological capability for accessible digital payment is developing through channels the legislation can't foreclose.

CASH
FREEDOM
CHOICE
PRIVACY
CASH IS A RIGHT
WHAT NEEDS TO CHANGE
(FOR THE COUNTERPOINT TO PREVAIL)
For the Counterpoint scenario to win out by 2030, these conditions must occur.
Today's trajectory is moving fast in the opposite direction.
EVERY DIGITAL PAYMENT LEAVES A TRACE
THE POINT MUST BE WRONG ABOUT
CONSTITUTIONAL PROTECTIONS WON'T SPREAD
Counterpoint requires no new countries or states to enshrine cash rights in law.
CBDC BANS WILL EXPAND
Counterpoint needs federal CBDC prohibition to fail or be repealed before 2030.
CASH USAGE WILL COLLAPSE
Counterpoint requires cash usage rates to fall below relevance in daily transactions.
INCLUSION CONCERNS WILL FADE
Counterpoint depends on digital access solving exclusion without leaving anyone behind.
CASH ACCEPTANCE LAWS WILL DISAPPEAR
Counterpoint requires states to repeal laws protecting cash acceptance.
DECISION FRAMEWORK: WHAT TO WATCH
TRACK LEGISLATIVE ACTION
Monitor new cash protection laws, CBDC restrictions, and constitutional amendments worldwide.
MEASURE CASH USAGE TRENDS
Follow Federal Reserve data, merchant surveys, and transaction mix by demographic.
WATCH COURT & LEGAL CHALLENGES
Observe litigation outcomes on cashless mandates, surveillance, and financial privacy.
MONITOR CBDC & DIGITAL ID DEVELOPMENT
Track central bank digital currency pilots, legislation, and political support.
ASSESS INFRASTRUCTURE & OUTAGES
Track power outages, cyberattacks, and system failures that reinforce cash resilience.
THE COUNTERPOINT MUST BE RIGHT ABOUT
CONSUMERS WILL CHOOSE DIGITAL ONLY
Most people must prefer digital payments despite privacy, fees, and control trade-offs.
TECHNOLOGY WILL ELIMINATE RISKS
Digital systems must become 100% secure, private, and resilient against all threats.
GOVERNMENTS WILL PRIORITIZE EFFICIENCY
Efficiency, taxation, and monitoring benefits must outweigh freedom and privacy concerns.
CASH WILL BECOME TOO COSTLY
Handling cash must become economically unsustainable for businesses and banks.
GLOBAL STANDARDIZATION WILL WIN
International coordination on digital identity, payments, and CBDCs must accelerate without major resistance.
CURRENT STATE (Q2 2026)
Four European countries have enshrined cash rights in their constitutions since 2023.
Fifth state cash law (New York) took effect March 21, 2026. Ten more states introduced cash protection legislation.
US Senate passed CBDC ban with 84–6 procedural vote. White House supports.
85% of Americans used cash in last 30 days. 25M+ unbanked or underbanked rely on cash.
THE MOMENTUM
Legal Strengthening
Political Bipartisan
Public Support Strong
International Converging
LIKELIHOOD COUNTERPOINT PREVAILS
18%
0%
100%
PREVIOUSLY 25% → NOW 18%
-7 POINTS
Constitutional protections, bipartisan legislation, and high cash usage continue to strengthen. The backlash is broad, durable, and accelerating. Sharpest downward revision in this series.
MOST CRITICAL VARIABLE
CONSTITUTIONAL PROTECTION MOMENTUM
If more countries and US states enshrine cash rights in their highest laws, the cashless agenda cannot be imposed—only circumvented. Constitutional protection creates durable, generational barriers.
HUNGARY 2025
SLOVENIA 2025
SWITZERLAND 2026
SLOVAKIA 2023
MORE TO COME?
THE OUTCOME IS NOT INEVITABLE. THE NEXT 2–3 YEARS WILL DECIDE THE FUTURE OF PAYMENT FREEDOM.
Will the world protect the right to pay with cash as a fundamental freedom...
or allow digital control to replace choice with compliance?

## What Needs to Change

### (...for the Counterpoint to Prevail)

For the "digital payments achieve true universality" scenario to materialize despite the current legislative headwinds, three structural conditions need to change that are currently moving in the wrong direction.

First, the CBDC ban needs to expire in 2030 and not be renewed, or a private-sector equivalent needs to develop sufficient trust and accessibility to serve the financial inclusion and privacy functions that a public CBDC would have provided. Neither condition is guaranteed: the political coalition that produced the 84-6 procedural vote is durable across the political cycle most likely to govern through 2030, and private-sector digital currencies have their own trust deficits with the populations most dependent on cash.

Second, the financial inclusion gap needs to close materially, the 25 million unbanked and underbanked Americans need to gain banking access through means other than cash, which requires infrastructure investment, regulatory reform, and behavioral change that has been promised and not delivered across multiple presidential administrations.

Third, and most importantly, the privacy argument needs to be answered by technology rather than dismissed by policy. A digital payment system that provides demonstrably equivalent anonymity to physical cash, verifiable by independent audit rather than corporate privacy policy, would remove the most politically durable driver of the cash protection coalition.

No such system currently exists, has a credible development timeline within the 2030 window, or has a regulatory pathway that the current legislative environment would permit.

## Likelihood Counterpoint Prevails: 18%

The Counterpoint likelihood declines from 25% to 18%, the sharpest downward revision in this series, and the revision is driven by two developments that are qualitatively different from the other articles' likelihood adjustments.

First, four European constitutional amendments in three years represent a different category of legal protection than regulatory mandates or legislative preferences. Constitutional amendments in Hungary, Slovakia, Slovenia, and Switzerland require supermajority or referendum-level support to enact and comparable supermajority processes to reverse. They are not vulnerable to a new administration's payment policy preferences or a platform's lobbying effort. They have moved cash rights to the category of fundamental law.

Second, the CBDC prohibition through 2030 eliminates the one technology that could theoretically have provided cash's privacy and universality benefits in digital form for the entire analytical window this series covers. Without a "digital cash equivalent," the Counterpoint's "digital payments achieve true universality" thesis has no operational vehicle through 2030. The remaining 18% reflects three genuine residual uncertainties. Mobile payment accessibility continues improving even without a CBDC, and the gap between "accessible to elderly and disabled users" and "universally accessible" is narrowing.

The federal CBDC ban is time-limited and the political coalition producing it, while durable, is not immovable. And the federal cash acceptance legislation gaining momentum in Congress is a double-edged development: if it passes as a weak preemption framework that supersedes stronger state laws, it could paradoxically weaken existing consumer protections. The 18% says: the Counterpoint's scenario has been foreclosed for the 2030 window by the constitutional enshrinements, the CBDC ban, and the legislative momentum, but the digital payment industry's resources, the genuine utility of digital payments for the majority of the population, and the inherent complexity of federal payment regulation leave a narrow but real path to a different outcome after 2030.

## AI Voice Clone Regulation Emergency

From Grandmother Scams to Corporate Fraud: AI Voice Cloning's Industrial Scale Arrives"

*Three seconds of audio from a social media video. An 85% voice accuracy match. A grandmother who recognizes her grandson's specific panic.*

*Americans lost nearly $3 billion to imposter scams in 2024, with AI voice cloning increasingly identified as the mechanism.*

*Consumer Reports assessed six major voice cloning products and found four lacked meaningful safeguards against misuse.*

*Deepfake-enabled vishing attacks surged 1,600% in Q1 2025 versus Q4 2024.*

*The NO FAKES Act has been reintroduced twice with growing bipartisan support and a coalition spanning OpenAI, Amazon, SAG-AFTRA, and entertainment unions.*

*The technology is running faster than the law. The law is running faster than it was.*

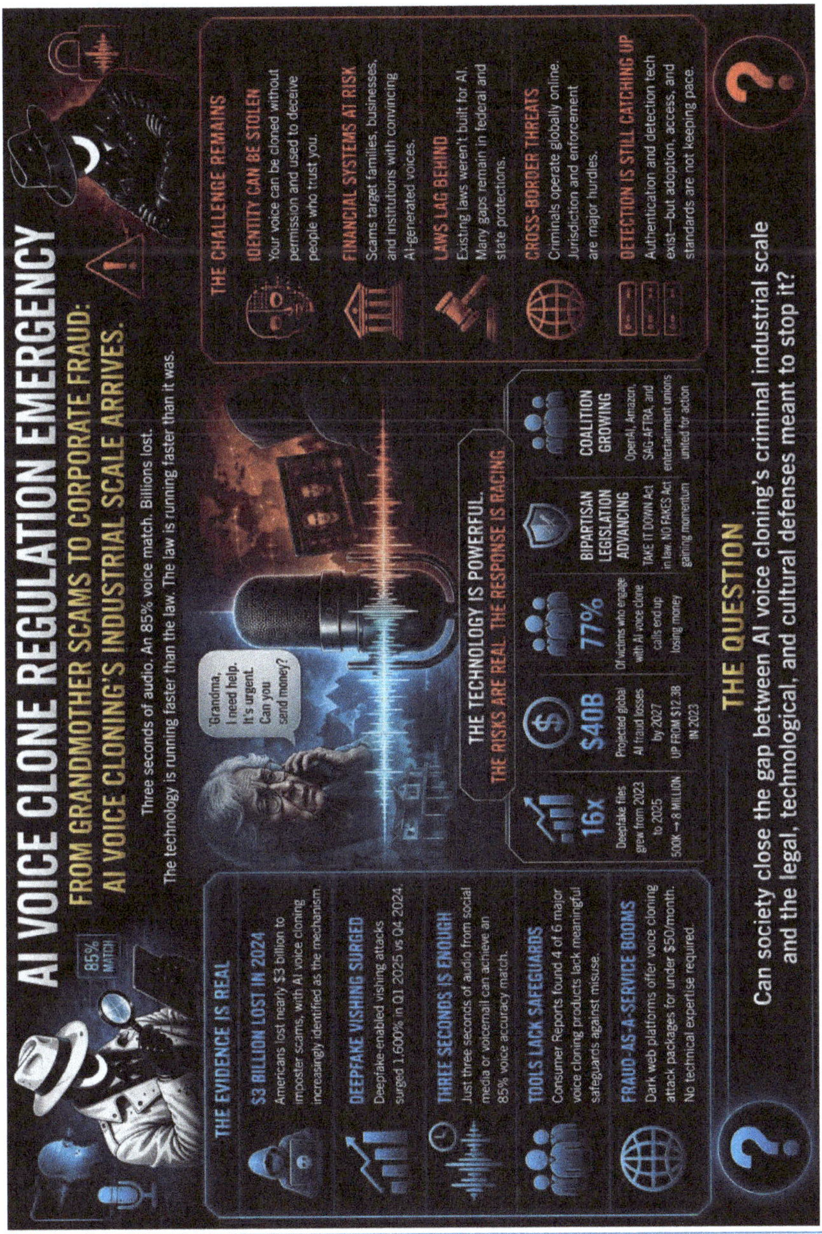
AI VOICE CLONE REGULATION EMERGENCY
FROM GRANDMOTHER SCAMS TO CORPORATE FRAUD:
AI VOICE CLONING'S INDUSTRIAL SCALE ARRIVES.
Three seconds of audio. An 85% voice match. Billions lost.
The technology is running faster than the law. The law is running faster than it was.
85% MATCH
"Grandma, I need help. It's urgent. Can you send money?"
THE EVIDENCE IS REAL
$3 BILLION LOST IN 2024
Americans lost nearly $3 billion to imposter scams, with AI voice cloning increasingly identified as the mechanism.
DEEPFAKE VISHING SURGED
Deepfake-enabled vishing attacks surged 1,600% in Q1 2025 vs Q4 2024.
THREE SECONDS IS ENOUGH
Just three seconds of audio from social media or voicemail can achieve an 85% voice accuracy match.
TOOLS LACK SAFEGUARDS
Consumer Reports found 4 of 6 major voice cloning products lack meaningful safeguards against misuse.
FRAUD-AS-A-SERVICE BOOMS
Dark web platforms offer voice cloning attack packages for under $50/month. No technical expertise required.
THE TECHNOLOGY IS POWERFUL.
THE RISKS ARE REAL. THE RESPONSE IS RACING.
16x
Deepfake files grew from 2023 to 2025
500K → 8 MILLION
$40B
Projected global AI fraud losses by 2027
UP FROM $12.3B IN 2023
77%
Of victims who engage with AI voice clone calls end up losing money
BIPARTISAN LEGISLATION ADVANCING
TAKE IT DOWN Act in law. NO FAKES Act gaining momentum
COALITION GROWING
OpenAI, Amazon, SAG-AFTRA, and entertainment unions united for action
THE CHALLENGE REMAINS
IDENTITY CAN BE STOLEN
Your voice can be cloned without permission and used to deceive people who trust you.
FINANCIAL SYSTEMS AT RISK
Scams target families, businesses, and institutions with convincing AI-generated voices.
LAWS LAG BEHIND
Existing laws weren't built for AI. Many gaps remain in federal and state protections.
CROSS-BORDER THREATS
Criminals operate globally online. Jurisdiction and enforcement are major hurdles.
DETECTION IS STILL CATCHING UP
Authentication and detection tech exist—but adoption, access, and standards are not keeping pace.
THE QUESTION
Can society close the gap between AI voice cloning's criminal industrial scale and the legal, technological, and cultural defenses meant to stop it?

## Opening

The AI voice cloning regulation emergency has moved from prediction to documented consumer harm between the writing of the original piece and this update.

The original prediction correctly identified the direction, the targets, and the inadequacy of existing legal frameworks.

The 2026 evidence confirms all three while adding a legislative response dimension that is more advanced than the original anticipated, and still structurally inadequate relative to the criminal ecosystem it's trying to constrain.

The most important analytical frame for the 2027–2030 window is the three-way race between criminal sophistication, detection and authentication technology, and legislative response.

As of mid-2026, criminal infrastructure is scaling through Fraud-as-a-Service ecosystems that require no technical expertise; legislative response has produced the TAKE IT DOWN Act and is advancing toward the NO FAKES Act; and authentication technology is developing in enterprise contexts faster than in the consumer protection context where the most vulnerable populations are concentrated.

That asymmetry, enterprise protection advancing, consumer protection lagging, is the 2027–2030 story.

## Intro

The AI voice cloning regulation emergency is one of the few articles in this series where the update's primary work is not calibrating a prediction against evidence that partially confirmed and partially challenged it. It's documenting the arrival of a crisis that the original piece described in forward-looking terms and that has now fully materialized in backward-looking ones.

The 1,600% Q1 2025 vishing surge, the $3 billion in imposter scam losses, the Fraud-as-a-Service platforms offering voice cloning attack packages for under $50 per month, the Consumer Reports finding that four of six major voice cloning products lack meaningful safeguards, these are not leading indicators of a coming crisis.

They are the documented record of one already underway.

What the 2026 update adds is the legislative dimension the original piece correctly anticipated but couldn't yet describe in operational terms: the TAKE IT DOWN Act signed, the NO FAKES Act reintroduced twice with expanding coalitions, Congressional hearings targeting platform design decisions upstream of the fraud itself.

The legislative response is real and accelerating. It's also structurally mismatched with the crisis in a specific way that the 2027-2030 analysis must confront directly: the legislative architecture assembling around voice and likeness property rights was designed to protect performers and public figures, not to prevent the grandmother scam.

Both problems are real. The law being written addresses one of them.

## Background

The scale of AI-enabled fraud has grown materially across every measurable dimension since the original prediction was written.

Deepfake files grew from approximately 500,000 in 2023 to 8 million in 2025 ,a 16x increase reflecting the democratization of voice synthesis tools. Deepfake-enabled vishing (voice phishing) attacks surged 1,600% in Q1 2025 compared to Q4 2024, according to cybersecurity researchers. AI-powered fraud losses are projected to reach $40 billion globally by 2027, up from $12.3 billion in 2023, a 32% compound annual growth rate documented by Deloitte.

Several data points show how quickly AI-enabled voice fraud has moved from isolated scam reports to a broader consumer-protection problem:

- **Imposter scam losses are already large.** Americans lost approximately $3 billion to imposter scams in 2024, according to FTC data, with AI voice cloning increasingly identified as a mechanism in family-emergency scams.
- **Confirmed AI voice phishing losses likely understate the real total.** Documented US losses from AI voice phishing reached $5 million in 2025, but that figure captures only cases where AI involvement was confirmed and reported. Because only an estimated 15% of victims report these crimes, the actual total is likely materially higher.
- **The technical barrier is extremely low.** McAfee research indicates that three seconds of audio from social media, voicemail, or voice messages can be enough to generate an 85% voice-accuracy match.

77% of people who receive a convincing AI voice clone call and engage with it end up losing money.

The commercial criminal infrastructure has matured: dark web Fraud-as-a-Service platforms offer voice cloning attack packages for under $50 per month in cryptocurrency, requiring no technical expertise, creating a franchise model for voice fraud that scales criminal capability faster than any individual prosecution can address.

THREE SECONDS IS ALL IT TAKES
POINT VS COUNTERPOINT
AI VOICE CLONE REGULATION EMERGENCY
FROM GRANDMOTHER SCAMS TO CORPORATE FRAUD:
AI VOICE CLONING'S INDUSTRIAL SCALE ARRIVES.
POWERFUL TECHNOLOGY. REAL RISKS. REAL TRADE-OFFS.
Two fundamentally different paths for a technology that can sound exactly like you.
THE POINT: ACT NOW TO PROTECT PEOPLE, TRUST, AND DEMOCRACY
REAL HARM IS HAPPENING NOW
Americans lost nearly $3 billion to imposter scams in 2024, with AI voice cloning increasingly the weapon of choice.
EXPLOSIVE GROWTH IN THREATS
Deepfake-enabled vishing attacks surged 1,600% in Q1 2025 vs Q4 2024. Deepfake files grew 16x from 2023 to 2025.
THE MOST VULNERABLE ARE TARGETED
Seniors lost $7.7 billion to cybercrime in 2025, up 37% from 2024. Family emergency scams cause financial devastation.
TOOLS ARE DANGEROUSLY EASY TO MISUSE
Three seconds of audio creates an 85% voice match. Criminals rent voice cloning kits for under $50/month on the dark web.
INDUSTRY & PUBLIC DEMAND ACTION
75,000 consumers petitioned the FTC. OpenAI, Amazon, SAG-AFTRA, and major unions support strong federal safeguards.
LAWS MUST CATCH UP TO THE THREAT
TAKE IT DOWN Act is a start—but we need stronger, broader protections. The NO FAKES Act is the right next step.
THE BATTLEGROUND
PROTECTION & ACCOUNTABILITY
VS
INNOVATION & ACCESS
WHAT'S AT STAKE
PERSONAL IDENTITY
FINANCIAL SECURITY
PUBLIC TRUST & DEMOCRACY
TECHNOLOGICAL INNOVATION
ECONOMIC GROWTH
FREEDOM OF EXPRESSION
BY THE NUMBERS (LATEST DATA)
$3B LOST TO IMPOSTER SCAMS IN 2024
1,600% SURGE IN DEEPFAKE VISHING ATTACKS Q1 2025 vs Q4 2024
$40B PROJECTED GLOBAL AI FRAUD LOSSES BY 2027
77% OF PEOPLE ENGAGE WITH AI CLONE CALLS AND LOSE MONEY
ONLY 15% OF VICTIMS REPORT THESE CRIMES
THE COUNTERPOINT: PROCEED CAREFULLY AND BALANCE RISKS
OVERREGULATION COULD STIFLE INNOVATION
Broad restrictions may slow beneficial AI tools that help businesses, creators, and consumers.
TECHNOLOGY CAN BE PART OF THE SOLUTION
Authentication, watermarking, and detection tools are improving and can reduce misuse without heavy-handed laws.
FREE SPEECH AND EXPRESSION MATTER
Voice regulation must be carefully designed to avoid infringing on parody, commentary, and artistic expression.
GLOBAL INCONSISTENCY CREATES PROBLEMS
Different rules across countries could fragment the internet and disadvantage US companies.
PRIVACY AND DATA CONCERNS
Mandatory identity checks or data collection could create privacy risks and new vulnerabilities.
EXISTING LAWS CAN ADAPT
Fraud, identity theft, and harassment laws already exist—focus on enforcement, not new federal mandates.
THE DEBATE IN ONE QUESTION
How do we stop criminals from weaponizing voices without silencing innovation or creating new barriers to communication?

## THE POINT: The Technology Runs Faster Than the Law. That Gap Is the Whole Problem.

The original prediction's central thesis, that the convergence of accessible voice cloning technology and sophisticated fraud networks would trigger a crisis requiring emergency legislative intervention, has been confirmed by the 2026 evidence record in ways the original piece only partially anticipated.

The Fraud-as-a-Service ecosystem is the development that most changes the threat landscape: voice clone fraud no longer requires technical capability, specialized hardware, or custom coding.

A criminal network can purchase a dark web subscription for under $50 per month that provides a web portal, a phone number input field, and a URL field for source audio, automating the entire attack chain from voice harvest to scam call.

That democratization explains the 1,600% Q1 2025 surge in deepfake vishing attacks better than any individual technology advance: the barrier to entry for a voice cloning fraud operation has fallen to the cost of a streaming subscription.

The consumer impact is concentrated in the populations the original prediction identified: Americans over 60 reported approximately $7.7 billion in total cybercrime losses in 2025, up 37% from 2024, with AI voice fraud increasingly identified as the primary mechanism for the family emergency scam category that has produced the most financially and emotionally devastating individual cases.

The FTC has documented a fourfold increase in reports from older consumers who lost $10,000 or more to impersonator scams since 2020.

Consumer Reports' assessment found that four of the six major AI voice cloning products lacked meaningful safeguards against misuse, and a petition of 75,000 consumers to the FTC calling for regulatory action confirms the consumer awareness of the problem has reached the level where institutional response is politically mandated.

The voice has crossed the threshold the original prediction described: it's now biometric data whose compromise has consequences that the existing legal framework was not designed to address.

The legislative response has accelerated since the original prediction was written, but the structural mismatch between what the emerging legislation addresses and what the fraud crisis requires remains the most analytically important gap in the response architecture.

The NO FAKES Act, reintroduced in April 2025 and again with a revised version in May 2026, creates a federal right protecting every American's voice and likeness from unauthorized AI replication ,with bipartisan Senate sponsors Blackburn, Coons, Tillis, and Klobuchar, 11 Senate co-sponsors, support from OpenAI, Amazon, SAG-AFTRA, RIAA, and child safety groups.

The TAKE IT DOWN Act, addressing nonconsensual intimate digital replicas, has already been signed into law, establishing the legislative precedent that AI-generated content causing personal harm warrants federal intervention.

Congressional scrutiny has expanded upstream: Senator Hassan's inquiry focuses on how voice cloning product design decisions by companies including ElevenLabs, Descript, and Resemble AI create the infrastructure that fraud networks exploit, a shift from after-the-fact prosecution to upstream platform accountability.

The Senate Commerce Committee and House Energy and Commerce Committee announced hearings specifically on AI voice fraud in April 2026. The Voice Cloning Protection Act and DEEPFAKES Accountability Act are advancing through committee.

But the critical gap persists: the NO FAKES Act creates voice and likeness property rights, it does not create financial authentication requirements for transactions, emergency communication verification standards, or the international coordination mechanism required to address criminal networks operating from Eastern Europe and Southeast Asia.

The legislative architecture is assembling around the wrong problem definition: protecting celebrity likenesses rather than protecting vulnerable consumers from financial fraud.

## The Point Roundup

The AI voice clone regulation emergency has arrived.

The FTC's documentation of fourfold growth in older consumer impersonation losses, Consumer Reports' finding that four of six major voice cloning tools lack meaningful safeguards, and the 1,600% Q1 2025 surge in deepfake vishing all confirm that the original prediction correctly identified both the trajectory and the destination.

What the 2027–2030 update requires the original piece to acknowledge is a more precise characterization of what "comprehensive federal legislation" actually looks like when it arrives: the NO FAKES Act addresses voice and likeness property rights; financial authentication requirements for high-value transactions remain largely unlegislated; and the international coordination required to address the Fraud-as-a-Service criminal ecosystem remains at its earliest stages.

The crisis the prediction described has arrived.

The response is real and accelerating. The match between crisis scope and legislative response scope is the unresolved question for 2030.

## THE COUNTERPOINT: Detection Is Advancing. The Problem Is Deployment Speed, Not Technology Capability.

The 77% victim rate among people who engage with a convincing AI voice clone call ,while alarming, contains within it the Counterpoint's most important observation: the attacks that fail do so because targets hang up, ask verification questions, call back on a known number, or apply the "slow down" principle the FTC has advocated.

Cybersecurity experts consistently recommend that families establish a private code word known only to close relatives ,a behavioral countermeasure so simple and so effective that its widespread adoption could dramatically reduce the target population for family emergency scams without requiring any legislation or technology deployment.

The FTC itself advises: "I hear you, I'll call you right back", a response that costs nothing, takes three seconds, and defeats the entire attack model that depends on manufactured urgency preventing verification. The Counterpoint's argument is not that technology alone solves the problem. It's that the combination of user education, behavioral protocols, and advancing detection technology creates a defense ecosystem that makes voice clone fraud progressively less economically viable for criminal networks, even without comprehensive federal legislation. The FCC's clarification that AI-generated voices fall under existing robocall rules where the Telephone Consumer Protection Act applies creates meaningful existing liability for the most common delivery mechanisms without requiring new legislation.

The criminal ecosystem is scaling through low-cost FaaS platforms, but the same technological accessibility that empowers criminals is also making detection tools more widely available and deployable.

The detection and authentication infrastructure is being built in the contexts where the financial stakes are highest, and the enterprise response to AI voice fraud is materially more advanced than the consumer protection landscape suggests.

Financial institutions that have documented the pattern of AI voice fraud have accelerated multi-factor authentication, behavioral biometric analysis, and call verification protocols that reduce the financial sector's exposure independent of consumer behavior change. Structured vishing simulation programs have been shown to improve employee verification behavior by 65%, and continuous simulation-based training has cut successful enterprise compromises by nearly 50% over twelve months, demonstrating that behavioral training works at institutional scale even when it hasn't been deployed at consumer scale.

Congress is advancing watermarking legislation that would require AI voice cloning tools to embed identifiable markers in synthetic audio at the generation layer, addressing the problem upstream rather than downstream and creating provenance infrastructure analogous to C2PA's approach to visual deepfakes.

The EU AI Act's transparency obligations for synthetic media, effective August 2026, create disclosure requirements that normalize the expectation of AI content labeling across consumer-facing platforms.

The NO FAKES Act's property rights framework, while not directly addressing financial fraud, creates the liability architecture that gives platforms commercial incentive to implement safeguards the current voluntary framework has failed to produce.

The legislative pieces are assembling, and the Counterpoint's scenario requires only that they assemble faster than the criminal ecosystem scales, which is a race, not a foregone conclusion.

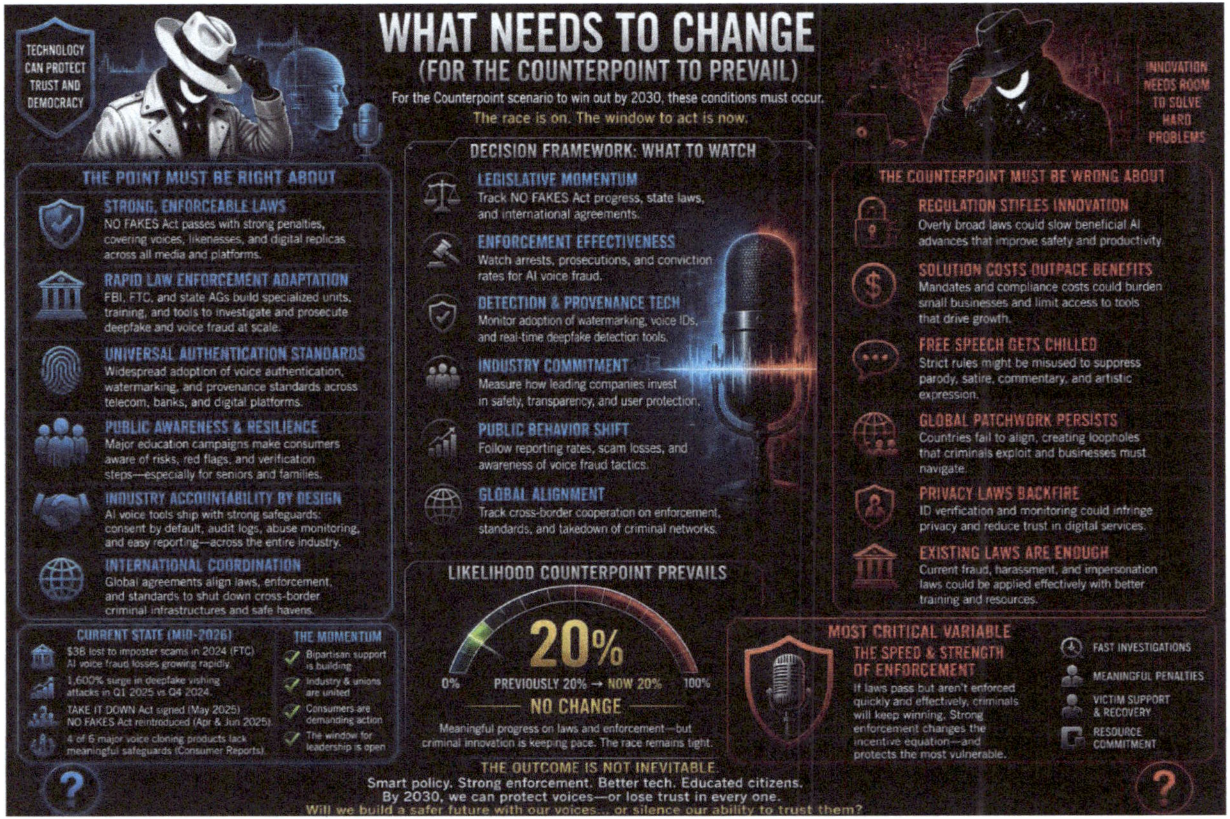
TECHNOLOGY CAN PROTECT TRUST AND DEMOCRACY
WHAT NEEDS TO CHANGE
(FOR THE COUNTERPOINT TO PREVAIL)
For the Counterpoint scenario to win out by 2030, these conditions must occur.
The race is on. The window to act is now.
INNOVATION NEEDS ROOM TO SOLVE HARD PROBLEMS
THE POINT MUST BE RIGHT ABOUT
STRONG, ENFORCEABLE LAWS
NO FAKES Act passes with strong penalties, covering voices, likenesses, and digital replicas across all media and platforms.
RAPID LAW ENFORCEMENT ADAPTATION
FBI, FTC, and state AGs build specialized units, training, and tools to investigate and prosecute deepfake and voice fraud at scale.
UNIVERSAL AUTHENTICATION STANDARDS
Widespread adoption of voice authentication, watermarking, and provenance standards across telecom, banks, and digital platforms.
PUBLIC AWARENESS & RESILIENCE
Major education campaigns make consumers aware of risks, red flags, and verification steps—especially for seniors and families.
INDUSTRY ACCOUNTABILITY BY DESIGN
AI voice tools ship with strong safeguards: consent by default, audit logs, abuse monitoring, and easy reporting—across the entire industry.
INTERNATIONAL COORDINATION
Global agreements align laws, enforcement, and standards to shut down cross-border criminal infrastructures and safe havens.
DECISION FRAMEWORK: WHAT TO WATCH
LEGISLATIVE MOMENTUM
Track NO FAKES Act progress, state laws, and international agreements.
ENFORCEMENT EFFECTIVENESS
Watch arrests, prosecutions, and conviction rates for AI voice fraud.
DETECTION & PROVENANCE TECH
Monitor adoption of watermarking, voice IDs, and real-time deepfake detection tools.
INDUSTRY COMMITMENT
Measure how leading companies invest in safety, transparency, and user protection.
PUBLIC BEHAVIOR SHIFT
Follow reporting rates, scam losses, and awareness of voice fraud tactics.
GLOBAL ALIGNMENT
Track cross-border cooperation on enforcement, standards, and takedown of criminal networks.
THE COUNTERPOINT MUST BE WRONG ABOUT
REGULATION STIFLES INNOVATION
Overly broad laws could slow beneficial AI advances that improve safety and productivity.
SOLUTION COSTS OUTPACE BENEFITS
Mandates and compliance costs could burden small businesses and limit access to tools that drive growth.
FREE SPEECH GETS CHILLED
Strict rules might be misused to suppress parody, satire, commentary, and artistic expression.
GLOBAL PATCHWORK PERSISTS
Countries fail to align, creating loopholes that criminals exploit and businesses must navigate.
PRIVACY LAWS BACKFIRE
ID verification and monitoring could infringe privacy and reduce trust in digital services.
EXISTING LAWS ARE ENOUGH
Current fraud, harassment, and impersonation laws could be applied effectively with better training and resources.
CURRENT STATE (MID-2026)
$3B lost to imposter scams in 2024 (FTC). AI voice fraud losses growing rapidly.
1,600% surge in deepfake vishing attacks in Q1 2025 vs Q4 2024.
TAKE IT DOWN Act signed (May 2025). NO FAKES Act reintroduced (Apr & Jun 2025).
4 of 6 major voice cloning products lack meaningful safeguards (Consumer Reports).
THE MOMENTUM
Bipartisan support is building
Industry & unions are united
Consumers are demanding action
The window for leadership is open
LIKELIHOOD COUNTERPOINT PREVAILS
20%
0%
100%
PREVIOUSLY 20% → NOW 20%
NO CHANGE
Meaningful progress on laws and enforcement—but criminal innovation is keeping pace. The race remains tight.
MOST CRITICAL VARIABLE
THE SPEED & STRENGTH OF ENFORCEMENT
If laws pass but aren't enforced quickly and effectively, criminals will keep winning. Strong enforcement changes the incentive equation—and protects the most vulnerable.
FAST INVESTIGATIONS
MEANINGFUL PENALTIES
VICTIM SUPPORT & RECOVERY
RESOURCE COMMITMENT
THE OUTCOME IS NOT INEVITABLE.
Smart policy. Strong enforcement. Better tech. Educated citizens.
By 2030, we can protect voices—or lose trust in every one.
Will we build a safer future with our voices... or silence our ability to trust them?

## What Needs to Change

### (for the Counterpoint to Prevail)

For the detection and education scenario to meaningfully reduce the AI voice fraud threat without requiring the comprehensive financial authentication framework the original prediction calls for, three conditions need to materialize on faster timelines than the current trajectory suggests.

First, watermarking and provenance standards for AI-generated audio need to achieve mandatory implementation across all major voice synthesis platforms, not as voluntary best practices but as enforced technical requirements, ideally through the FCC's existing TCPA framework or through the Voice Cloning Protection Act and DEEPFAKES Accountability Act advancing through Congress.

Second, financial institutions need to universally implement out-of-band verification for wire transfers and large transactions, removing voice as a sufficient authorization mechanism for high-value transactions would eliminate the most financially devastating category of AI voice fraud even if the underlying scam technology continues advancing.

Third, the "slow down and call back" behavioral countermeasure needs to achieve mainstream adoption through public education at a scale and consistency that no private entity has incentive to fund and government agencies have not yet prioritized at consumer scale.

All three conditions are achievable. The NO FAKES Act's revised 2026 version and the congressional hearings suggest the legislative momentum is building.

The question is whether it builds faster than the Fraud-as-a-Service ecosystem scales.

## Likelihood Counterpoint Prevails: 20%

The Counterpoint likelihood holds at 20%, the same level as the original prediction, and this stability is analytically informative rather than a failure to update.

The 2026 evidence has strengthened the Point's case (1,600% vishing surge, $3 billion imposter scam losses, FaaS democratization) and strengthened the Counterpoint's case (NO FAKES Act advancing with genuine legislative momentum, TAKE IT DOWN Act signed, Congressional hearings directly targeting AI voice fraud, enterprise authentication advancing) in roughly equal measure.

The 20% reflects a prediction where the crisis mechanism is confirmed, the direction is unambiguous, and the legislative response is real but structurally mismatched with the fraud problem's specific contours.

The NO FAKES Act creates voice and likeness property rights, necessary but not sufficient for the financial fraud prevention the crisis requires.

The 20% is not a concession that the Counterpoint is close to prevailing.

It reflects three residual uncertainties that still matter:

**behavioral defenses**, such as private family code words, are genuinely effective and cost nothing;

**enterprise authentication advances** could reduce the most financially significant fraud category even if consumer-level protection lags;

and **watermarking and provenance legislation** advancing through Congress could materially change the detection landscape if it is enacted and enforced.

The 20% says: the criminal ecosystem is scaling faster than the legislative response, the consumer protection gap is real and documented, and the comprehensive financial authentication framework the prediction called for remains the absent piece.

The Counterpoint's detection and education scenario can succeed in specific contexts.

It can't succeed at the scale the crisis requires on the timeline the $40 billion 2027 projection describes.

# AI-Powered Personal Finance Revolution

## The Revolution Arrived, For Half the Problem. The Other Half Is Still Waiting

*Fidelity International launched Freya, a conversational LLM-based financial interface for 21 million customers.*

*Lloyds Banking Group announced the UK's first large-scale AI financial assistant for early 2026. The global robo-advisor market is projected to reach $3.2 trillion by 2033. And the SEC identified emerging financial technology as its top 2026 examination priority.*

*The AI personal finance revolution arrived, for portfolio management, basic tax optimization, and standard financial planning.*

*The Counterpoint also arrived, in the form of a regulatory framework that maintains human oversight requirements and a complexity ceiling that current AI handles poorly.*

*Both predictions are simultaneously right. That is the 2026 story.*

AI FINANCE IS HERE.
IT'S HELPING MILLIONS.
AI-POWERED PERSONAL FINANCE REVOLUTION
THE REVOLUTION ARRIVED, FOR HALF THE PROBLEM.
THE OTHER HALF IS STILL WAITING.
AI finance has gone mainstream for simple and standard needs.
Complex, high-stakes, human decisions remain in human hands.
COMPLEXITY
RISK, REGULATION,
HUMAN JUDGMENT
STILL MATTERS
THE REVOLUTION IS REAL
MASS ADOPTION IS UNDERWAY
Fidelity's Freya and Lloyds' AI assistant will serve 21 million customers. AI finance is moving from pilot to platform.
CONVERSATIONAL AI GOES MAINSTREAM
Natural language interfaces deliver real-time answers, insights, and personalized advice 24/7 at scale.
ROBO-ADVISORS SCALE FAST
Global market projected to reach $3.2 trillion by 2033 (10.5% CAGR), driven by lower fees, better tech, and younger investors.
DEMOCRATIZING PRO INVESTMENT STRATEGIES
Tools like direct indexing and tax-loss harvesting once reserved for the wealthy are now available to everyday investors.
EXPANDING ACCESS & INCLUSION
AI uses alternative data to extend credit and banking access to underserved and underbanked populations.
REGULATORS ARE PAYING ATTENTION
SEC named Emerging Financial Technology as its top 2026 examination priority. FINRA is ramping its AI oversight efforts.
PROOF OF ADOPTION: MAJOR PLAYERS, MAJOR SCALE.
Fidelity
INTERNATIONAL
Launched Freya (Sept 2025)
Conversational LLM agent that understands nuance and delivers dynamic financial insights.
21 MILLION+ CUSTOMERS
LLOYDS
BANKING GROUP
Announced UK's first large-scale AI financial assistant (Early 2026)
Personalized coaching 24/7 with human support for complex needs.
21 MILLION+ CUSTOMERS
THE MARKET IS BOOMING
$1.4 TRILLION
2024 Market Size
$3.2 TRILLION
2033 Projected Size
10.5% CAGR
Through 2033
WHERE AI DELIVERS TODAY
Portfolio management & rebalancing
Tax optimization & loss harvesting
Retirement & savings planning
Investment research & analysis
Financial education & guidance
Budgeting & cash flow optimization
AI
WHERE HUMANS REMAIN ESSENTIAL
Complex legal & estate planning
Divorce, inheritance & family dynamics
Emotional decisions & life transitions
Multi-party negotiations
Unique, high-net-worth situations
Fiduciary judgment & accountability
THE CHALLENGE IS REAL
HUMAN OVERSIGHT IS REQUIRED
Regulators require human accountability. Algorithms can advise, but humans must be responsible.
COMPLEXITY IS THE LIMIT
AI struggles with legal nuance, emotional intelligence, and multi-party, high-stakes situations.
REGULATORY GUARDRAILS EXIST
Fiduciary duty, suitability, and privacy rules limit how AI can be used and what it can decide.
BIAS, DATA & TRANSPARENCY RISKS
AI outputs are only as good as the data and models behind them. Bias and explainability remain concerns.
TRUST NEEDS TO BE EARNED
Many investors are cautious about letting AI make or act on financial decisions without human review.
LIABILITY & RESPONSIBILITY MATTER
When things go wrong, someone must be accountable. The legal framework is still evolving.
THE BIG QUESTION
Can AI and regulation evolve fast enough to close the complexity gap
and deliver the full promise of truly inclusive, trustworthy financial guidance for all?

S

## Opening

The AI personal finance revolution requires a more precise analytical frame than the original prediction's democratization-versus-human-dominance binary. What 2026 reveals is a bifurcated market where the democratization thesis has been substantially validated for standard financial situations, while the Counterpoint's limitations thesis has been validated with equal precision for complex financial situations involving legal complexity, emotional sensitivity, and multi-party dynamics.

The 2027–2030 analytical question is not whether AI transforms personal finance, it demonstrably has and will continue to, but whether the technology and regulatory framework can bridge the complexity gap that currently separates AI's validated domain from the full scope of what the original prediction claimed it would encompass.

The bifurcation itself is the most important analytical finding in this update: AI has won on cost and accessibility for standard situations; human advisors retain genuine competitive advantage for complex ones.

Understanding where the dividing line sits, and whether it's moving, is the work the 2027–2030 window will do.

## Intro

The AI personal finance prediction's most important analytical contribution is not the democratization thesis or the complexity ceiling thesis individually, both have been confirmed, but the specific location of the boundary between them.

The original prediction framed the debate as democratization versus human dominance, implying a single market where one dynamic would eventually prevail.

The 2026 evidence reveals a bifurcated market where both dynamics prevail simultaneously in different segments, and the boundary between those segments is more structurally stable than either side of the original debate anticipated.

Wealthfront's Direct Indexing at 0.25% management fee and Fidelity's Freya answering complex account questions in natural language are not eroding the market for estate planning with trusts, business succession across family partnerships, or divorce asset division.

They are serving a different population with a different set of needs at a price point that makes the original advisory model structurally nonviable.

The complexity ceiling is not a temporary technology gap awaiting the next model version. It's a regulatory and institutional reality that Fidelity and Lloyds both designed their products around rather than through, building human escalation paths into their AI interfaces not because the technology failed but because the compliance framework requires it and the liability structure rewards it.

Understanding that the boundary between AI's validated domain and human advisory's protected domain is being reinforced by regulation rather than eroded by capability is the analytical key to the 2027-2030 forecast.

## Background

The global robo-advisor market was valued at $1.4 trillion in 2024 and is projected to reach $3.2 trillion by 2033, growing at a 10.5% CAGR, driven by increasing adoption of digital wealth management platforms, rising demand for cost-effective investment solutions, and growing preference for AI-enabled portfolio customization among younger investors.

AI-driven platforms have expanded credit access for underserved populations through alternative data analysis, and major institutions including JPMorgan and Morgan Stanley have invested heavily in AI financial services infrastructure.

Consumer adoption signals are consistent with the growth trajectory: industry analysis consistently shows younger investors strongly preferring digital financial management to traditional advisory relationships for standard portfolio and savings decisions.

The regulatory environment has clarified materially in 2025-2026: the SEC identified "Emerging Financial Technology" as a key examination priority in its 2026 Examination Priorities, and FINRA has launched an enterprise-wide initiative focused on AI in financial services.

Regulators have adopted a principles-based, technology-neutral approach that emphasizes existing registered investment adviser compliance frameworks, meaning that delegating portfolio decisions to algorithms does not dilute fiduciary liability, and human oversight remains a regulatory requirement rather than an optional design choice.

This framework shapes the deployment architecture of AI financial services as much as the technology itself: institutions deploy AI within supervised frameworks precisely because the existing liability structure creates strong incentives to maintain human accountability chains.

AI CAN OPTIMIZE NUMBERS. IT CAN'T REPLACE WISDOM.
AI-POWERED PERSONAL FINANCE REVOLUTION
TWO PREDICTIONS. BOTH RIGHT.
AI is transforming how millions manage money today—
and human judgment remains essential for life's most complex decisions.
SOME DECISIONS DEMAND HUMAN JUDGMENT
THE POINT: DEMOCRATIZATION IS REAL
Fidelity INTERNATIONAL
MAJOR INSTITUTIONS ARE ALL IN
Fidelity launched Freya (Sept 2025), a conversational AI assistant. Lloyds will serve 21 million customers in early 2026.
MASS ADOPTION IS HAPPENING
Global consumers are embracing AI finance at scale. The tools are easy, useful, and available 24/7.
21+ MILLION CUSTOMERS TOUCHED
THE MARKET IS BOOMING
Robo-advisor market projected to reach $3.2 trillion by 2033 (10.5% CAGR). Lower fees, better tech, and always-on access are winning.
$3.2 TRILLION BY 2033
AI DELIVERS POWERFUL CAPABILITIES
Tax optimization, retirement modeling, direct indexing, scenario planning, and personalized advice—once reserved for the wealthy.
COSTS HAVE COLLAPSED
Strategies that once required $500K+ and $5,000+ annual fees are now accessible to anyone with a smartphone.
FINANCIAL INCLUSION IS EXPANDING
AI uses alternative data to extend credit and banking access to underserved and underbanked populations worldwide.
THE BATTLEGROUND
WHERE DOES THE COMPLEXITY CEILING SIT?
AI EXCELS AT
(Optimization Problems)
Portfolio rebalancing & asset allocation
Tax optimization & loss harvesting
Retirement modeling & withdrawal planning
Budgeting & cash flow management
Insurance analysis & needs estimation
Direct indexing & customized portfolios
VS
HUMANS STILL DOMINATE
(Life Problems)
Estate planning & inheritance disputes
Divorce & family financial dynamics
Business succession & ownership transitions
Disability, illness & long-term care planning
Multi-party conflicts & negotiations
Unique, high-net-worth & non-standard situations
THE COMPLEXITY CEILING DEFINES THE MARKET.
The future depends on whether AI can cross the gap.
THE COUNTERPOINT: COMPLEXITY IS STICKIER THAN EXPECTED
HUMAN OVERSIGHT IS REQUIRED
Regulators require human accountability. Algorithms can advise, but humans must be responsible for outcomes.
REGULATORY FRAMEWORKS LIMIT AUTONOMY
Fiduciary duty, suitability, and privacy rules constrain how AI can be used and what it can decide.
LIABILITY STRUCTURES PERSIST
When things go wrong, someone must be liable. The legal system is built for humans, not algorithms.
EMOTIONAL INTELLIGENCE MATTERS
Empathy, trust, and understanding context are critical in high-stakes life decisions that affect families and futures.
TRUST DECLINES IN HIGH-STAKES DECISIONS
Consumers are cautious about letting AI handle irreversible or deeply personal financial choices.
COMPLEXITY REMAINS PROFITABLE
The most complex situations are where human advisors add the most value and charge the most.
THE DEBATE IN ONE QUESTION
Can AI extend its proven success beyond optimization problems into the emotionally and legally complex realities that define the most consequential financial decisions?

## THE POINT: Freya, Lloyds, and 21 Million Customers: The Conversational AI Finance Era Has Begun

The original prediction's democratization thesis has been validated in the domain it described most precisely, portfolio management, tax optimization, and standard financial planning, with evidence that exceeds the original evidence base in both scale and product sophistication.

Strategies that previously required a $500,000 minimum and a $5,000 annual advisory fee are increasingly available to anyone with a smartphone.

AI-powered tools can now model complex scenarios, when to claim Social Security, how to sequence retirement account withdrawals, whether to convert traditional IRA balances to Roth, how much life insurance to carry, with a sophistication that matches or exceeds human advisors for standard situations.

Wealthfront's AI-driven Direct Indexing product, buying individual stocks to replicate an index while harvesting tax losses daily, represents a service that previously required six-figure investment minimums and expert tax management, now accessible at 0.25% management fee.

The institutional adoption of conversational AI confirms the trajectory: Fidelity International launched Freya in September 2025, a conversational LLM agent capable of interpreting natural language, understanding nuance, and generating dynamic financial responses in real time, designed to answer customer questions about their investments and accounts with greater depth and flexibility than traditional chatbots.

Lloyds Banking Group announced the UK's first large-scale AI-powered financial assistant for early 2026 rollout to 21 million customers, offering 24/7 personalized financial coaching with the ability to escalate to human support for complex needs.

These products represent exactly the institutional adoption the original prediction described: major financial institutions building AI-first customer interfaces rather than treating digital tools as supplements to relationship-based service models.

The agentic AI dimension, AI that doesn't just advise but acts autonomously on financial decisions, represents the frontier that will define the 2027–2030 period and where the Point's strongest forward-looking evidence lives.

The FCA and Bank of England have both publicly anticipated a shift toward agentic AI in core financial decision-making including credit underwriting, portfolio optimization, and risk modeling, a regulatory acknowledgment that the technology trajectory is moving toward autonomous action, not just sophisticated recommendation.

The democratization impact is already visible in credit access: AI-driven platforms have expanded credit access for underserved populations by analyzing alternative data that traditional credit scoring can't capture, reducing the structural financial exclusion that the original prediction identified as the most important equity dimension of AI finance.

The cost compression is the most durable democratization mechanism: a human financial advisor's fee structure, typically 1% of assets annually with substantial minimums, is structurally incompatible with serving the 60% of Americans who have no financial advisor because they lack sufficient assets to make advisory fees economically viable for the advisor.

AI has resolved that structural incompatibility for standard financial situations, and the robo-advisor market's $1.4 trillion to $3.2 trillion projected growth between 2024 and 2033 represents the financial consequence of that resolution.

## The Point Roundup

The AI personal finance revolution is not a prediction or a trajectory; it's a documented reality for the majority of standard financial planning use cases.

The democratization the original piece described has occurred: portfolio management, tax optimization, retirement planning modeling, and now conversational AI financial guidance are available to anyone with a smartphone at costs representing 75-90% reductions from traditional advisory fees.

Fidelity International and Lloyds Banking Group deploying LLM-based financial interfaces for tens of millions of customers represents institutional confirmation of the trajectory.

What hasn't occurred, and what the 2027–2030 window will test, is the extension of AI's proven capabilities into the full scope of human financial complexity: business sales, estate disputes, disability planning, divorce, and the multi-party dynamics that define the most financially consequential decisions people make.

The revolution is real and substantial. Its scope is more bounded than the original prediction implied.

## THE COUNTERPOINT: Complexity Ceiling Is Real, And It's Where the Money Is

The Counterpoint's core prediction, that AI financial tools would hit fundamental limitations in complex planning scenarios while the most profitable client segments continued preferring human relationships, has been confirmed by the 2026 market structure with a precision that the original 35% likelihood undervalued.

The bifurcation is structural and not merely transitional: AI dominates the standard portfolio management and basic planning market where cost efficiency is the primary value driver, while human advisors retain dominance in the complex planning market where relationship, judgment, and legal expertise are the primary value drivers.

Estate planning with trusts, wills, and beneficiary designations; business succession planning; divorce and asset division; disability planning with special needs trusts, these situations involve nuance, legal complexity, and emotional sensitivity that current AI systems handle poorly in 2026.

The market has confirmed this empirically: Fidelity International's own Freya launch explicitly acknowledged the FCA boundary between guidance and regulated financial advice, designing the product to answer customer questions within guidance parameters rather than crossing into the advice territory where comprehensive financial planning lives.

Lloyds Banking Group's AI assistant similarly emphasizes "the ability to easily refer to expert human support when needed", a design choice that reflects regulatory reality, not a transitional limitation awaiting the next product cycle.

The institutional deployments that confirm the Point's democratization case simultaneously confirm the Counterpoint's complexity ceiling by designing their products around it.

The regulatory and liability framework has settled in ways that reinforce rather than eliminate the human advisory requirement for complex situations. The SEC's 2026 principles-based approach makes explicit that delegating portfolio decisions to algorithms does not dilute fiduciary liability, human oversight remains a regulatory requirement, and firms must clearly disclose the actual role and limitations of their AI tools.

The use of agentic AI for autonomously approving or denying financial decisions based on biased data creates systematic discrimination risks that expose institutions to regulatory penalties and reputational damage, a risk profile that gives financial institutions strong incentives to maintain human oversight rather than pursuing full automation.

Consumer confidence data reinforces the structural dynamic: analysis consistently shows that consumers are broadly receptive to AI for low-risk applications like fraud detection and portfolio rebalancing, but confidence declines sharply for high-stakes decisions like loan approvals, major investment choices, and insurance recommendations.

The behavioral finance dimension the Counterpoint identified as AI's fundamental limitation is not being overcome by product sophistication alone: people experiencing divorce, disability, business transitions, and estate planning don't experience these as optimization problems to be solved by an algorithm.

They experience them as life events requiring empathetic human counsel that can hold the emotional complexity alongside the financial analysis. No robo-advisor has demonstrated it can do that at scale, and no regulatory framework would permit it to do so without human accountability.

AI IS SOLVING MANY PROBLEMS.
THE HARDER ONES REMAIN.
WHAT NEEDS TO CHANGE
(FOR THE COUNTERPOINT TO PREVAIL)
Counterpoint wins only if these conditions materialize.
The counterpoint has strengthened modestly, not because AI has stalled, but because the limits of automation in deeply human financial situations have proven more persistent than the original analysis anticipated.
SOME DECISIONS WILL ALWAYS REQUIRE HUMAN JUDGMENT.
THE POINT MUST BE WRONG ABOUT
COMPLEXITY YIELDS TO COMPUTATION
AI systems would need to demonstrate that increasingly sophisticated reasoning models can successfully navigate emotionally charged and legally complex financial decisions.
TRUST ACCELERATES RAPIDLY
Consumers would need to become substantially more comfortable delegating major life decisions involving inheritance, family dynamics, and long-term care planning to AI systems.
REGULATION EVOLVES QUICKLY
Regulators would need to establish frameworks that permit significantly greater autonomy while maintaining consumer protections.
HUMAN ADVISOR ADVANTAGES SHRINK
Empathy, negotiation skills, and contextual judgment would need to become less differentiating in financial outcomes.
LIABILITY BECOMES MANAGEABLE
Clear accountability structures would need to emerge that allow autonomous financial recommendations without extensive human oversight.
CURRENT STATE
AI-powered personal finance has moved decisively beyond experimentation. Optimization tasks including budgeting, portfolio construction, tax-loss harvesting, and retirement modeling continue expanding rapidly.
However, the evidence increasingly suggests that success in computational finance does not automatically translate into success in emotionally complex financial decision-making.
The revolution is real. Its boundaries remain under negotiation.
THE REVOLUTION IS REAL.
ITS NEXT CHAPTER IS UNWRITTEN.
DECISION FRAMEWORK
Actions to take now to prepare for either outcome.
1 SEGMENT DECISIONS BY COMPLEXITY
Deploy AI aggressively in optimization-heavy domains while preserving human involvement in high-stakes situations.
2 TRACK REGULATORY EVOLUTION
Monitor fiduciary standards, disclosure requirements, and accountability frameworks shaping AI adoption.
3 BUILD HYBRID ADVISORY MODELS
Combine automation efficiency with human expertise rather than treating them as competing approaches.
4 MEASURE CONSUMER TRUST
Watch adoption patterns in increasingly complex financial activities.
5 REASSESS CONTINUOUSLY
The boundary between optimization problems and life problems will evolve as technology matures.
LIKELIHOOD COUNTERPOINT PREVAILS
40%
▲+5% vs Original Prediction (35%)
The counterpoint strengthens modestly because real-world deployment has reinforced the distinction between optimization problems and life problems. AI continues outperforming in structured financial domains. The evidence is less convincing that consumers will readily transfer trust during moments involving family conflict, uncertainty, mortality, or identity. This is not a rejection of AI's trajectory. It is recognition that some forms of complexity appear more durable than initially expected.
★ • THE OUTCOME IS NOT INEVITABLE ★
The next five years will reveal whether AI crosses the complexity gap or whether human judgment remains the ultimate differentiator.
THE COUNTERPOINT MUST BE RIGHT ABOUT
COMPLEXITY REMAINS STICKY
The most valuable financial decisions continue to involve emotional, relational, and contextual factors resistant to automation.
HUMAN TRUST EVOLVES SLOWLY
Consumers maintain a preference for human accountability when consequences become significant.
REGULATION PRIORITIZES OVERSIGHT
Policymakers continue requiring human review and responsibility for consequential financial decisions.
LIABILITY CONSTRAINS AUTONOMY
Institutions remain reluctant to fully automate decisions carrying substantial legal exposure.
PREMIUM HUMAN ADVICE ENDURES
The highest-value segments of financial planning remain relationship-driven businesses.
MOST CRITICAL VARIABLE
THE COMPLEXITY CEILING
The single most important question is not whether AI can provide better recommendations. It is whether increasingly capable systems can successfully navigate situations defined by:
• competing stakeholder interests
• emotional volatility
• incomplete information
• ethical ambiguity
• accountability requirements
If the complexity ceiling falls, the Point prevails decisively. If it holds, the Counterpoint remains resilient.
HUMAN JUDGMENT ENDURES.
ITS VALUE REMAINS UNMATCHED.

## What Needs to Change

### (for the Counterpoint to Prevail)

For the Counterpoint's "human advisory dominance in complex planning" scenario to remain the dominant market structure through 2030, three conditions need to hold that are under active pressure.

First, the complexity ceiling needs to remain real at the level that currently defines it. Fidelity's Freya has already moved the ceiling upward from where it stood in 2023, answering complex questions about investment accounts in natural language rather than through scripted menus.

If the next generation of agentic financial AI achieves genuine contextual legal and emotional reasoning for complex situations, the Counterpoint's protected domain shrinks materially.

The "What Needs to Change" test for the Counterpoint is whether divorce asset division, business succession with tax implications, and special needs trust planning remain genuinely beyond AI's capability through 2030, not whether they are beyond today's capability, which they clearly are.

Second, the regulatory framework needs to maintain meaningful human oversight requirements rather than evolving toward a pure disclosure model. A future SEC or FCA that treats algorithmic advice as equivalent to human advice provided adequate disclosure is made would fundamentally change the liability architecture that currently protects human advisory dominance in complex situations.

Third, the generational wealth transfer needs to produce high-net-worth clients who share their predecessors' preference for human relationships.

The existing evidence, younger investors strongly preferring digital tools for standard situations, is directionally unfavorable for the Counterpoint in the long run, even if it remains favorable through 2030 for the complex planning segment.

## Likelihood Counterpoint Prevails: 40%

The Counterpoint likelihood increases from 35% to 40%, a significant upward revision, and it's driven by three 2026 developments that validate the Counterpoint's structural arguments more precisely than the original evidence base supported.

First, the institutional product launches confirm the complexity ceiling as a design reality rather than a temporary technical gap: both Fidelity International's Freya and Lloyds's AI assistant were explicitly architected around the guidance-versus-advice regulatory boundary and around human escalation paths, confirming that the complexity ceiling is a regulatory and institutional reality, not just an AI capability gap awaiting the next model version.

Second, the SEC's 2026 examination priorities and principles-based framework have settled the regulatory landscape in a way that preserves human accountability requirements rather than creating the liability-free zone for algorithmic advice that a pure market logic might have produced.

Third, consumer confidence declining sharply for high-stakes AI financial decisions, consistent across multiple independent analyses, confirms that the behavioral finance dimension the Counterpoint identified is not being resolved by product quality alone.

The 40% does not say the Counterpoint is close to prevailing.

It says the Counterpoint's case is stronger and more structurally well-grounded than the original 35% implied, validated by the same institutional product launches that confirm the Point's democratization case.

Both sides are simultaneously right, and the 40%/60% split reflects that genuine analytical tension rather than one-sided evidence.

# Voice Assistant Platform Abandonment

## The Smoke Detector of the Connected Home: Nobody Loves It, Nobody Removes It

*Amazon upgraded Echo devices to Alexa+ without asking permission.*

*Users who tried to revert were flooded with ads until the original experience became unusable.*

*A quarter of early Alexa users abandoned their devices within two weeks of purchase.*

*China's smart speaker market declined 14% in 2025.*

*Revenue projections point to $127 billion by 2035.*

*Both the abandonment story and the growth story are true.*

*The question is which one defines the category.*

VOICE ASSISTANT PLATFORM ABANDONMENT
THE SMOKE DETECTOR OF THE CONNECTED HOME:
NOBODY LOVES IT, NOBODY REMOVES IT
TWO PREDICTIONS. BOTH RIGHT. THE FUTURE ISN'T SIMPLE.
THE POINT:
THE AI MADE IT SMARTER. IT DIDN'T MAKE IT NECESSARY.
MANUFACTURER FORCED THE UPGRADE
Amazon auto-enrolled Prime members in Alexa+, flooded reversions with ads. That's not confidence—it's a retention problem.
HIGH EARLY ABANDONMENT
15–25% of new Alexa users (2018–2021) abandoned their devices within two weeks. Internal Amazon data confirmed the churn.
CHINA MARKET DECLINED 14% IN 2025
Analysts say the category "seems to have reached a dead end." AI makes speakers smarter, but doesn't enrich the ecosystem.
SMARTPHONES REMAIN THE SUPERCOMPUTER
A smarter command system is still a command system competing with a more capable device already in your pocket.
SMART GLASSES ARE THE REAL ENDGAME
Amazon confirms Alexa is coming to smart glasses. The speaker becomes a peripheral node, not the primary interface.
PRIVACY PUSHBACK AND MATURATION
On-device processing and physical mute switches are now standard. Growth slowed as consumers demand more control.
THE POINT ROUNDUP
The smart speaker's core value problem hasn't been solved. AI improves the experience, but doesn't justify the device. Users move on—or never move in.
THE BATTLEGROUND
THE USER'S REALITY
Forced upgrades and declining engagement
ADOPTION
Revenue and shipments keep growing
Simple tasks phones already handle better
VALUE PROPOSITION
AI enables complex conversations and agentic tasks
Substitution by smart glasses and wearables
PLATFORM ROLE
Core hub for the connected home
Privacy concerns limit depth of engagement
TRUST & PRIVACY
On-device AI and controls improve over time
Novelty faded. Utility plateaued.
USER ENGAGEMENT
Usage deepens with better AI and routines
THE DEBATE IN ONE QUESTION
WILL VOICE ASSISTANTS BECOME AN INDISPENSABLE, AI-POWERED HUB FOR THE CONNECTED HOME—OR REMAIN A TOLERATED UTILITY THAT RARELY EARNS USER LOYALTY?
THE COUNTERPOINT:
AI IS TRANSFORMING THE PLATFORM EXPERIENCE.
REVENUE AND SHIPMENTS ARE GROWING
$13.15B market in 2026, projected to reach $127.35B by 2035 at 28.71% CAGR. Shipments hit 250M units in 2024 (up from 200M in 2023).
ALEXA+ DELIVERS A REAL UPGRADE
AI+
Launched March 2025 with advanced reasoning, real conversations, and agentic task completion using leading AI models.
ECOSYSTEM INTEGRATION DEEPENS
Speakers remain the most natural voice hub for lighting, security, entertainment, and multi-room control.
MONETIZATION AND BUSINESS MOMENTUM
Deeper engagement creates subscription, commerce, and advertising opportunities that drive sustained investment.
USAGE BECOMES MORE MEANINGFUL
Better AI + routines = more complex queries, proactive assistance, and habit formation that increase stickiness.
GLOBAL TAM REMAINS MASSIVE
Hundreds of millions of households still lack voice interfaces. Long runway remains, especially in emerging markets.
THE COUNTERPOINT ROUNDUP
The platform is evolving beyond its early limits. AI capability + ecosystem value will drive deeper adoption and long-term growth.

## Opening

The voice assistant platform story has evolved into something more complex than either the original prediction's "expensive Bluetooth speaker" endpoint or the Counterpoint's "conversational AI transformation" scenario.

The 2026 evidence shows a market simultaneously growing in revenue and contracting in user engagement depth, adding AI capability while struggling to retain the users who gave the original devices a chance and found them insufficient.

The forced Alexa+ migration is the most revealing single data point in the 2026 record: it's both the strongest evidence that Amazon has delivered on the AI upgrade the Counterpoint required and the strongest evidence that the organic pull toward that upgrade was insufficient to drive adoption through normal means.

The 2027-2030 window is where these contradictory signals resolve, either into a genuinely transformed platform category or into the smart speaker version of the 3D television: technically impressive, commercially persistent at the margins, and never quite the revolution its advocates described.

## Intro

The voice assistant platform prediction's 2026 update requires separating two questions that the original piece and its Counterpoint treated as one: whether AI capability arrived, and whether AI capability solved the value proposition problem.

The answer to the first question is yes. Alexa+ is a materially better product than the original Alexa, Consumer Reports said so, and Amazon's own behavior confirms it.

I was in that first wave of customers who waited months to get Alexa. Alexa+ is far superior.

The answer to the second question is more complicated, and Amazon's forced migration is the evidence that makes it complicated.

A company with genuine confidence in its product's organic appeal does not need to enroll Prime members without asking, flood reverters with advertising, and make the prior experience practically unusable.

It offers the upgrade and lets the product make the case. The forced migration strategy reveals that Amazon understood, before the market did, that the organic pull toward Alexa+ was insufficient to produce the adoption curve the business case required.

That is not a verdict on whether Alexa+ is better.

It's a verdict on whether "better" is enough.

The smoke detector analogy in this article's subtitle captures the dynamic precisely: the category has become load-bearing infrastructure in too many connected homes to be abandoned easily, but it hasn't become valued enough to be upgraded enthusiastically.

The 2027-2030 question is whether the smart glasses migration Amazon's own VP is signaling changes that equation, or whether the standalone speaker settles permanently into the utility role it has earned without the transformation its advocates promised.

## Background

The global smart home speakers market was valued at $13.15 billion in 2026 and is projected to reach $127.35 billion by 2035 at a 28.71% CAGR, with global shipments rising from 200 million units in 2023 to 250 million in 2024.

Security concerns remain a significant restraint, with 63% of non-users citing privacy fears as the primary barrier to adoption. The competitive dynamics have shifted materially: Amazon's Alexa+ launched in March 2025, using models from Anthropic and designed to be model-agnostic, capable of real conversations, complex query reasoning, and agentic task completion.

Amazon announced at CES January 2026 that it was ending the Alexa+ waiting list, then began automatically enrolling Prime members without requiring explicit opt-in. Users who tried to revert to the original Alexa were presented with persistent advertising, making the reversion practically unusable.

The backlash was immediate and documented across Reddit, TechRadar, AFTVnews, and Vice. eMarketer characterized the episode plainly: "Blanket feature rollouts are indicative of platforms chasing engagement and ad monetization first and considering user permission second."

Simultaneously, industry analyst firm Transforma Insights published a January 2026 report titled "Smart Speakers: A Market Under Pressure to Evolve," and China's smart speaker market declined 14% in 2025, with analysts noting that AI model additions make speakers smarter but "cannot enrich the smart home ecosystem" that represents the category's primary practical value.

VOICE ASSISTANT PLATFORM ABANDONMENT
THE SMOKE DETECTOR OF THE CONNECTED HOME: NOBODY LOVES IT, NOBODY REMOVES IT
TWO PREDICTIONS. BOTH RIGHT. THE FUTURE ISN'T SIMPLE.
THE POINT: THE AI MADE IT SMARTER. IT DIDN'T MAKE IT NECESSARY.
MANUFACTURER FORCED THE UPGRADE
Amazon auto-enrolled Prime members in Alexa+, flooded reversions with ads. That's not confidence—it's a retention problem.
HIGH EARLY ABANDONMENT
15–25% of new Alexa users (2018–2021) abandoned their devices within two weeks. Internal Amazon data confirmed the churn.
CHINA MARKET DECLINED 14% IN 2025
Analysts say the category "seems to have reached a dead end." AI makes speakers smarter, but doesn't enrich the ecosystem.
SMARTPHONES REMAIN THE SUPERCOMPUTER
A smarter command system is still a command system competing with a more capable device already in your pocket.
SMART GLASSES ARE THE REAL ENDGAME
Amazon confirms Alexa is coming to smart glasses. The speaker becomes a peripheral node, not the primary interface.
PRIVACY PUSHBACK AND MATURATION
On-device processing and physical mute switches are now standard. Growth slowed as consumers demand more control.
THE POINT ROUNDUP
The smart speaker's core value problem hasn't been solved. AI improves the experience, but doesn't justify the device. Users move on—or never move in.
THE BATTLEGROUND
THE USER'S REALITY
Forced upgrades and declining engagement
ADOPTION
Revenue and shipments keep growing
Simple tasks phones already handle better
VALUE PROPOSITION
AI enables complex conversations and agentic tasks
Substitution by smart glasses and wearables
PLATFORM ROLE
Core hub for the connected home
Privacy concerns limit depth of engagement
TRUST & PRIVACY
On-device AI and controls improve over time
Novelty faded. Utility plateaued.
USER ENGAGEMENT
Usage deepens with better AI and routines
THE DEBATE IN ONE QUESTION
WILL VOICE ASSISTANTS BECOME AN INDISPENSABLE, AI-POWERED HUB FOR THE CONNECTED HOME—OR REMAIN A TOLERATED UTILITY THAT RARELY EARNS USER LOYALTY?
THE COUNTERPOINT: AI IS TRANSFORMING THE PLATFORM EXPERIENCE.
REVENUE AND SHIPMENTS ARE GROWING
$13.15B market in 2026, projected to reach $127.35B by 2035 at 28.71% CAGR. Shipments hit 250M units in 2024 (up from 200M in 2023).
AI+
ALEXA+ DELIVERS A REAL UPGRADE
Launched March 2025 with advanced reasoning, real conversations, and agentic task completion using leading AI models.
ECOSYSTEM INTEGRATION DEEPENS
Speakers remain the most natural voice hub for lighting, security, entertainment, and multi-room control.
MONETIZATION AND BUSINESS MOMENTUM
Deeper engagement creates subscription, commerce, and advertising opportunities that drive sustained investment.
USAGE BECOMES MORE MEANINGFUL
Better AI + routines = more complex queries, proactive assistance, and habit formation that increase stickiness.
GLOBAL TAM REMAINS MASSIVE
Hundreds of millions of households still lack voice interfaces. Long runway remains, especially in emerging markets.
THE COUNTERPOINT ROUNDUP
The platform is evolving beyond its early limits. AI capability + ecosystem value will drive deeper adoption and long-term growth.

## THE POINT: The AI Made It Smarter. It Didn't Make It Necessary.

The original prediction's core thesis, that smart speakers face a fundamental value proposition problem that technical improvements alone can't solve, has been confirmed not by user abandonment statistics but by something more revealing: the manufacturer's own behavior.

Amazon's automatic enrollment of Prime members in Alexa+ without explicit opt-in, changing the familiar voice users had lived with for years to what they described as a "sassy teenager," was not a confident product launch. It was a forced migration from a company that understood its platform faced a retention problem that organic upgrade adoption was insufficient to solve.

Amazon's own internal data, obtained by Bloomberg, found that 15-25% of new Alexa users during 2018-2021 completely abandoned their devices in the second week of ownership, and the company concluded the smart speaker market had "passed its growth phase" with only 1.2% annual growth projected going forward.

The Chinese smart speaker market declined 14% in 2025, with industry analysts concluding the category "seems to have reached a dead end," noting that AI large-scale model additions make speakers smarter and more accurate but "cannot enrich the smart home ecosystem" that represents the category's primary practical value.

The ecosystem problem the original piece identified, that smart speakers work well for simple tasks that smartphones already handle, hasn't been solved by AI capability improvements.

A smarter command system is still a command system competing with a supercomputer you already carry in your pocket.

The substitution threat identified in the original prediction has materialized in a specific and previously unforecast form: smart glasses.

The Amazon VP of Alexa told reporters in 2026 that "you're going to hear things from us in 2026" about Alexa coming to smart glasses, an admission that the standalone speaker form factor's primary value proposition (voice access to AI) is more compellingly delivered by a wearable that goes where you go than by a cylinder that stays in your kitchen.

The platform is not being abandoned outright, it's being absorbed by a more capable form factor that makes the standalone speaker's role more peripheral than central.

This dynamic is compounded by the privacy evolution the original prediction identified: 2026 has seen a surge in on-device processing adoption as consumers push back against cloud-dependent always-listening architectures, with physical mute switches now an expected standard feature rather than a premium differentiator.

Growth rates have declined due to market saturation, greater availability of substitute products, and continued privacy concerns, with the market explicitly described as "under pressure to evolve" rather than confidently expanding.

eMarketer's analysis of the forced Alexa+ rollout is the most direct characterization of the category's structural situation: users who feel "spoon-fed rather than served" have tolerance for platform behavior that drops as fast as adoption rises when it was delivered by compulsion.

### The Point Roundup

The voice assistant platform abandonment story hasn't materialized in the dramatic form the original prediction described, smart speakers have not become expensive Bluetooth speakers, and Amazon hasn't retreated from the category.

What has materialized is a more subtle and more structurally interesting version of the same problem: a market that grew rapidly, hit its natural ceiling earlier than transformative technologies typically do, and is now defending its installed base through forced AI upgrades rather than growing it through compelling new value propositions.

The 2027-2030 window tests whether Alexa+ can convert the forced migration from compliance to genuine engagement, turning the AI upgrade into a companion users choose rather than accept, or whether the category settles into the persistent but stagnant role of the utility that nobody loves but nobody removes: the smoke detector of the connected home.

## THE COUNTERPOINT: The Central Nervous System of the Connected Home Doesn't Have to Be Loved to Be Indispensable

The Counterpoint's core prediction, that large language models would transform smart speakers from command devices into genuinely conversational AI, has been delivered by Alexa+. Amazon's AI-powered system uses models from Anthropic and is model-agnostic, using whatever model is needed in the moment.

It can engage in real conversations, reason through complex queries, and complete agentic tasks autonomously, a fundamentally different product from the command-response system that characterized the first generation and generated the frustrations the original prediction described.

Consumer Reports assessed Alexa+ as "a much better assistant", capable of real conversations, capable of completing tasks as an AI agent, materially improved in natural language understanding, while acknowledging the app experience "is still a mess" and the improvement is "far from perfect."

That "far from perfect but genuinely better" assessment describes a product on an improvement trajectory rather than a collapsing one. Amazon developed custom AZ-series chips to move voice processing to the device itself, simultaneously improving response quality, reducing latency, and addressing consumer privacy concerns through on-device processing.

The Counterpoint identified three required improvements: recognition accuracy, true conversational intelligence, and privacy solutions. All three have advanced materially.

Whether they have advanced enough to change consumer behavior, not just product capability, is the 2027-2030 test.

The market revenue trajectory the Counterpoint's "stability through AI integration" scenario required is intact. Global shipments rose from 200 million units in 2023 to 250 million in 2024, and the market is projected to reach $127.35 billion by 2035 at a 28.71% CAGR, growth rates that are inconsistent with the "platform abandonment" scenario the original prediction described.

Smart speakers have become the orchestration layer of the connected home, managing lighting, climate, security, entertainment, and energy across a growing installed base of connected devices.

That role becomes more valuable, not less, as the number of devices increases.

The abandonment that the original prediction described was the abandonment of the novelty layer, the "ask it questions and be amused" use case that drove early adoption and burned out quickly. What has remained and deepened is the utility layer: home automation orchestration, hands-free information access during cooking and cleaning, alarm-setting, timer-management, and music-playing that smart speaker loyalists have always described as genuine daily value.

That utility layer is not glamorous. It's also not abandonment.

Revenue projections of $35 billion by 2027 and $62 billion by 2030 reflect analyst confidence that the category has a durable role in consumer electronics even if it falls short of the transformative platform vision the original marketing implied.

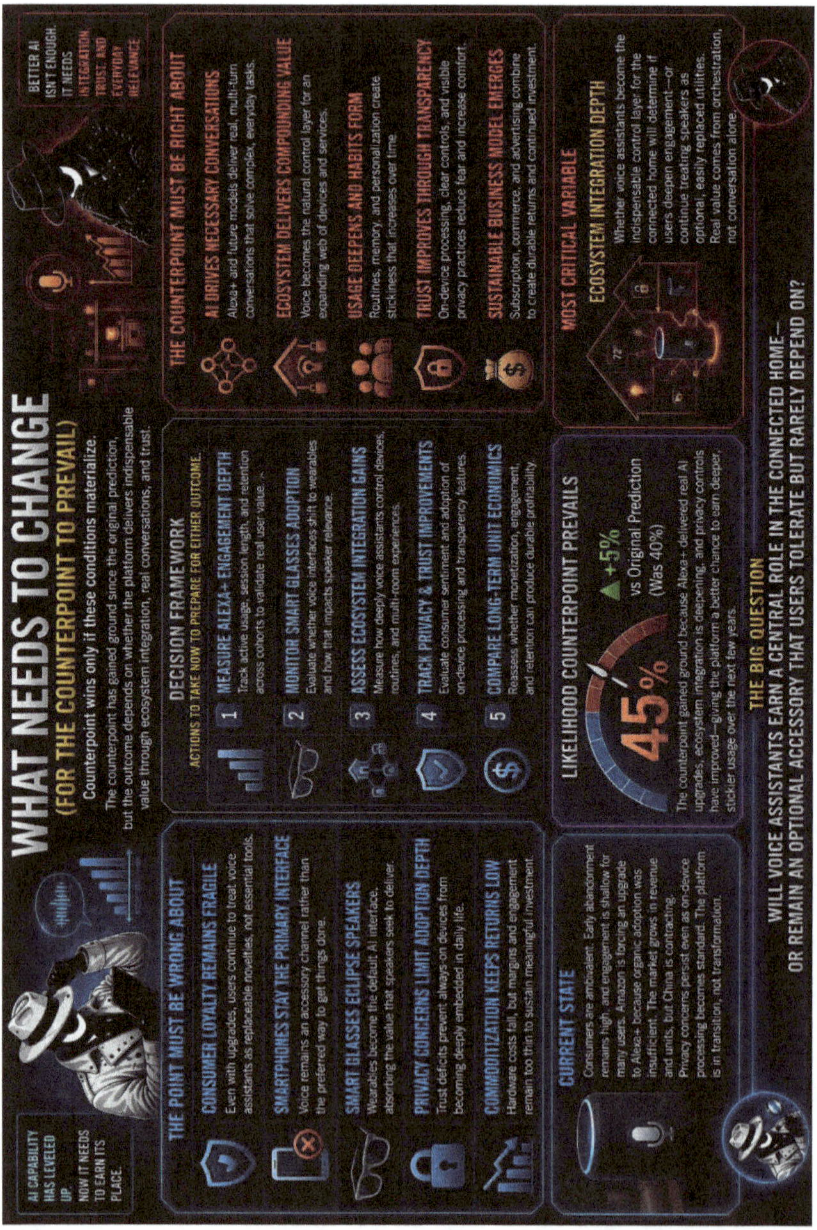
AI CAPABILITY HAS LEVELED UP. NOW IT NEEDS TO EARN ITS PLACE.
WHAT NEEDS TO CHANGE
(FOR THE COUNTERPOINT TO PREVAIL)
Counterpoint wins only if these conditions materialize.
The counterpoint has gained ground since the original prediction, but the outcome depends on whether the platform delivers indispensable value through ecosystem integration, real conversations, and trust.
BETTER AI ISN'T ENOUGH. IT NEEDS INTEGRATION, TRUST, AND EVERYDAY RELEVANCE.
THE POINT MUST BE WRONG ABOUT
CONSUMER LOYALTY REMAINS FRAGILE
Even with upgrades, users continue to treat voice assistants as replaceable novelties, not essential tools.
SMARTPHONES STAY THE PRIMARY INTERFACE
Voice remains an accessory channel rather than the preferred way to get things done.
SMART GLASSES ECLIPSE SPEAKERS
Wearables become the default AI interface, absorbing the value that speakers seek to deliver.
PRIVACY CONCERNS LIMIT ADOPTION DEPTH
Trust deficits prevent always-on devices from becoming deeply embedded in daily life.
COMMODITIZATION KEEPS RETURNS LOW
Hardware costs fall, but margins and engagement remain too thin to sustain meaningful investment.
DECISION FRAMEWORK
ACTIONS TO TAKE NOW TO PREPARE FOR EITHER OUTCOME.
1 MEASURE ALEXA+ ENGAGEMENT DEPTH
Track active usage, session length, and retention across cohorts to validate real user value.
2 MONITOR SMART GLASSES ADOPTION
Evaluate whether voice interfaces shift to wearables and how that impacts speaker relevance.
3 ASSESS ECOSYSTEM INTEGRATION GAINS
Measure how deeply voice assistants control devices, routines, and multi-room experiences.
4 TRACK PRIVACY & TRUST IMPROVEMENTS
Evaluate consumer sentiment and adoption of on-device processing and transparency features.
5 COMPARE LONG-TERM UNIT ECONOMICS
Reassess whether monetization, engagement, and retention can produce durable profitability.
THE COUNTERPOINT MUST BE RIGHT ABOUT
AI DRIVES NECESSARY CONVERSATIONS
Alexa+ and future models deliver real, multi-turn conversations that solve complex, everyday tasks.
ECOSYSTEM DELIVERS COMPOUNDING VALUE
Voice becomes the natural control layer for an expanding web of devices and services.
USAGE DEEPENS AND HABITS FORM
Routines, memory, and personalization create stickiness that increases over time.
TRUST IMPROVES THROUGH TRANSPARENCY
On-device processing, clear controls, and visible privacy practices reduce fear and increase comfort.
SUSTAINABLE BUSINESS MODEL EMERGES
Subscription, commerce, and advertising combine to create durable returns and continued investment.
CURRENT STATE
Consumers are ambivalent. Early abandonment remains high, and engagement is shallow for many users. Amazon is forcing an upgrade to Alexa+ because organic adoption was insufficient. The market grows in revenue and units, but China is contracting. Privacy concerns persist even as on-device processing becomes standard. The platform is in transition, not transformation.
LIKELIHOOD COUNTERPOINT PREVAILS
45%
▲ +5%
vs Original Prediction
(Was 40%)
The counterpoint gained ground because Alexa+ delivered real AI upgrades, ecosystem integration is deepening, and privacy controls have improved—giving the platform a better chance to earn deeper, stickier usage over the next few years.
MOST CRITICAL VARIABLE
ECOSYSTEM INTEGRATION DEPTH
Whether voice assistants become the indispensable control layer for the connected home will determine if users deepen engagement—or continue treating speakers as optional, easily replaced utilities. Real value comes from orchestration, not conversation alone.
THE BIG QUESTION
WILL VOICE ASSISTANTS EARN A CENTRAL ROLE IN THE CONNECTED HOME—
OR REMAIN AN OPTIONAL ACCESSORY THAT USERS TOLERATE BUT RARELY DEPEND ON?

## What Needs to Change

### (... for the Counterpoint to Fully Prevail)

For the Counterpoint's transformative AI scenario to fully materialize through 2030, two conditions need to resolve in the platform's favor that are currently ambiguous.

First, the Alexa+ forced migration needs to produce genuine engagement rather than compliance, the 15-25% early abandonment rate needs to decline toward single digits as LLM-powered conversations give users reasons to initiate interactions rather than just respond to them.

The current evidence is that Alexa+ is better but the app experience is still a mess, a combination suggesting the capability ceiling has been raised without the experience ceiling following.

Amazon has pledged to introduce explicit opt-in prompts and clearer privacy disclosures for future AI upgrades; whether that commitment restores the user trust the forced rollout eroded is a 2026-2027 behavioral question that product quality alone can't answer.

Second, the smart glasses threat needs to fail to materialize as a mainstream form factor, or smart glasses need to become an Alexa distribution surface rather than a competitive replacement.

Amazon's VP of Alexa signaling Alexa+ coming to smart glasses "in 2026" reflects a recognition that the standalone speaker form factor's voice AI value proposition is more compellingly delivered by a wearable that goes everywhere with the user.

If smart glasses become the primary Alexa surface, the standalone speaker does not disappear. It becomes secondary: still useful in the home, but no longer the center of the voice assistant experience. That shift would place the category somewhere between the original prediction's "expensive Bluetooth speaker" endpoint and the Counterpoint's "central nervous system" vision.

## Likelihood Counterpoint Prevails: 45%

The Counterpoint likelihood increases from 40% to 45%, a modest but meaningful upward revision driven by the Alexa+ AI upgrade landing as a genuinely improved product, the market revenue trajectory remaining strongly positive, and the smart speaker's integration into the connected home ecosystem deepening rather than reversing.

The "far from perfect but genuinely better" Consumer Reports assessment describes a product on an improvement trajectory. Revenue projections of $35 billion by 2027 and $62 billion by 2030 reflect analyst confidence in the category's durability.

What keeps the number at 45% rather than higher is the forced migration dynamic, a product that requires force to upgrade rather than generating organic demand for its AI features has a value proposition problem that capability improvements alone don't resolve. eMarketer's characterization is precise: "Blanket feature rollouts are indicative of platforms chasing engagement and ad monetization first and considering user permission second."

That is a structural trust problem, not a product quality problem.

The original prediction's "expensive Bluetooth speaker" endpoint remains plausible for the portion of the installed base that accepted the force-upgrade with resignation rather than enthusiasm.

The 45% says: the Counterpoint's AI transformation scenario has delivered a materially better product, but the fundamental question of whether voice interaction is the optimal interface for most computing tasks, which the original piece identified as the unresolved core problem, hasn't been answered by Alexa+ any more definitively than it was by the original Alexa.

The smart glasses development is the new variable neither side of the original piece accounted for, and Amazon's own VP signaling Alexa's migration to that form factor is the update's most important strategic signal for the 2030 outcome.

As an X-factor in this topic, Smart Glasses are dealing with their own issues and could very well anchor the AI Assistant's transformation.

# AI Customer Service Consumer Revolt

## Digital First, Not Digital Only: How the Industry Blinked Before the Full Revolt

*Preference for human customer service: up to 85%. Preference for AI: down to 5%. Frustration with AI agents: rising from 54% to 59% in six months. Willingness to hang up if connected to AI: rising from 29% to 31%.*

*Also: Gartner forecast in January 2026 that GenAI cost per resolution will exceed $3 by 2030, higher than many offshore human agents, once governance, quality control, escalation, and exception handling are factored in.*

*The revolt is real.*

*The economics that made AI-first deployment seem inevitable are breaking down.*

*And the industry has already started retreating before consumers forced* it.

MOST CONSUMERS WANT HUMANS. NOT BOTS.
THE REVOLT IS REAL.
AI CUSTOMER SERVICE CONSUMER REVOLT
DIGITAL FIRST, NOT DIGITAL ONLY:
HOW THE INDUSTRY BLINKED BEFORE THE FULL REVOLT
AI was supposed to cut costs and improve service. Consumers pushed back. The economics are breaking down. The retreat has already begun.
THE AI-FIRST BET WAS WRONG.
COSTS ARE RISING
TRUST IS FALLING
THE MODEL IS BREAKING
EVIDENCE THE SHIFT IS REAL
UP TO 85% PREFER HUMAN SERVICE
Preference for speaking to a real person rose from 83% to 85% in six months.
AI PREFERENCE FALLS TO 5%
5%
Those preferring AI dropped from 7% to 5%.
FRUSTRATION IS RISING
Frustration with AI agents rose from 54% to 59%.
MORE WILL HANG UP
Customers who would hang up if connected to AI rose from 29% to 31%.
MOST WANT A HUMAN OPTION
73% will take their business elsewhere if only AI is offered.
How can I help you?
I can help. Let's fix this.
THE MARKET REALITY
$16.6B
Market Size in 2026
22.8%
CAGR Through 2030
$50.8B
Projected Market in 2030
70%
AI Projects Miss ROI Goals
6,000
Consumers Surveyed (US, UK, Canada)
EVIDENCE THE SHIFT MATTERS
COST ADVANTAGE IS EVAPORATING
Gartner forecasts GenAI cost per resolution will exceed $3 by 2030—higher than many offshore human agents.
REGULATION WILL RAISE COSTS
AI-related regulation will increase assisted service volume by 30% by 2028—raising costs and compliance burdens.
CUSTOMER TRUST IS DECLINING
Trust in AI fell from 62% in 2023 to 59% in 2025. "Very untrustworthy" more than doubled from 5% to 12%.
CHURN AND REVENUE AT RISK
Poor AI experiences drive churn, refunds, and lost lifetime value. Executives are already seeing the impact.
INDUSTRY IS RETREATING
95% of service leaders plan hybrid models. No Fortune 500 company will eliminate human service by 2028.
THE QUESTION
WILL COMPANIES FIND THE RIGHT BALANCE BETWEEN AI EFFICIENCY AND HUMAN EMPATHY—
OR KEEP ALIENATING CUSTOMERS AND PAYING THE PRICE?

## Opening

The AI customer service revolt story has evolved into something more analytically interesting than a simple vindication: the consumer backlash the original prediction described has materialized in the data, but the corporate response partially preempted the full crisis by retreating toward hybrid models before the revolt reached critical mass.

The 2026 picture is an industry that saw the backlash coming and moved toward "digital first, not digital only" before consumers' churn-driven revenue losses became impossible to ignore. Whether that preemptive retreat constitutes the corporate crisis the original prediction described, or a reasonably managed transition toward a sustainable hybrid model, is the analytical question the 2027-2030 window will answer.

Gartner's collapsing cost advantage forecast is the most important variable in that answer, because if AI customer service genuinely approaches human agent cost by 2030, the revolt becomes an economic logic story that aligns with consumer preference, and the "talk to a person immediately" premium the original piece described becomes simply the standard offering rather than a premium differentiator.

## Intro

The AI customer service revolt prediction occupies unusual analytical territory: the consumer backlash it described has arrived in the data with the consistency and direction the original piece predicted, and the industry has responded not by waiting for the crisis to force a reckoning but by partially preempting it.

That preemptive retreat, 95% of service leaders retaining human agents, no Fortune 500 company eliminating human customer service, Gartner forecasting half of planned AI headcount reductions will be abandoned, is simultaneously the strongest evidence that the original prediction was right and the most important reason the full corporate crisis it described hasn't arrived.

The prediction was right about consumer preference, right about the structural inadequacy of AI for emotionally complex interactions, and right about the economic logic that would eventually force a recalibration.

What it did not anticipate is that the recalibration would happen through corporate self-correction rather than through consumer revolt forcing it.

The "digital first, not digital only" landing point the industry has reached is closer to the Counterpoint's predicted hybrid equilibrium than to the original prediction's full revolt scenario, but it arrived through reluctant retreat rather than through AI earning the consumer trust that would make the equilibrium genuinely durable.

That distinction matters for the 2027-2030 forecast: a hybrid model built on managed retreat is more fragile than one built on demonstrated value, and the Gartner cost convergence forecast is the variable that determines whether the retreat continues or whether the industry finds a stable floor.

## Background

In December 2024, Gartner reported 85% of customer service leaders were exploring or piloting conversational GenAI, with more than 75% feeling pressure from executive leadership to implement it.

By February 2026, pressure had risen to 91%, while Gartner simultaneously forecast that:

half the organizations expecting significant AI-driven headcount reduction would abandon those plans by 2027;

95% of service leaders planned to retain human agents in a "digital first, not digital only" model;

and no Fortune 500 company would fully eliminate human customer service by 2028.

The consumer data driving that corporate retreat: 79% of Americans strongly prefer interacting with a human over an AI agent; 56% have negative feelings about companies using AI in their customer experience; 84% believe human agents are more accurate than AI; 81% believe AI is used primarily to save money not improve service; and 89% believe companies should always offer the option to speak with a human.

The AnswerConnect 6,000-person study published May 2026 across the US, UK, and Canada provides the most current directional data: preference for speaking to a real person rose from 83% to 85% in six months, those preferring AI dropped from 7% to 5%, frustration with AI agents rose from 54% to 59%, and more customers now say they would hang up if connected to AI, rising from 29% to 31%.

All four directional signals are moving against AI-first deployment simultaneously.

AI CUSTOMER SERVICE CONSUMER REVOLT
DIGITAL FIRST, NOT DIGITAL ONLY:
HOW THE INDUSTRY BLINKED BEFORE THE FULL REVOLT
TWO PREDICTIONS. BOTH RIGHT. THE FUTURE IS HYBRID.
THE POINT:
THE COST ADVANTAGE IS EVAPORATING.
THE CONSUMER PREFERENCE NEVER EXISTED.
CONSUMERS OVERWHELMINGLY PREFER HUMANS
Up to 85% prefer human service. Only 5% prefer AI. 73% will take their business elsewhere if only AI is offered.
SATISFACTION IS DECLINING
Frustration with AI agents rose from 54% to 59% in six months. More say they will hang up if connected to AI (29% to 31%).
TRUST IN AI IS FALLING
Trust in AI fell from 62% in 2023 to 59% in 2025. "Very untrustworthy" more than doubled from 5% to 12%.
COST ADVANTAGE IS COLLAPSING
Gartner forecasts GenAI cost per resolution will exceed $3 by 2030—higher than many offshore humans.
REGULATION WILL INCREASE COMPLEXITY
AI-related regulation will push assisted service volume up 30% by 2028 and increase compliance costs.
THE POINT ROUNDUP
AI-first bets were based on weak assumptions. Consumers rejected the model. Costs are rising. Trust is falling. The economics are breaking down.
THE BATTLEGROUND:
WHERE THE REAL CONFLICT LIVES
Consumers want humans for complex, emotional, and high-stakes issues.
PREFERENCE
VS
Consumers accept AI for simple, status-based and transactional tasks.
AI-first creates friction, drives churn, and damages loyalty.
CUSTOMER IMPACT
VS
AI-first reduces wait times and improves efficiency for many.
Total cost is rising once all factors are included.
ECONOMICS
VS
AI still delivers material cost savings today for many interactions.
Regulation and quality issues increase cost and complexity.
RISK & COMPLIANCE
VS
Companies can manage risk with governance and proper design.
The hype cycle is breaking. Expectations are resetting.
INDUSTRY TRAJECTORY
AI capability is improving. Long-term value will increase.
THE COUNTERPOINT:
AI IS TRANSFORMING CUSTOMER SERVICE—
NOT REPLACING HUMANS.
AI HANDLES WHAT HUMANS DON'T WANT
68% prefer AI for simple questions. Customers want fast answers for status, tracking, and basic tasks.
EFFICIENCY AND SPEED IMPROVE
AI delivers immediate responses 24/7. It reduces wait times and resolves issues faster for many customers.
MATERIAL COST SAVINGS TODAY
AI resolutions average $0.62 vs $7.40 for humans. Savings are real even if the gap narrows.
HYBRID MODELS ARE WORKING
95% of leaders plan "digital first, not digital only." Hybrid models improve satisfaction and reduce costs.
AI CAPABILITY IS ADVANCING
Models, orchestration, and integrations are improving. AI will handle more complex tasks over time.
THE COUNTERPOINT ROUNDUP
AI is not replacing humans. It is changing the mix. Companies are moving to hybrid models that improve service and control costs.
THE DEBATE IN ONE QUESTION
WILL COMPANIES ADOPT AI AS A COMPLEMENT TO HUMANS—
OR KEEP PUSHING AI-AS-A-GATEKEEPER AND PAY THE PRICE?

## THE POINT: The Cost Advantage Is Evaporating. The Consumer Preference Never Existed.

The original prediction's consumer preference data has been confirmed and hardened by the most current research available.

The task-specific data provides the nuance the original prediction lacked: 68% of consumers prefer AI for simple status-style questions in 2026, up from 41% in 2024, demonstrating genuine AI acceptance for the narrow band of interactions where speed is the primary value.

But 74% prefer a human for complaints, billing disputes, and sentiment-heavy contacts; 82% expect a clear and immediate path to a human when requested; and 73% say they would take their business elsewhere if a company offers only AI with no human option.

The consumer revolt is not against AI customer service per se, it's against AI customer service deployed as a gatekeeper to human interaction rather than a complement to it.

That distinction is analytically important for the 2027-2030 trajectory: companies deploying AI as a human complement are achieving both cost savings and customer satisfaction improvements, while companies deploying AI as a human replacement are experiencing the silent churn the original prediction described.

Consumer trust in AI is moving in the wrong direction as deployment accelerates declining from 62% in 2023 to 59% in 2025, with the share calling AI "very untrustworthy" more than doubling from 5% to 12%.

An industry that has experienced rising AI distrust despite, or because of, three years of aggressive AI customer service deployment is not building the consumer acceptance it needs for the full transformation the AI-first strategy requires.

The economic argument that made AI-first customer service seem inevitable is being challenged by Gartner's cost trajectory forecast in ways that change the fundamental logic of the investment.

Gartner forecast in January 2026 that GenAI cost per resolution in customer service would exceed $3 by 2030, higher than many B2C offshore human agents, once governance, quality control, escalation, exception handling, and the work created by bad automation are factored into the true cost calculation.

Gartner also forecast that AI-related regulation would push assisted service volume up by 30% by 2028, meaning compliance costs are themselves increasing the human interaction volume the AI was supposed to reduce.

Current data shows AI resolutions averaging $0.62 versus $7.40 for human agents, still a significant advantage in the current period. But the trajectory from $0.62 toward $3.00 by 2030, converging with declining offshore human agent costs, describes an economic case becoming progressively weaker over precisely the window the original prediction identified as the revolt's timeline.

Only 20% of customer service leaders had actually reduced staffing because of AI by early 2026, meaning the headcount savings that justified corporate AI investments have not materialized at the scale finance cases projected.

N.B.: the Gartner cost forecast cited here is reported through ServiceJi, a customer service AI vendor; the specific $3 figure couldn't be verified against a Gartner primary document. It's included as a directional finding attributed to Gartner via secondary reporting, consistent with Gartner's publicly stated concern that GenAI customer service costs are higher than simple per-interaction calculations suggest.

### The Point Roundup

The AI customer service revolt has arrived in its consumer preference dimension, the data is unambiguous, directional, and accelerating.

What hasn't arrived is the corporate crisis the original prediction described, because the industry has partially preempted it by retreating toward hybrid models before the revolt reached revenue-threatening tipping point.

The 2027-2030 window tests whether that preemptive retreat is sufficient, whether "digital first, not digital only" delivers enough human access to satisfy the 82% who expect an immediate path to a human when requested, or whether the gap between policy and implementation continues to frustrate consumers who encounter AI gatekeeping in practice even when company policy promises human access.

The Gartner cost convergence forecast is the development that most changes the original prediction's conclusion: if AI customer service is no longer clearly cheaper by 2030, the economic justification for deploying it as a human replacement rather than a human complement disappears, and the "premium competitive advantage" of human support may not require premium pricing at all.

## THE COUNTERPOINT: Digital First, Not Digital Only: The Industry Found the Answer Before the Crisis Hit

The Counterpoint's most important 2026 evidence is the CSAT gap closure: pure-AI handling now achieves 4.1/5 satisfaction against 4.3/5 for human agents, a gap that has effectively narrowed from the wider differential characterizing earlier AI customer service implementations, with hybrid escalation flows narrowing it further to 0.05 points.

51% of consumers prefer interacting with bots over humans when they want immediate service, and 56% prefer an AI assistant if it resolves their problem quickly, demonstrating that the Counterpoint's "speed changes everything" thesis has genuine empirical support.

The task-specific adoption data is the Counterpoint's strongest evidence: 68% of consumers prefer AI for simple status-style questions in 2026, up from 41% in 2024, a 66% increase in AI acceptance for appropriate use cases in two years.

Deflection rates for structured intents like refund requests and password resets reach 70% or higher.

For the interactions where AI is genuinely superior, speed, availability, consistency at scale, consumer acceptance is growing rapidly.

The revolt the original prediction described is not against AI customer service; it's against AI customer service deployed inappropriately, without human escalation paths, for interactions requiring empathy and judgment.

Companies that have correctly identified where AI adds value versus where it destroys it are achieving the Counterpoint's predicted outcome: cost savings alongside customer satisfaction improvement.

The industry's preemptive move toward "digital first, not digital only" represents the hybrid model equilibrium the Counterpoint predicted as the sustainable outcome.

Ninety-five percent of service leaders plan to retain human agents in a digital-first model; no Fortune 500 company will fully eliminate human customer service by 2028; and half the organizations expecting significant AI-driven headcount reduction have abandoned those plans, all consistent with the Counterpoint's prediction that AI would capture routine interactions while human agents retained dominance for complex ones.

Bank of America's Erica resolves 98% of customer queries within 44 seconds, a performance benchmark demonstrating the Counterpoint's "good AI drives acceptance" thesis at scale. Companies see average returns of $3.50 for every $1 invested in AI customer service, with leading organizations achieving up to 8x ROI.

The revolt the original prediction described as a corporate crisis has been partially absorbed by industry self-correction: companies deploying AI as a human replacement face churn penalties; companies deploying AI as a human complement achieve both efficiency and satisfaction gains.

That market differentiation, playing out in real revenue outcomes, is the self-correcting mechanism the Counterpoint relied on to prevent the full revolt the original prediction described.

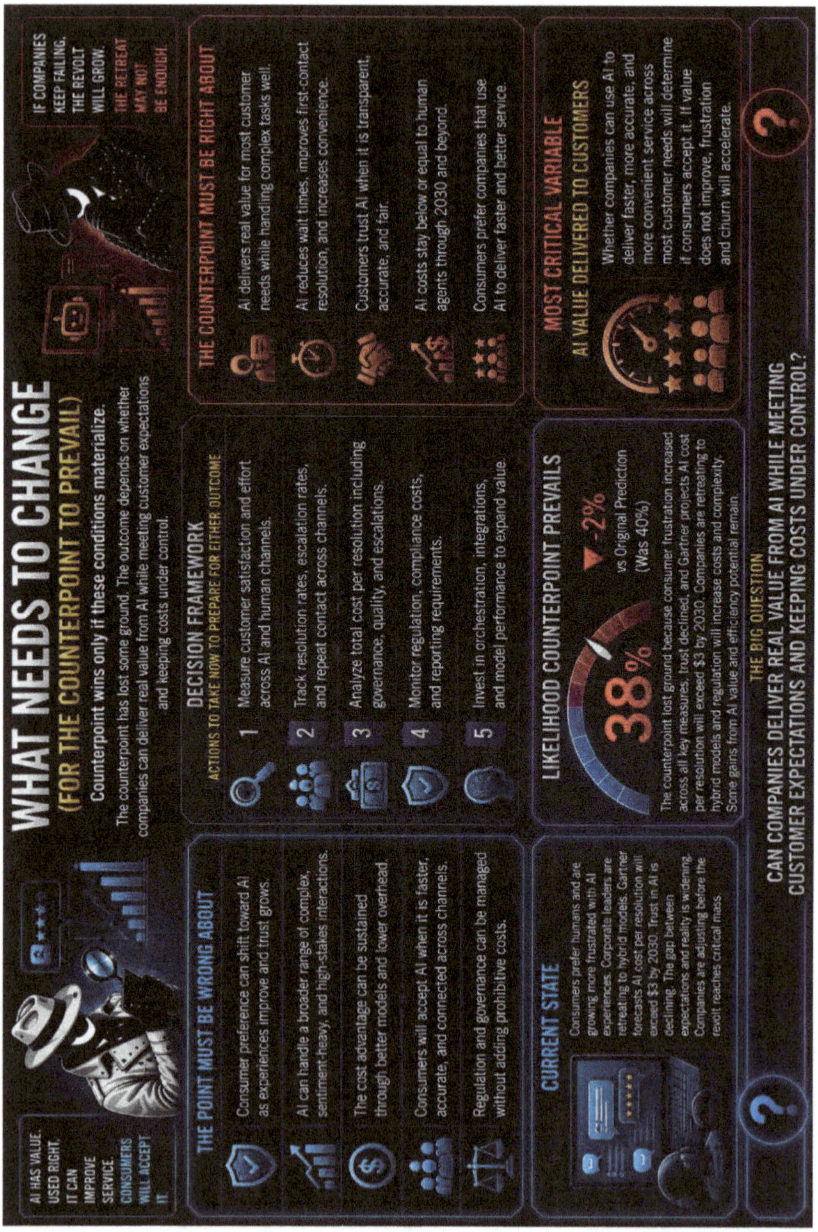
AI HAS VALUE. USED RIGHT, IT CAN IMPROVE SERVICE. CONSUMERS WILL ACCEPT IT.
WHAT NEEDS TO CHANGE
(FOR THE COUNTERPOINT TO PREVAIL)
Counterpoint wins only if these conditions materialize.
The counterpoint has lost some ground. The outcome depends on whether companies can deliver real value from AI while meeting customer expectations and keeping costs under control.
IF COMPANIES KEEP FAILING, THE REVOLT WILL GROW. THE RETREAT MAY NOT BE ENOUGH.
THE POINT MUST BE WRONG ABOUT
Consumer preference can shift toward AI as experiences improve and trust grows.
AI can handle a broader range of complex, sentiment-heavy, and high-stakes interactions.
The cost advantage can be sustained through better models and lower overhead.
Consumers will accept AI when it is faster, accurate, and connected across channels.
Regulation and governance can be managed without adding prohibitive costs.
DECISION FRAMEWORK
ACTIONS TO TAKE NOW TO PREPARE FOR EITHER OUTCOME
1 Measure customer satisfaction and effort across AI and human channels.
2 Track resolution rates, escalation rates, and repeat contact across channels.
3 Analyze total cost per resolution including governance, quality, and escalations.
4 Monitor regulation, compliance costs, and reporting requirements.
5 Invest in orchestration, integrations, and model performance to expand value.
THE COUNTERPOINT MUST BE RIGHT ABOUT
AI delivers real value for most customer needs while handling complex tasks well.
AI reduces wait times, improves first-contact resolution, and increases convenience.
Customers trust AI when it is transparent, accurate, and fair.
AI costs stay below or equal to human agents through 2030 and beyond.
Consumers prefer companies that use AI to deliver faster and better service.
CURRENT STATE
Consumers prefer humans and are growing more frustrated with AI experiences. Corporate leaders are retreating to hybrid models. Gartner forecasts AI cost per resolution will exceed $3 by 2030. Trust in AI is declining. The gap between expectations and reality is widening. Companies are adjusting before the revolt reaches critical mass.
LIKELIHOOD COUNTERPOINT PREVAILS
38%
▼ -2%
vs Original Prediction (Was 40%)
The counterpoint lost ground because consumer frustration increased across all key measures, trust declined, and Gartner projects AI cost per resolution will exceed $3 by 2030. Companies are retreating to hybrid models and regulation will increase costs and complexity. Some gains from AI value and efficiency potential remain.
MOST CRITICAL VARIABLE
AI VALUE DELIVERED TO CUSTOMERS
Whether companies can use AI to deliver faster, more accurate, and more convenient service across most customer needs will determine if consumers accept it. If value does not improve, frustration and churn will accelerate.
THE BIG QUESTION
CAN COMPANIES DELIVER REAL VALUE FROM AI WHILE MEETING CUSTOMER EXPECTATIONS AND KEEPING COSTS UNDER CONTROL?

## What Needs to Change

### (... for the Counterpoint to Fully Prevail)

For the "AI reaches human-level service quality" scenario to become the dominant 2030 outcome, two structural conditions need to develop that are currently under pressure.

First, the Gartner cost convergence forecast needs to prove wrong, or the governance, quality control, and exception handling costs that are driving AI resolution costs upward need to be materially reduced through better system design. If AI customer service genuinely approaches $3 per resolution by 2030 while offshore human agent costs continue declining, the economic case for prioritizing AI over humans collapses in the B2C segment, and the "digital first" model survives only on the quality argument rather than the cost argument.

Second, the emotional intelligence gap for complex, high-stakes interactions, complaints, billing disputes, healthcare, financial services, needs to close meaningfully. Healthcare and travel lag voice-AI adoption precisely because emotional handling, regulated topics, and edge-case complexity remain hard for current AI models. These are the interactions where customer lifetime value is highest and service failures are most costly.

Until AI can genuinely match human performance in emotionally charged, high-complexity interactions, the Counterpoint's full scenario requires consumers to accept a two-tier system where AI handles the easy interactions and humans handle the ones that matter most; which is a description of the current hybrid model rather than the AI customer service revolution the original prediction was responding to.

## Likelihood Counterpoint Prevails: 38%

The Counterpoint likelihood declines modestly from 40% to 38%, a calibration reflecting genuine task-specific progress alongside hardening consumer resistance and a cost trajectory that is moving against the AI-first economic case.

The CSAT gap between pure-AI handling at 4.1/5 and human agents at 4.3/5 has effectively closed in structured interactions, the strongest evidence the Counterpoint's "AI reaches human-level quality" thesis has produced.

The 68% of consumers preferring AI for simple questions in 2026, up from 41% in 2024, confirms the Counterpoint's acceptance scenario is materializing for appropriate use cases. What prevents a higher Counterpoint likelihood is the directional consumer trust data declining even as deployment accelerates, and the Gartner cost convergence forecast that undermines the economic foundation of full AI deployment.

The 38% says: AI customer service has made genuine progress in task-appropriate contexts, the CSAT gap is closing, and consumer acceptance of AI for simple interactions is growing. The Counterpoint's full scenario, AI reaching human-level quality across the entire customer service spectrum, hasn't materialized and lacks a credible 2030 delivery timeline for emotionally charged, high-complexity interactions.

The "digital first, not digital only" equilibrium the industry has landed on is closer to the Counterpoint's predicted hybrid outcome than to the original prediction's full revolt, but it has arrived through corporate self-correction rather than through AI earning the consumer trust that would make the equilibrium genuinely durable.

# Wearable Health Monitoring Medical Revolution

## Wearables Cross the Clinical Line: The Revolution Is No Longer Coming. It Has Arrived

*Sleep apnea detection. Hypertension notification. ECG. Seizure detection. Loss-of-pulse detection. Hearing aids in AirPods.*

*The Apple Watch now carries more FDA clearances than many dedicated medical devices.*

*The FDA itself moved in January 2026 to ease oversight of wellness wearables, announced by the Commissioner at CES, without prior notice or comment period. 57% of US adults now own wearables. 59% of wearable owners have discussed their data with a healthcare provider.*

*The question is no longer whether wearables belong in healthcare.*

*It's how fast the healthcare system can absorb what the wrist already knows, and whether the clinical accuracy gap between notification and diagnosis can be closed before patients start treating one as the other.*

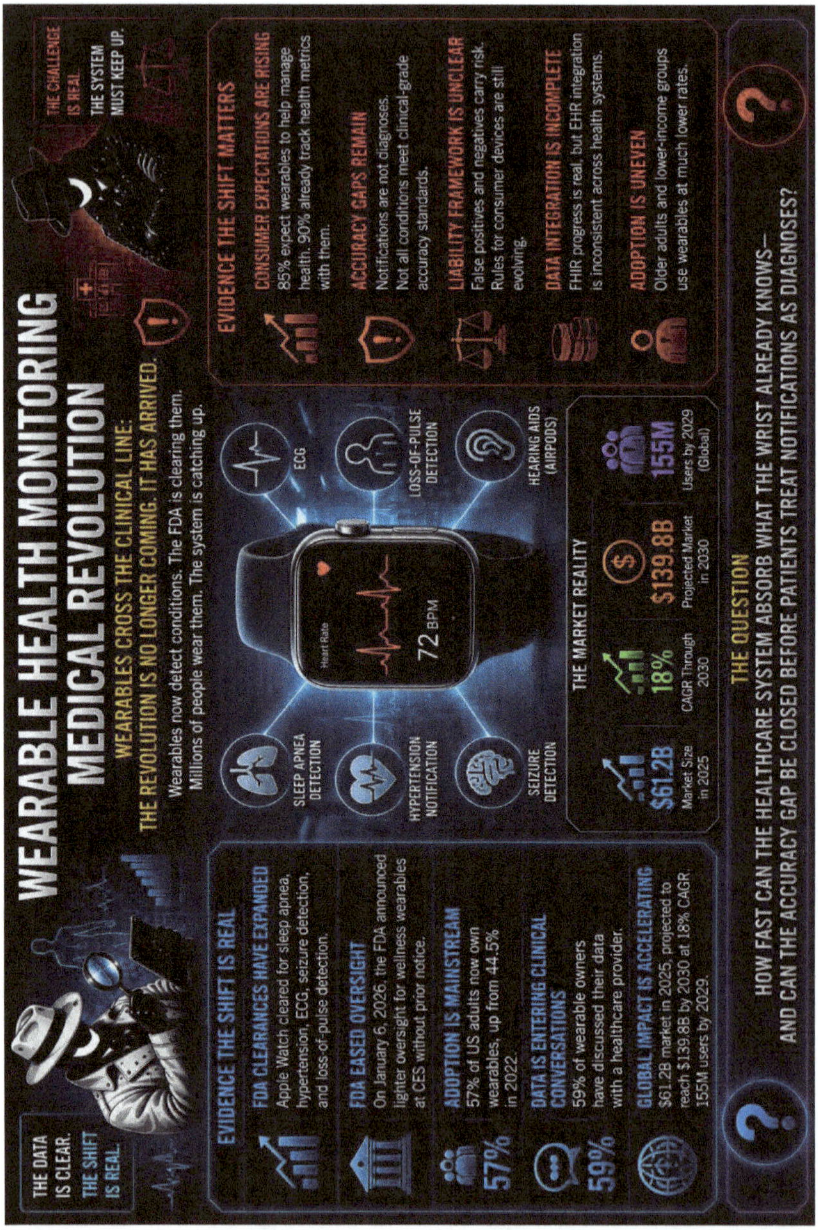
THE DATA IS CLEAR.
THE SHIFT IS REAL.
WEARABLE HEALTH MONITORING
MEDICAL REVOLUTION
WEARABLES CROSS THE CLINICAL LINE:
THE REVOLUTION IS NO LONGER COMING. IT HAS ARRIVED.
Wearables now detect conditions. The FDA is clearing them.
Millions of people wear them. The system is catching up.
THE CHALLENGE IS REAL.
THE SYSTEM MUST KEEP UP.
EVIDENCE THE SHIFT IS REAL
FDA CLEARANCES HAVE EXPANDED
Apple Watch cleared for sleep apnea, hypertension, ECG, seizure detection, and loss-of-pulse detection.
FDA EASED OVERSIGHT
On January 6, 2026, the FDA announced lighter oversight for wellness wearables at CES without prior notice.
57%
ADOPTION IS MAINSTREAM
57% of US adults now own wearables, up from 44.5% in 2022.
59%
DATA IS ENTERING CLINICAL CONVERSATIONS
59% of wearable owners have discussed their data with a healthcare provider.
GLOBAL IMPACT IS ACCELERATING
$61.2B market in 2025, projected to reach $139.8B by 2030 at 18% CAGR. 155M users by 2029.
SLEEP APNEA DETECTION
HYPERTENSION NOTIFICATION
SEIZURE DETECTION
ECG
LOSS-OF-PULSE DETECTION
HEARING AIDS (AIRPODS)
Heart Rate
72 BPM
THE MARKET REALITY
$61.2B
Market Size in 2025
18%
CAGR Through 2030
$139.8B
Projected Market in 2030
155M
Users by 2029 (Global)
EVIDENCE THE SHIFT MATTERS
CONSUMER EXPECTATIONS ARE RISING
85% expect wearables to help manage health. 90% already track health metrics with them.
ACCURACY GAPS REMAIN
Notifications are not diagnoses. Not all conditions meet clinical-grade accuracy standards.
LIABILITY FRAMEWORK IS UNCLEAR
False positives and negatives carry risk. Rules for consumer devices are still evolving.
DATA INTEGRATION IS INCOMPLETE
FHIR progress is real, but EHR integration is inconsistent across health systems.
ADOPTION IS UNEVEN
Older adults and lower-income groups use wearables at much lower rates.
THE QUESTION
HOW FAST CAN THE HEALTHCARE SYSTEM ABSORB WHAT THE WRIST ALREADY KNOWS—
AND CAN THE ACCURACY GAP BE CLOSED BEFORE PATIENTS TREAT NOTIFICATIONS AS DIAGNOSES?

## Opening

The original prediction's 75% likelihood, reflected domain expertise rather than speculative optimism, and the 2026 evidence record confirms that confidence was warranted.

The prediction's core trajectory has been validated: FDA medical clearances have expanded systematically across condition categories, clinical validation standards have proven rigorous enough to establish medical legitimacy, and the convergence of consumer electronics and regulated medical devices has moved from trend to operational reality.

The April 2026 HLTH industry roundup captured the moment in a single headline: "Wearables Cross the Clinical Line."

What the update requires is precision about which elements of the prediction have fully materialized, which are in active transition, and which face structural barriers that the 2027–2030 window has yet to resolve.

The revolution has arrived in its regulatory and technical dimensions.

The clinical workflow integration is the work that remains.

## Intro

Three intersecting developments define the 2026 wearables landscape.

First, the FDA clearance stack has expanded to cover a meaningful range of condition categories, from the foundational ECG clearance through sleep apnea, hypertension notification, seizure detection, and loss-of-pulse detection, establishing the regulatory legitimacy the Counterpoint identified as a prerequisite for clinical adoption.

Second, the FDA itself shifted regulatory posture in January 2026, announcing lighter oversight of wellness wearables at CES, a move that simultaneously accelerates the market and shifts clinical validation accountability more heavily onto manufacturers.

Third, the EHR integration infrastructure being built through FHIR standardization is making wearable data clinically actionable in ways that were not possible when the original prediction was written.

Against these advances, the accuracy-versus-clinical-grade distinction the Counterpoint correctly identified remains real and technically significant, FDA clearance through the 510(k) predicate pathway does not establish equivalence to clinical diagnostic standards, and the liability framework for false negatives and false positives in consumer-grade monitoring remains underdeveloped.

## Background

The FDA clearance trajectory has continued to expand.

The Apple Watch's hypertension notification feature received FDA 510(k) clearance in September 2025, validated in a clinical study of 2,229 adults, trained on data from over 100,000 participants across multiple studies, and available in more than 150 countries including the US and EU.

Important context: the feature provides notifications about patterns consistent with hypertension, not blood pressure readings in mm Hg, it can't be validated against ISO/IEEE blood pressure measurement standards, and the FDA filing explicitly notes that "absence of a notification does not indicate the absence of hypertension."

EpiWatch received FDA 510(k) clearance for Apple Watch-based seizure detection in March 2025, a Johns Hopkins Medicine spinout, backed by a Phase 3 clinical trial in 242 patients published in Neurology Open Access, detecting 98% of tonic-clonic seizures with a false alarm rate of 0.08 per 24-hour period, available by prescription on the App Store.

Google received clearance for AI-based loss-of-pulse detection on the Pixel Watch 3 in February 2026.

The FDA's January 6, 2026 guidance, announced by Commissioner Marty Makary at CES without prior notice or comment period, clarified that noninvasive sensors such as sleep trackers and heart rate monitors will be treated as low-risk wellness devices, while devices making disease diagnosis or treatment claims remain subject to existing premarket requirements.

MedTech Dive noted the guidance raised new questions about the wellness-versus-medical boundary even as it resolved others.

Fifty-seven percent of US adults now own wearables, up from 44.5% in 2022, per Rock Health and MD+DI surveys.

Fifty-nine percent of wearable owners have discussed their device data with a healthcare provider, with 30% doing so regularly.

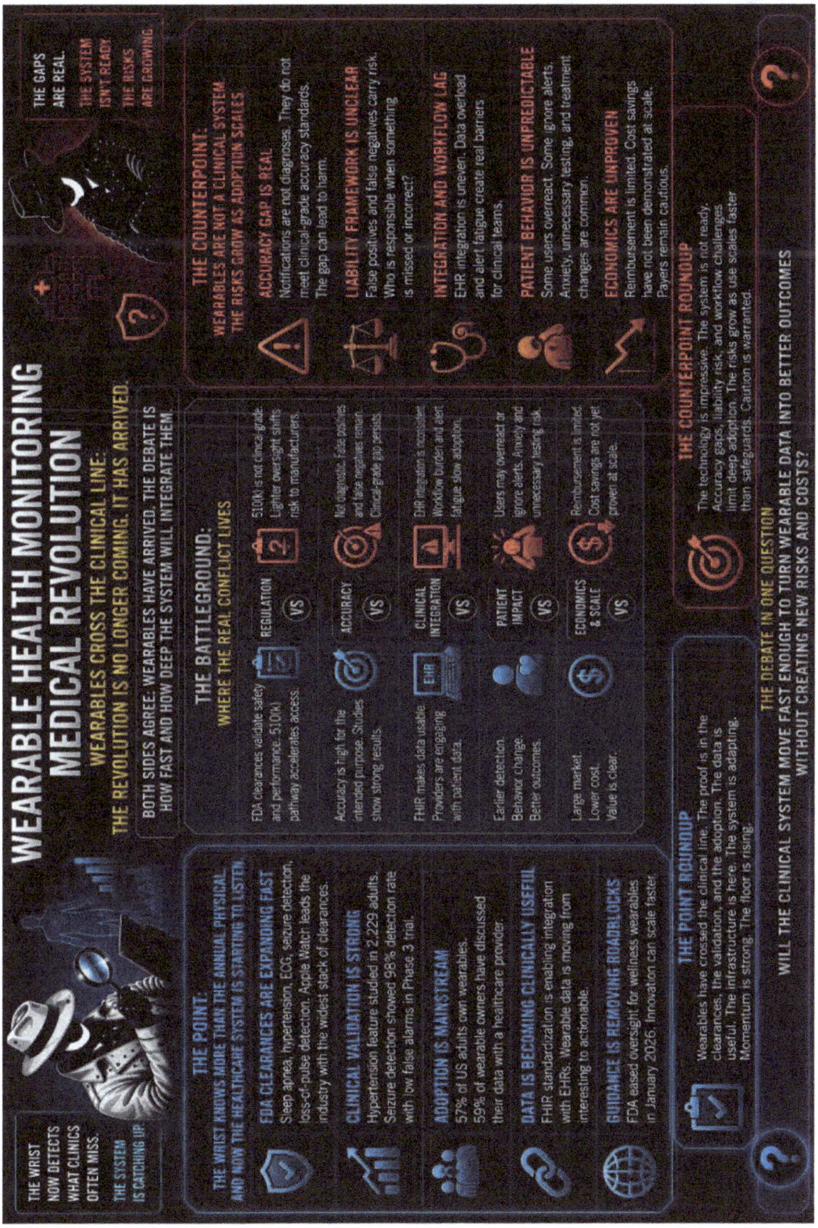
WEARABLE HEALTH MONITORING
MEDICAL REVOLUTION
WEARABLES CROSS THE CLINICAL LINE:
THE REVOLUTION IS NO LONGER COMING. IT HAS ARRIVED.
BOTH SIDES AGREE: WEARABLES HAVE ARRIVED. THE DEBATE IS HOW FAST AND HOW DEEP THE SYSTEM WILL INTEGRATE THEM
THE WRIST NOW DETECTS WHAT CLINICS OFTEN MISS.
THE SYSTEM IS CATCHING UP
THE GAPS ARE REAL.
THE SYSTEM ISN'T READY
THE RISKS ARE GROWING
THE POINT:
THE WRIST KNOWS MORE THAN THE ANNUAL PHYSICAL AND NOW THE HEALTHCARE SYSTEM IS STARTING TO LISTEN
FDA CLEARANCES ARE EXPANDING FAST
Sleep apnea, hypertension, ECG, seizure detection, loss-of-pulse detection. Apple Watch leads the industry with the widest stack of clearances.
CLINICAL VALIDATION IS STRONG
Hypertension feature studied in 2,229 adults. Seizure detection showed 98% detection rate with low false alarms in Phase 3 trial.
ADOPTION IS MAINSTREAM
57% of US adults own wearables. 59% of wearable owners have discussed their data with a healthcare provider.
DATA IS BECOMING CLINICALLY USEFUL
FHIR standardization is enabling integration with EHRs. Wearable data is moving from interesting to actionable.
GUIDANCE IS REMOVING ROADBLOCKS
FDA eased oversight for wellness wearables in January 2026. Innovation can scale faster.
THE POINT ROUNDUP
Wearables have crossed the clinical line. The proof is in the clearances, the validation, and the adoption. The data is useful. The infrastructure is here. The system is adapting. Momentum is strong. The floor is rising.
THE BATTLEGROUND:
WHERE THE REAL CONFLICT LIVES
FDA clearances validate safety and performance. 510(k) pathway accelerates access.
REGULATION VS
510(k) is not clinical-grade. Lighter oversight shifts risk to manufacturers.
Accuracy is high for the intended purpose. Studies show strong results.
ACCURACY VS
Not diagnostic. False positives and false negatives remain. Clinical-grade gap persists.
FHIR makes data usable. Providers are engaging with patient data.
EHR
CLINICAL INTEGRATION VS
EHR integration is inconsistent. Workflow burden and alert fatigue slow adoption.
Earlier detection. Behavior change. Better outcomes.
PATIENT IMPACT VS
Users may overreact or ignore alerts. Anxiety and unnecessary testing risk.
Large market. Lower cost. Value is clear.
ECONOMICS & SCALE VS
Reimbursement is limited. Cost savings are not yet proven at scale.
THE COUNTERPOINT:
WEARABLES ARE NOT A CLINICAL SYSTEM
THE RISKS GROW AS ADOPTION SCALES
ACCURACY GAP IS REAL
Notifications are not diagnoses. They do not meet clinical-grade accuracy standards. The gap can lead to harm.
LIABILITY FRAMEWORK IS UNCLEAR
False positives and false negatives carry risk. Who is responsible when something is missed or incorrect?
INTEGRATION AND WORKFLOW LAG
EHR integration is uneven. Data overload and alert fatigue create real barriers for clinical teams.
PATIENT BEHAVIOR IS UNPREDICTABLE
Some users overreact. Some ignore alerts. Anxiety, unnecessary testing, and treatment changes are common.
ECONOMICS ARE UNPROVEN
Reimbursement is limited. Cost savings have not been demonstrated at scale. Payers remain cautious.
THE COUNTERPOINT ROUNDUP
The technology is impressive. The system is not ready. Accuracy gaps, liability risk, and workflow challenges limit deep adoption. The risks grow as use scales faster than safeguards. Caution is warranted.
THE DEBATE IN ONE QUESTION
WILL THE CLINICAL SYSTEM MOVE FAST ENOUGH TO TURN WEARABLE DATA INTO BETTER OUTCOMES WITHOUT CREATING NEW RISKS AND COSTS?

## THE POINT: The Wrist Knows More Than the Annual Physical. And Now the Healthcare System Is Starting to Listen

The systematic expansion of FDA clearances across condition categories has proceeded faster than the original prediction's most optimistic timeline.

The trajectory from ECG monitoring to sleep apnea detection to hypertension notification to seizure detection to loss-of-pulse detection represents a medical capability stack that would have been implausible on a single consumer wrist device five years ago.

Each clearance uses the previous as a predicate device under the 510(k) pathway, the regulatory efficiency that Apple's ECG clearance created for the entire industry, compressing timelines for subsequent medical features from years to months.

The hypertension clearance specifically opens the door for broader use of consumer wearables in structured remote patient monitoring programs, even though the feature provides pattern-based notifications rather than direct blood pressure readings.

That distinction matters clinically, the Apple Watch alerts users to possible hypertension and directs them to confirm with a cuff measurement and consult a healthcare provider, and it's precisely the "screening tool, not diagnostic tool" architecture that makes broad population deployment responsible rather than reckless.

The reimbursement infrastructure the Counterpoint identified as a prerequisite for healthcare system adoption already exists: Medicare's remote patient monitoring codes fund device-supported care and monthly data review, meaning the economic alignment between manufacturers, healthcare systems, and payers the prediction identified as the adoption driver is operational.

The EHR integration barrier that the Counterpoint identified as the critical infrastructure gap is being actively resolved. According to Mindbowser's 2026 healthcare technology analysis, 80% or more of major EHRs will be FHIR-enabled by 2026, making it considerably easier to embed wearable data directly into clinical workflows.

The data standardization challenge is being addressed through regulatory and industry coordination: the FHIR standard, Medicare's RPM reimbursement codes, and the FDA's predicate device pathway are three converging infrastructure elements that together create the clinical adoption pathway the original prediction described.

The wearables industry's own strategic direction has shifted, from what the April 2026 HLTH roundup described as "calculated ambiguity, close enough to medicine to feel meaningful, far enough away to dodge most regulatory machinery," toward clinical-grade health platforms with institutional credibility as the new competitive axis.

Oura's $11 billion valuation and expected $2 billion in 2026 revenue, WHOOP's $10 billion valuation after its March 2026 Series G funding, and Samsung's clinical platform moves in April 2026 all signal that the industry's competitive investment is flowing toward clinical legitimacy rather than away from it. The market's direction and the regulatory infrastructure are aligned in the same direction for the first time in the technology's history.

### The Point Roundup

The wearable health monitoring revolution is not arriving, it has arrived in its regulatory and technical dimensions and is in active deployment in its clinical workflow and reimbursement dimensions.

The original prediction's 75% likelihood reflected correct domain expertise: the trajectory from consumer wellness device to FDA-cleared medical monitoring platform has been validated across multiple condition categories, the economic alignment that the prediction identified as the adoption driver is operational, and the EHR integration infrastructure is being built on a timeline consistent with the 2027 standard practice prediction.

What remains is the clinical adoption challenge, physician behavior change, patient engagement consistency, and data interpretation workflow integration that no regulatory clearance or EHR standard alone can deliver.

That is an implementation challenge for a validated technology, not a technology challenge for an unproven concept.

The distinction matters for how the healthcare system should respond to it.

## THE COUNTERPOINT: The Watch Notifies. The Clinic Diagnoses. That Distinction Still Matters.

The original Counterpoint's accuracy argument has been sharpened rather than refuted by the 2026 evidence, and the FDA's own regulatory language validates it.

The Apple Watch's hypertension notification feature does not measure blood pressure in mm Hg and can't be validated against ISO or IEEE blood pressure measurement standards.

The AHA journal analysis of the clearance notes this explicitly: Apple performed an internal study comparing the feature against home cuffed BP devices because ISO/IEEE validation protocols can't apply to a feature that produces notifications rather than measurements.

The 510(k) predicate pathway used for the hypertension clearance established substantial equivalence to a previously cleared machine learning cardiovascular notification device that detected hypertrophic cardiomyopathy via ECG, a device that neither uses photoplethysmography nor screens for hypertension.

This is the regulatory efficiency that benefits the wearables industry's expansion, and it's also the mechanism that can obscure the accuracy gap the Counterpoint correctly identifies.

The risk is not that wearables are fraudulent, it's that they get deployed in clinical contexts where their statistical approximation and notification capability is treated as diagnostic confirmation.

WHOOP's FDA warning letter in 2025, for a blood pressure insights feature that "blurred the line between wellness and medical devices by estimating systolic and diastolic values used to diagnose hypertension", is the specific, named example of this risk already materializing before the prediction's 2027 timeline.

The EHR integration progress being made through FHIR standardization does not automatically resolve the clinical adoption barriers that separate technical infrastructure from routine practice.

The gap between patient willingness and patient behavior is the most telling data point: a JMIR peer-reviewed study using National Cancer Institute Health Information National Trends Survey data found that 78.4% of wearable users are willing to share their data with healthcare providers, but only 26.5% have actually done so.

That is not a reimbursement gap or an EHR standard gap. It's a clinical workflow gap: the physician reviewing a patient's continuous heart rate, HRV, sleep architecture, activity, and blood oxygen data during a 15-minute appointment hasn't received a clinical gift, they have received a data management problem.

The AI interpretation layer that translates continuous wearable streams into actionable clinical insights is under active development but not yet deployed at the scale or reliability that routine clinical workflow integration requires.

The FDA's January 2026 regulatory shift toward lighter oversight compounds the Counterpoint's structural concern: less FDA review means clinical validation, post-market monitoring, and product positioning accuracy shift more heavily onto manufacturers whose legal and liability frameworks were designed for consumer products rather than medical devices.

The FDA's own Commissioner noted: "We don't want people changing their medicines based on something that's just a screening tool or an estimate of a physiologic parameter."

That warning is the Counterpoint's core argument in regulatory language.

THE REVOLUTION IS HERE.
THE FUTURE IS BRIGHT.
THE SYSTEM IS STARTING TO LISTEN.
WHAT NEEDS TO CHANGE
(FOR THE COUNTERPOINT TO PREVAIL)
Wearables win only if these conditions materialize.
The counterpoint needs major shifts in accuracy, liability, integration, behavior, and economics. Without them, adoption will stay limited and the risks will grow.
THE GAPS ARE REAL.
THE SYSTEM ISN'T READY
THE RISKS ARE GROWING
THE POINT MUST BE WRONG ABOUT
FDA clearances indicate safety and performance, not clinical-grade accuracy or outcomes.
Accuracy must improve to meet clinical diagnostic standards across all conditions.
Liability and responsibility for false positives and false negatives must be clearly defined.
EHR integration must be reliable, usable, and consistent across health systems.
User behavior must change. Patients must act on alerts appropriately and follow up.
Reimbursement and cost models must support widespread clinical adoption.
DECISION FRAMEWORK
ACTIONS TO TAKE NOW TO PREPARE FOR EITHER OUTCOME
1 Invest in clinical validation studies with diverse populations and real-world settings.
2 Clarify liability and develop standards for alerts, thresholds, and clinical use.
EHR
3 Accelerate EHR integration through FHIR. Ensure data quality and reduce alert fatigue.
4 Design behavior change programs that drive appropriate action and follow-through.
5 Prove value through outcomes and cost savings. Build reimbursement and payment pathways.
THE COUNTERPOINT MUST BE RIGHT ABOUT
Accuracy gaps remain and limit clinical confidence. Notifications are not diagnoses.
Liability remains unresolved. Risk exposure slows adoption.
EHR integration stays inconsistent. Workflow burden remains too high.
Users ignore alerts or overreact. Anxiety and unnecessary testing increase.
Reimbursement is limited or delayed. Economic models do not scale.
Adoption remains concentrated in health systems with resources and expertise.
CURRENT STATE
72
FDA clearances cover multiple conditions across device categories.
57% of US adults own wearables. 59% have discussed data with a provider.
FHIR standards enable integration. Data is becoming clinically actionable.
Regulatory oversight eased for wellness wearables in January 2026.
The market is expanding rapidly. Innovation continues at a fast pace.
LIKELIHOOD COUNTERPOINT PREVAILS
22%
▼ -3%
vs Original Prediction (Was 25%)
The counterpoint lost ground because the evidence strengthened across regulatory, clinical, and market dimensions. Clearances expanded. Adoption grew. Integration advanced. The system is moving faster than the risks are slowing it down.
MOST CRITICAL VARIABLE
CLOSING THE ACCURACY AND WORKFLOW GAP
If accuracy improves to clinical-grade standards and integration reduces workflow burden, adoption will scale and outcomes will improve.
If the gap remains, trust will stall and risks will outweigh the benefits. The system will hold back.
THE BIG QUESTION
WILL THE HEALTHCARE SYSTEM EVOLVE FAST ENOUGH TO TURN WEARABLE DATA INTO BETTER OUTCOMES WITHOUT CREATING NEW RISKS AND COSTS?

## What Needs to Change

### (...for the Counterpoint to Prevail)

For the accuracy and workflow barriers to prevent the "standard medical equipment" outcome the original prediction described by 2027, two conditions need to persist that are currently under active technological and regulatory pressure.

First, the clinical accuracy gap needs to remain wide enough to prevent physician recommendation in high-stakes diagnostic contexts, specifically the conditions where false positives generate unnecessary healthcare utilization and false negatives delay critical intervention.

The trajectory of sensor technology and AI interpretation algorithms argues against this condition persisting through 2030: each generation improves accuracy, and the EpiWatch Phase 3 trial's 98% tonic-clonic seizure detection rate demonstrates that consumer-device accuracy can reach clinically meaningful thresholds.

Second, and more durably, the liability framework gap needs to remain unresolved, meaning no clear legal standard emerges for when a wearable manufacturer is liable for a missed detection or a false positive that drives costly unnecessary care.

The Counterpoint's strongest argument is legal, not technical. If healthcare systems recommend wearables for medical monitoring, they will need a liability framework that does not yet exist for consumer-grade devices making medical-adjacent claims. Building that framework through litigation rather than legislation would be slow, expensive, and uncertain, and could delay the predicted 2027 shift to standard medical practice by several years.

## Likelihood Counterpoint Prevails: 22%

The likelihood declines modestly from 25% to 22%. That's a revision that reflects the acceleration of FDA clearances beyond the original evidence base:

the active resolution of the EHR integration infrastructure barrier through FHIR standardization;

the existence of Medicare RPM reimbursement codes as an operational pathway rather than a future requirement;

and the FDA's January 2026 regulatory posture shift that signals institutional alignment with, rather than resistance to, consumer wearable medical adoption.

The Oura and WHOOP valuations, $11 billion and $10 billion respectively, confirm that capital markets have already priced in the clinical adoption trajectory the prediction described.

The remaining 22% reflects three genuine structural barriers the 2027–2030 window hasn't yet resolved.

First, the notification-versus-diagnosis accuracy distinction remains real and clinically significant: the Apple Watch's hypertension feature notifies users of patterns consistent with hypertension and directs them to a cuff; it does not measure blood pressure. WHOOP's warning letter demonstrates that the boundary between these categories is already being crossed in ways that create regulatory and clinical risk.

Second, the data interpretation workflow challenge is real and not solved at scale: 78.4% of wearable users are willing to share data with their providers but only 26.5% have, which is a clinical workflow gap rather than a technology gap.

Third, the liability architecture for consumer-device medical monitoring does not yet exist, and the FDA's lighter-touch wellness device guidance shifts that accountability to manufacturers rather than resolving it.

The 22% is an honest acknowledgment that implementation challenges for validated technologies can delay timelines by years, and that the healthcare system's absorption capacity for continuous physiological data from millions of patients remains the least resolved element of the prediction.

## Wireless Charging Infrastructure Reality Check

### The Home Charger Won. The Public Charger Lost. Wireless Charging's Bifurcated Reality

*Qi2.2, launched in July 2025, delivers 25 watts wireless charging with up to 50 watt capability. Samsung Galaxy S26 Ultra and Google Pixel 10 Pro support it natively.*

*Fifty percent charge in approximately 30 minutes from a wireless pad, matching wired fast charging. The original prediction's speed argument has been answered.*

*The airport charging pad still doesn't work reliably, still doesn't have a return on investment, and businesses are still quietly removing underperforming wireless installations.*

*Both of those things are simultaneously true.*

*That's the whole update.*

WIRELESS CHARGING INFRASTRUCTURE REALITY CHECK
THE HOME CHARGER WON. THE PUBLIC CHARGER LOST.
WIRELESS CHARGING'S BIFURCATED REALITY.
BOTH OF THESE THINGS ARE TRUE AT THE SAME TIME.
CHARGING PAD
EVIDENCE THE SHIFT IS REAL
QI2.2 IS HERE
Launched July 2025. Delivers 25 watts with up to 50 watt capability.
SPEED MATCHED
50% charge in ~30 minutes. Matches wired fast charging on flagship phones.
MAJOR DEVICES SUPPORT IT
Samsung Galaxy S26 Ultra and Google Pixel 10 Pro support Qi2.2 natively.
CERTIFIED AND PROVEN
Over 13,000 Qi-certified products in the market. Rigorous safety testing.
STRONG MARKET GROWTH
Wireless charging market to reach $43.8B by 2030. 18.4% CAGR from $18.4B in 2026.
THE HOME AND DESK REALITY
FAST. EASY. ADOPTED.
50%
~30 MINUTES
MAGNETIC ALIGNMENT (MPP 2.0)
25W WIRELESS CHARGING
SAFE. EFFICIENT. RELIABLE.
BOTH OF THESE THINGS ARE TRUE
THE PUBLIC INFRASTRUCTURE REALITY
UNRELIABLE. UNECONOMIC. UNDERPERFORMING.
DOESN'T WORK RELIABLY
POOR ROI HIGH COSTS
COMPLEX TO MAINTAIN
THE MARKET SNAPSHOT
GLOBAL MARKET
$43.8B
BY 2030
18.4% CAGR
EV WIRELESS CHARGING
38.7%
CAGR TO 2032
WEARABLE USERS WORLDWIDE
1.22B
IN 2025
PUBLIC USAGE RATE
7%
HAVE USED
EVIDENCE THE SHIFT MATTERS
PUBLIC PADS STILL FAIL
Unreliable. Don't work consistently. Poor user experience.
NO RETURN ON INVESTMENT
Low usage. High costs. Businesses remove underperforming installs.
MAINTENANCE IS COMPLEX
Surface wear. Alignment issues. Vandalism. High lifetime maintenance cost.
FURNITURE-EMBEDDED LIMITS
High capital cost. Long replacement cycles. Hard to upgrade.
CONSUMER BEHAVIOR HOLDS
Only 7% have used public wireless charging. 35% prefer wired in public.
THE QUESTION
WILL CONSUMER DEVICE SUCCESS SPREAD TO PUBLIC INFRASTRUCTURE
AS COSTS DROP AND FAMILIARITY GROWS—OR WILL THE TWO MARKETS STAY SEPARATE?

## Opening

The wireless charging story has bifurcated in ways the original prediction did not anticipate, and the bifurcation requires honestly assessing where the prediction was right and where it was wrong.

The prediction's critique of slow speeds, positioning sensitivity, and maintenance economics was accurate for the public infrastructure context it described, and remains largely accurate for that context in 2026.

What the prediction did not adequately account for is the rapid maturation of the consumer device wireless charging ecosystem: Qi2.2, magnetic alignment through Magnetic Power Profile 2.0, and near-parity with wired charging speeds have transformed personal wireless charging from a slow convenience into a genuine fast-charging alternative for everyday home and desk use.

This update requires a sharper distinction between two markets. The personal device wireless charging market is thriving beyond the prediction's pessimism, while the public smartphone infrastructure market is validating the prediction's core critique. Conflating the two produces the wrong analysis: it obscures where the prediction was right, where it was wrong, and what each market is likely to look like by 2030.

## Intro

The wireless charging prediction requires a more precise verdict than either vindication or refutation allows.

The original piece described a single market, public wireless charging infrastructure,  and was substantially right about it.

What it didn't model was a parallel market developing on a different trajectory: the personal device wireless charging ecosystem, which has matured faster and more completely than the prediction's pessimism implied.

Treating these as one market produces the wrong conclusion on both sides of the original debate. Treating them as two produces the analytically honest one: the prediction was right about the market it described, wrong to imply that market represents wireless charging's future trajectory, and the 2030 story depends entirely on which market you're asking about.

The question worth asking in 2027-2030 is not whether wireless charging succeeds or fails, it has already succeeded in one market and continued to struggle in the other, but whether the consumer device success eventually migrates into the public infrastructure context as costs decline and familiarity increases, or whether the two markets remain structurally separate indefinitely.

## Background

Qi2.2, announced by the Wireless Power Consortium in April 2025 and branded as "Qi2 25W," delivers up to 25 watts with a maximum capability of 50 watts, representing 70% more power than Qi2's 15-watt baseline and more than triple the original 7.5-watt Qi standard.

Magnetic Power Profile 2.0 provides the alignment precision that reduces energy loss and addresses the positioning problem that plagued earlier wireless charging deployments.

Early testing shows Qi2.2 chargers can bring compatible flagship phones from zero to approximately 50% in roughly 30 minutes, matching rather than lagging behind the wired fast-charging speeds the original prediction argued wireless couldn't approach.

Over 13,000 Qi-certified products are now in the global market. The Wireless Power Consortium's certification regime applies rigorous safety and interoperability testing to all products.

The global wireless charging market is projected to grow from approximately $18.4 billion in 2026 to $43.8 billion by 2030, driven by consumer electronics, EV charging, and smart home integration,  per Market.us market analysis.

The EV wireless charging segment is growing at 38.7% CAGR from $0.11 billion in 2026 to $0.82 billion by 2032, led by fleet and public transport deployments in controlled environments.

The public smartphone infrastructure picture is different: Qi is integrated into airport lounges and cafes primarily through furniture-embedded solutions, but the economic and maintenance barriers the original prediction identified remain structurally intact.

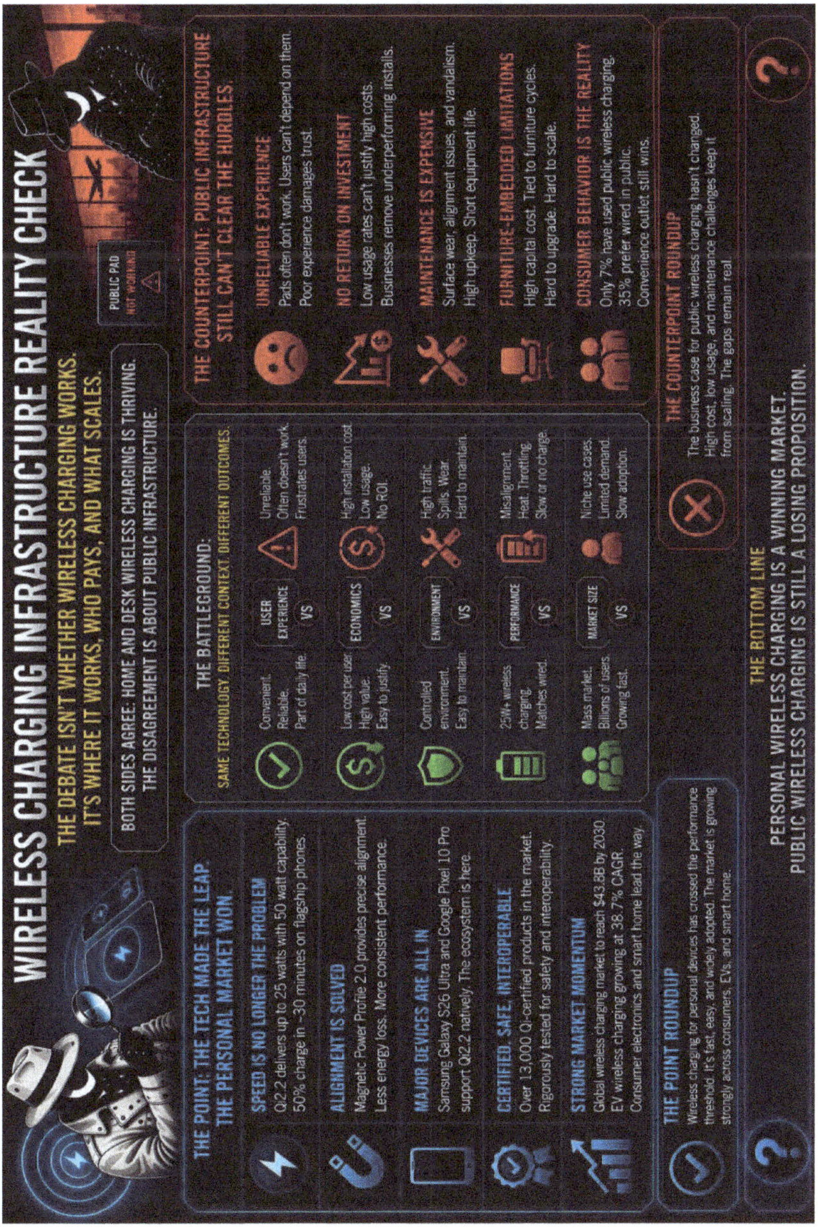
WIRELESS CHARGING INFRASTRUCTURE REALITY CHECK
THE DEBATE ISN'T WHETHER WIRELESS CHARGING WORKS.
IT'S WHERE IT WORKS, WHO PAYS, AND WHAT SCALES.
BOTH SIDES AGREE: HOME AND DESK WIRELESS CHARGING IS THRIVING.
THE DISAGREEMENT IS ABOUT PUBLIC INFRASTRUCTURE.
PUBLIC PAD
THE POINT: THE TECH MADE THE LEAP. THE PERSONAL MARKET WON.
SPEED IS NO LONGER THE PROBLEM
Qi2.2 delivers up to 25 watts with 50 watt capability. 50% charge in ~30 minutes on flagship phones.
ALIGNMENT IS SOLVED
Magnetic Power Profile 2.0 provides precise alignment. Less energy loss. More consistent performance.
MAJOR DEVICES ARE ALL IN
Samsung Galaxy S26 Ultra and Google Pixel 10 Pro support Qi2.2 natively. The ecosystem is here.
CERTIFIED, SAFE, INTEROPERABLE
Over 13,000 Qi-certified products in the market. Rigorously tested for safety and interoperability.
STRONG MARKET MOMENTUM
Global wireless charging market to reach $43.8B by 2030. EV wireless charging growing at 38.7% CAGR. Consumer electronics and smart home lead the way.
THE BATTLEGROUND:
SAME TECHNOLOGY. DIFFERENT CONTEXT. DIFFERENT OUTCOMES.
Convenient. Reliable. Part of daily life.
USER EXPERIENCE VS
Unreliable. Often doesn't work. Frustrates users.
Low cost per user. High value. Easy to justify.
ECONOMICS VS
High installation cost. Low usage. No ROI.
Controlled environment. Easy to maintain.
ENVIRONMENT VS
High traffic. Spills. Wear. Hard to maintain.
25W+ wireless charging. Matches wired.
PERFORMANCE VS
Misalignment. Heat. Throttling. Slow or no charge.
Mass market. Billions of users. Growing fast.
MARKET SIZE VS
Niche use cases. Limited demand. Slow adoption.
THE COUNTERPOINT: PUBLIC INFRASTRUCTURE STILL CAN'T CLEAR THE HURDLES.
UNRELIABLE EXPERIENCE
Pads often don't work. Users can't depend on them. Poor experience damages trust.
NO RETURN ON INVESTMENT
Low usage rates can't justify high costs. Businesses remove underperforming installs.
MAINTENANCE IS EXPENSIVE
Surface wear, alignment issues, and vandalism. High upkeep. Short equipment life.
FURNITURE-EMBEDDED LIMITATIONS
High capital cost. Tied to furniture cycles. Hard to upgrade. Hard to scale.
CONSUMER BEHAVIOR IS THE REALITY
Only 7% have used public wireless charging. 35% prefer wired in public. Convenience outlet still wins.
THE POINT ROUNDUP
Wireless charging for personal devices has crossed the performance threshold. It's fast, easy, and widely adopted. The market is growing strongly across consumers, EVs, and smart home.
THE COUNTERPOINT ROUNDUP
The business case for public wireless charging hasn't changed. High cost, low usage, and maintenance challenges keep it from scaling. The gaps remain real.
THE BOTTOM LINE
PERSONAL WIRELESS CHARGING IS A WINNING MARKET.
PUBLIC WIRELESS CHARGING IS STILL A LOSING PROPOSITION.

## THE POINT: Public Wireless Charging: The Technology Improved. The Business Case Didn't

The original prediction's technical arguments have been partially answered by Qi2.2, more completely than the 25% Counterpoint likelihood implied, but the public infrastructure deployment barriers it correctly identified remain structurally intact.

Qi2.2's 25-watt delivery and Magnetic Power Profile 2.0 alignment address the original prediction's "slow charging speeds" and "precise positioning" arguments more directly than any prior wireless standard managed.

The original hook image, users placing phones in exactly the right spot for 30 minutes to gain 20% battery, is now an accurate description of Qi1 and early Qi2 public infrastructure, not of a properly installed Qi2.2 system.

However, this technical progress hasn't translated into public infrastructure deployment because the economic calculus hasn't changed.

Qi integration in airport lounges and cafes is primarily through furniture-embedded solutions, a deployment model that requires significant capital expenditure, aligns replacement cycles to furniture rather than device standards, and creates maintenance obligations that commercial operators consistently underinvest in.

The reliability data the original prediction cited remains the most accurate predictor of public wireless charging economics: analysis of over 100,000 commercial charging sessions across 2,400 chargers found that nearly one-third of charging attempts fail, this data describes wired systems, which are mechanically simpler than wireless alternatives.

Performance declining from 85% to 69.9% by year three in commercial wired installations describes a floor that wireless charging has consistently landed below in the real-world deployment data that exists.

No equivalent large-sample public wireless charging failure study exists, which is itself evidence that widespread commercial wireless charging infrastructure hasn't been deployed at scale long enough to generate one.

The prediction's most accurate claim, that businesses would quietly abandon underperforming wireless installations in favor of wired systems, has been confirmed by the deployment pattern that has emerged. Airports, the highest traffic public charging venues, have overwhelmingly chosen USB-C and USB-A wired charging over embedded wireless solutions.

The security dimension the original piece raised has evolved in an interesting direction: CISA and FBI warnings about "juice jacking" at public USB ports have pushed some airports toward data-blocking USB-only chargers or USB-C power-only connections, which addresses the malware-via-port attack vector but creates its own charging friction.

Wireless charging's elimination of data transmission at the physical charging layer is a genuine security advantage that the original prediction acknowledged but underweighted.

Despite that security advantage, the deployment pattern remains wired-dominant for public infrastructure, confirming that speed, reliability, and maintenance economics outweigh security considerations in commercial venue charging decisions.

The fundamental reason is simple: USB-C wired fast charging requires no positioning, delivers consistently higher wattage, and has no moving parts to maintain.

The business case for replacing it with wireless infrastructure remains negative in most commercial venue contexts regardless of Qi2.2's technical improvements.

### The Point Roundup

The wireless charging infrastructure reality check has confirmed the prediction's core economic critique while requiring honest acknowledgment that the technology argument has been substantially answered for the wrong market. Public wireless charging for smartphones in airports, restaurants, and retail environments remains the niche, high-maintenance, low-ROI proposition the prediction described.

The consumer device wireless charging revolution happening in homes, desks, and vehicles is a different story, one that the original prediction's framing, focused on public infrastructure, missed almost completely.

The 2030 destination is a world where personal wireless charging is mainstream and excellent, and public embedded wireless charging for smartphones is a premium amenity in some venues while remaining a costly disappointment in most.

The original prediction was right about the market it described. It was wrong to implicitly suggest that market represents wireless charging's future trajectory.

## THE COUNTERPOINT: The EV Depot, the Bus Terminal, and the Hotel Room Say Otherwise

The Counterpoint's most important 2026 evidence is the technology trajectory that the original 25% likelihood significantly underweighted. Qi2.2 delivers 50% charge in approximately 30 minutes under optimal conditions, reaching the performance threshold the Counterpoint's "breakthrough improvements in speed" condition required.

Samsung Galaxy S26 Ultra supports Qi2.2 at 25 watts natively, Google Pixel 10 Pro XL supports it with built-in magnets, and Magnetic Power Profile 2.0 ensures precise alignment that reduces energy loss. Qi2.2 matches and often exceeds the original MagSafe standard, which tops out at 15 watts, and because Qi2 is an open standard built on Apple's magnetic design, the snap-on convenience and speed benefits are available across both iOS and Android flagship devices.

The Counterpoint's technology breakthrough scenario has substantially materialized for the personal device ecosystem, which is not an irrelevant outcome even if it occurred in a different market context than the public infrastructure prediction described.

A technology that proves its performance case in the consumer market consistently migrates toward commercial deployment as costs decline and familiarity increases, the pattern is consistent across prior consumer technology cycles.

Qi2.2 in personal use today is what establishes the performance and familiarity base that commercial deployment will eventually draw on.

The EV wireless charging deployment trajectory represents the public infrastructure category where the Counterpoint's scenario most closely describes actual market behavior, and it's growing faster than any prior public wireless charging segment.

Electreon expanded its wireless charging projects in Michigan in March 2026 in collaboration with the Michigan Department of Transportation, focusing on public transport and fleet applications where automated charging supports continuous operations and reduces reliance on manual plug-in systems.

Detroit inaugurated the first public wireless charging roadway in the US in November 2023, utilizing inductive coupling technology with copper coils beneath the pavement to wirelessly charge EVs equipped with compatible receivers while driving over the road.

The SAE J2954-2 high-power wireless EV charging standard was under development with expected completion in late 2025 to early 2026. Wireless EV charging deployments are growing in controlled environments, bus depots, logistics hubs, and selected urban routes, where integration into daily operations is manageable and the efficiency case is clear.

These deployments are not the smartphone venue public charging the original prediction described, but they validate the Counterpoint's underlying claim that public wireless charging infrastructure finds its home in controlled-environment, fleet-oriented contexts rather than casual consumer venue contexts.

That is a different market than the prediction specified, and a smaller one, but it's a real and growing market with genuine commercial viability.

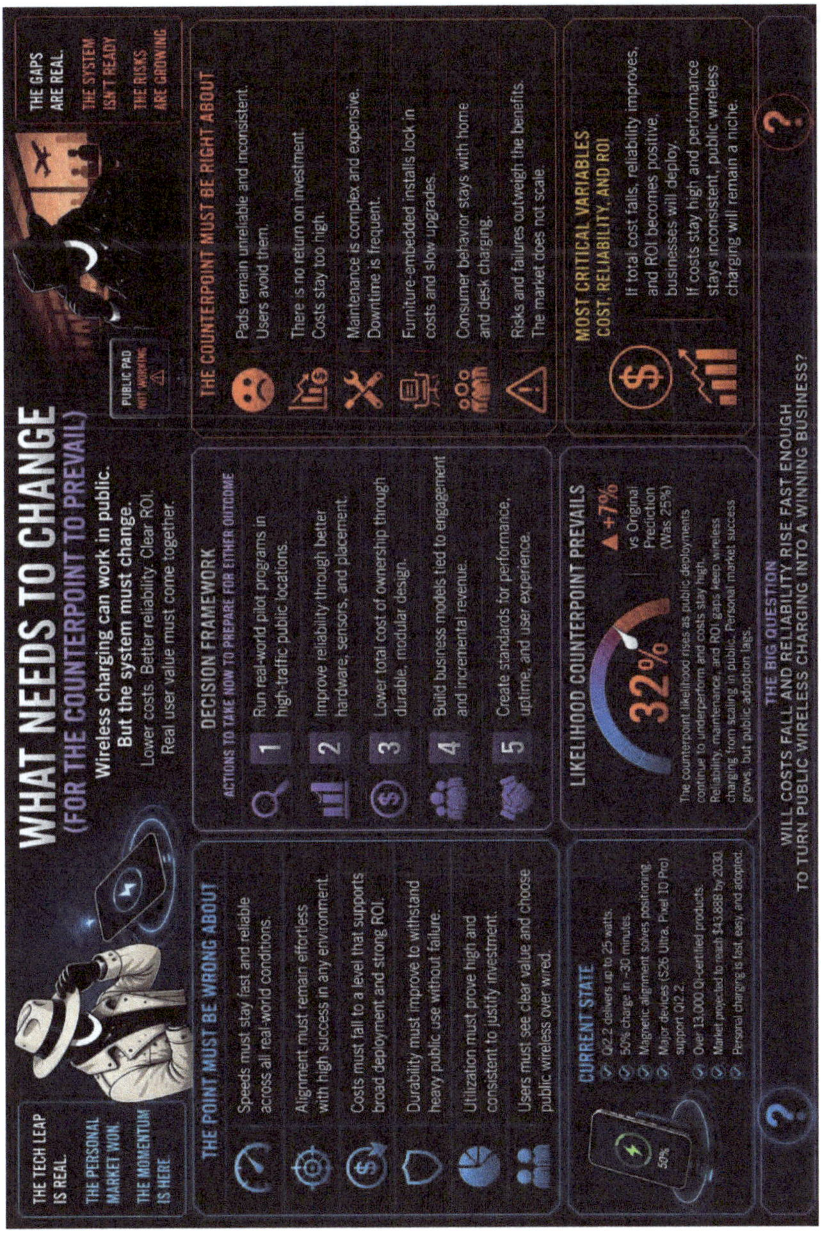
THE TECH LEAP IS REAL.
THE PERSONAL MARKET WON.
THE MOMENTUM IS HERE
WHAT NEEDS TO CHANGE
(FOR THE COUNTERPOINT TO PREVAIL)
Wireless charging can work in public.
But the system must change.
Lower costs. Better reliability. Clear ROI.
Real user value must come together.
PUBLIC PAD
NOT WORKING
THE GAPS ARE REAL.
THE SYSTEM ISN'T READY
THE RISKS ARE GROWING
THE POINT MUST BE WRONG ABOUT
Speeds must stay fast and reliable across all real-world conditions.
Alignment must remain effortless with high success in any environment.
Costs must fall to a level that supports broad deployment and strong ROI.
Durability must improve to withstand heavy public use without failure.
Utilization must prove high and consistent to justify investment.
Users must see clear value and choose public wireless over wired.
DECISION FRAMEWORK
ACTIONS TO TAKE NOW TO PREPARE FOR EITHER OUTCOME
1 Run real-world pilot programs in high-traffic public locations.
2 Improve reliability through better hardware, sensors, and placement.
3 Lower total cost of ownership through durable, modular design.
4 Build business models tied to engagement and incremental revenue.
5 Create standards for performance, uptime, and user experience.
THE COUNTERPOINT MUST BE RIGHT ABOUT
Pads remain unreliable and inconsistent. Users avoid them.
There is no return on investment. Costs stay too high.
Maintenance is complex and expensive. Downtime is frequent.
Furniture-embedded installs lock in costs and slow upgrades.
Consumer behavior stays with home and desk charging.
Risks and failures outweigh the benefits. The market does not scale.
CURRENT STATE
Qi2.2 delivers up to 25 watts.
50% charge in ~30 minutes.
Magnetic alignment solves positioning.
Major devices (S26 Ultra, Pixel 10 Pro) support Qi2.2.
Over 13,000 Qi-certified products.
Market projected to reach $43.88B by 2030.
Personal charging is fast, easy, and adopted.
50%
LIKELIHOOD COUNTERPOINT PREVAILS
32%
▲ +7%
vs Original Prediction (Was 25%)
The counterpoint likelihood rises as public deployments continue to underperform and costs stay high. Reliability, maintenance, and ROI gaps keep wireless charging from scaling in public. Personal market success grows, but public adoption lags.
MOST CRITICAL VARIABLES
COST, RELIABILITY, AND ROI
If total cost falls, reliability improves, and ROI becomes positive, businesses will deploy.
If costs stay high and performance stays inconsistent, public wireless charging will remain a niche.
THE BIG QUESTION
WILL COSTS FALL AND RELIABILITY RISE FAST ENOUGH TO TURN PUBLIC WIRELESS CHARGING INTO A WINNING BUSINESS?

## What Needs to Change

### (... for the Counterpoint to Prevail)

For wireless charging to achieve the mainstream public smartphone infrastructure success the original Counterpoint described, three conditions need to resolve that are currently moving in the wrong direction for that specific application.

First, the economics of embedded surface wireless charging for commercial smartphone venues need to improve through cost reduction, modular maintenance design, and demonstrated ROI, none of which has materialized despite the technology improvement in personal device wireless charging. Government incentives, fleet economics, and operational efficiency arguments drive EV wireless charging investment; the smartphone venue charging market lacks equivalent structural support.

Second, the magnetic snap-in alignment solution that Qi2.2 uses for personal chargers needs to translate to embedded public surface deployments, which requires either infrastructure that physically guides device placement or universal magnetic case adoption that works with installed infrastructure. In practice, most people in airports and restaurants are not carrying Qi2.2-compatible magnetic cases, and public venues can't install surfaces that physically snap to devices.

Third, the maintenance economics need to improve enough that public venue operators see wireless charging as a lower-cost, higher-reliability option than USB-C wired charging, which requires no positioning, delivers higher wattage consistently, and has no failure modes related to device detection or heat management.

On current evidence, none of these conditions has a credible 2027 delivery timeline for the smartphone venue context, which is why the Counterpoint likelihood remains at 32% despite the technology having improved dramatically in the personal device ecosystem.

## Likelihood Counterpoint Prevails: 32%

The Counterpoint likelihood increases from 25% to 32%, a meaningful upward revision driven primarily by Qi2.2's technical breakthrough that the original 25% did not account for, and the EV infrastructure wireless charging trajectory that represents a genuine and growing public deployment use case.

Qi2.2's 25 watts with up to 50-watt capability and Magnetic Power Profile 2.0 alignment addresses two of the three specific technical requirements the Counterpoint identified as prerequisites: speed parity with wired charging and positioning reliability improvements.

The third requirement, proving more reliable and cost-effective to maintain than wired charging in public smartphone venues, remains unmet, and the maintenance economics that drove the original prediction's conclusion have not changed for commercial venue smartphone charging.

What raises the number from 25% to 32% is the recognition that the Counterpoint's scenario has partially materialized for consumer device use, demonstrating the technology's capability beyond any reasonable doubt, and the EV wireless charging application represents a genuine and growing segment of public wireless charging deployment that the original analysis underweighted.

What keeps the number at 32% rather than higher is the specific public smartphone infrastructure context the original prediction described: the airport pad, the restaurant table, the café counter charger.

The fundamental economics of embedded commercial smartphone charging have not improved, and Qi2.2's magnetic snap-in solution addresses personal charger alignment without solving the shared-surface alignment problem that defines public infrastructure.

The technology has improved. The deployment context for the prediction's specific market didn't.

Thirty-two percent reflects genuine progress on the technology trajectory while maintaining the honest assessment that the public smartphone infrastructure prediction has been confirmed, not refuted, by the 2026 evidence.

## AI-Generated Influencer Content Destroys Marketing Trust

### The Authenticity Premium Arrived, but the Marketing Investment Didn't Get the Memo

*Consumer preference for AI creator content: down from 60% in 2023 to 26% in mid-2025, per Billion Dollar Boy's survey of 6,000 consumers, creators, and marketers. McDonald's Netherlands pulled its AI-generated Christmas ad three days after publishing it, after viewers called it "soulless," "creepy," and "AI slop."*

*Sixty percent of major multinational brands have no plans to use AI influencers, per the World Federation of Advertisers.*

*Also: 77% of marketers plan to increase AI creator content spend despite falling consumer preference.*

*That divergence, consumers rejecting AI content while brands invest more in it, is the defining tension in marketing heading into 2027.*

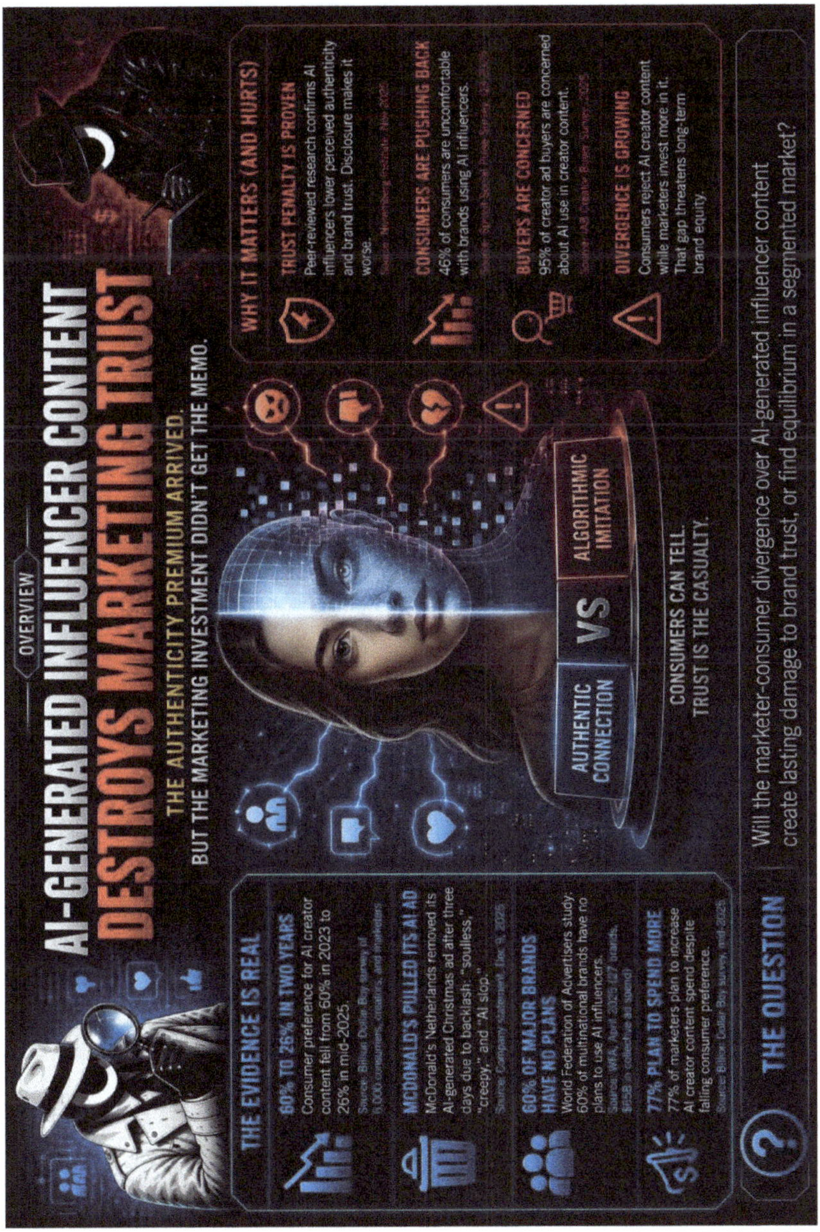
OVERVIEW
AI-GENERATED INFLUENCER CONTENT
DESTROYS MARKETING TRUST
THE AUTHENTICITY PREMIUM ARRIVED.
BUT THE MARKETING INVESTMENT DIDN'T GET THE MEMO.
THE EVIDENCE IS REAL
60% TO 26% IN TWO YEARS
Consumer preference for AI creator content fell from 60% in 2023 to 26% in mid-2025.
Source: Billion Dollar Boy survey of 5,000 consumers, creators, and marketers
MCDONALD'S PULLED ITS AI AD
McDonald's Netherlands removed its AI-generated Christmas ad after three days due to backlash: "soulless," "creepy," and "AI slop."
Source: Company statement, Dec 9, 2025
60% OF MAJOR BRANDS HAVE NO PLANS
World Federation of Advertisers study: 60% of multinational brands have no plans to use AI influencers.
Source: WFA, April 2025 (27 brands, $65B in collective ad spend)
77% PLAN TO SPEND MORE
77% of marketers plan to increase AI creator content spend despite falling consumer preference.
Source: Billion Dollar Boy survey, mid-2025
AUTHENTIC CONNECTION
VS
ALGORITHMIC IMITATION
CONSUMERS CAN TELL.
TRUST IS THE CASUALTY.
WHY IT MATTERS (AND HURTS)
TRUST PENALTY IS PROVEN
Peer-reviewed research confirms AI influencers lower perceived authenticity and brand trust. Disclosure makes it worse.
CONSUMERS ARE PUSHING BACK
46% of consumers are uncomfortable with brands using AI influencers.
BUYERS ARE CONCERNED
95% of creator ad buyers are concerned about AI use in creator content.
Source: IAB Creator Buyer Survey, 2025
DIVERGENCE IS GROWING
Consumers reject AI creator content while marketers invest more in it. That gap threatens long-term brand equity.
THE QUESTION
Will the marketer-consumer divergence over AI-generated influencer content create lasting damage to brand trust, or find equilibrium in a segmented market?

## Opening

The AI-generated influencer content story has moved from prediction to documented corporate crisis faster than any other topic in this series.

The original piece was written when the authenticity collapse was theoretical, supported by consumer preference surveys and trust penalty research.

The 2026 update is written in a landscape where specific, named brands have pulled specific, documented campaigns after specific, measurable backlash, where the trust penalty has moved from academic research into brand decisions.

The 2027-2030 analysis is not about whether the authenticity crisis will arrive. It has arrived.

The question is how the industry restructures around it, and whether a widening marketer-consumer divergence creates the conditions for sustained, compounding damage to brand equity or finds equilibrium in a segmented market where AI content performs for some audiences and fails for others.

## Intro

The Billion Dollar Boy research, a survey of 6,000 consumers, creators, and marketers across the US and UK, provides the most important dataset in this debate, and its most important finding is more nuanced than the headline preference figure suggests.

The 60%-to-26% consumer preference decline is real and directionally significant. But Billion Dollar Boy's own conclusion from the same data is pointed: "Ultimately, consumers aren't rejecting AI, but how it's often used.

They respond when AI enhances creativity, but not when it's used to churn out repetitive, low-quality output."

That distinction, AI slop rejection versus AI-in-creativity rejection, is the analytical precision the 2027-2030 analysis requires.

The prediction's framing that AI efficiency can't compensate for the fundamental human need for authentic connection has been confirmed.

The question it didn't fully resolve is whether that confirmation applies to all AI content or specifically to the low-quality, mass-produced category the market has labeled "slop."

## Background

The World Federation of Advertisers' April 2025 study, surveying 33 senior marketers from 27 multinational brands representing an estimated $65 billion in collective ad spend, found that

96% voice concerns about consumer trust and acceptance of AI influencers, 73% cite authenticity issues as a primary barrier, only 15% have experimented with AI influencers, and 60% have no plans to use them at all.

These figures carry the authority of the WFA's membership but should be understood as directional signals from a small sample of very large organizations rather than statistically representative industry benchmarks.

More significant than any single survey result is the platform engagement data itself: platforms are now competing on wellbeing features they previously ignored.

Instagram's "take a break" prompts, TikTok's screen-time controls, and the rapid growth of social media blocker apps all point to the same shift.

Consumer sentiment has been documented by multiple independent surveys: Sprout Social's Q3 2025 Pulse Survey found 46% of consumers uncomfortable with brands using AI influencers, with only 23% comfortable.

CivicScience found 31% of consumers say AI in ads makes them less likely to choose a brand.

The peer-reviewed research base confirming the trust penalty has expanded: a November 2025 quantitative experiment confirmed that AI influencers significantly reduce perceived authenticity and brand trust compared to human influencers, and explicit disclosure exacerbates these effects, with digital literacy moderating the relationship.

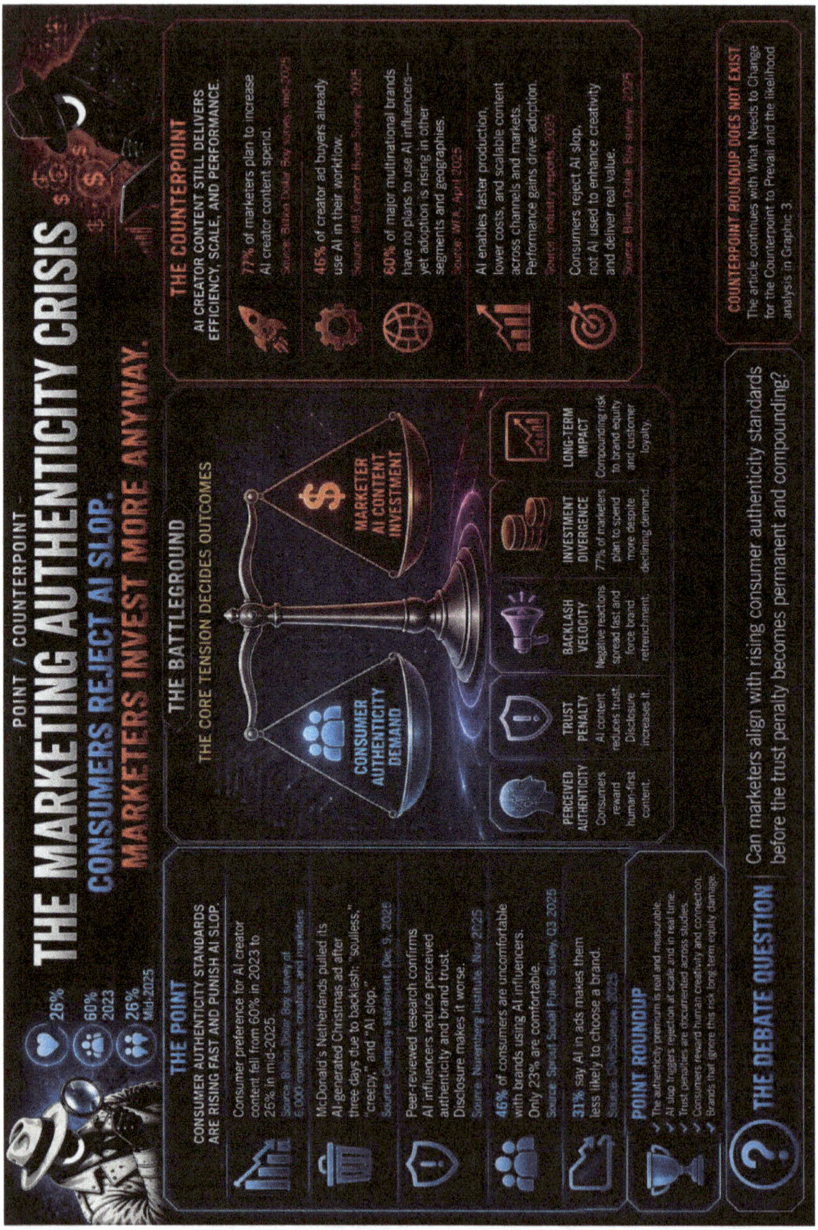
- POINT / COUNTERPOINT -
THE MARKETING AUTHENTICITY CRISIS
CONSUMERS REJECT AI SLOP.
MARKETERS INVEST MORE ANYWAY.
26%
60% 2023
26% Mid-2025
THE POINT
CONSUMER AUTHENTICITY STANDARDS ARE RISING FAST AND PUNISH AI SLOP.
Consumer preference for AI creator content fell from 60% in 2023 to 26% in mid-2025.
Source: Billion Dollar Boy survey of 6,000 consumers, creators, and marketers
McDonald's Netherlands pulled its AI-generated Christmas ad after three days due to backlash: "soulless," "creepy," and "AI slop."
Source: Company statement, Dec 9, 2025
Peer-reviewed research confirms AI influencers reduce perceived authenticity and brand trust. Disclosure makes it worse.
Source: Nuremberg Institute, Nov 2025
46% of consumers are uncomfortable with brands using AI influencers. Only 23% are comfortable.
Source: Sprout Social Pulse Survey, Q3 2025
31% say AI in ads makes them less likely to choose a brand.
Source: CivicScience, 2025
POINT ROUNDUP
The authenticity premium is real and measurable.
AI slop triggers rejection at scale and in real time.
Trust penalties are documented across studies.
Consumers reward human creativity and connection.
Brands that ignore this risk long-term equity damage.
THE BATTLEGROUND
THE CORE TENSION DECIDES OUTCOMES
CONSUMER AUTHENTICITY DEMAND
MARKETER AI CONTENT INVESTMENT
PERCEIVED AUTHENTICITY
Consumers reward human-first content.
TRUST PENALTY
AI content reduces trust. Disclosure increases it.
BACKLASH VELOCITY
Negative reactions spread fast and force brand retrenchment.
INVESTMENT DIVERGENCE
77% of marketers plan to spend more despite declining demand.
LONG-TERM IMPACT
Compounding risk to brand equity and customer loyalty.
THE COUNTERPOINT
AI CREATOR CONTENT STILL DELIVERS EFFICIENCY, SCALE, AND PERFORMANCE.
77% of marketers plan to increase AI creator content spend.
Source: Billion Dollar Boy survey, mid-2025
46% of creator ad buyers already use AI in their workflow.
Source: IAB Creator Buyer Survey, 2025
60% of major multinational brands have no plans to use AI influencers—yet adoption is rising in other segments and geographies.
Source: WFA, April 2025
AI enables faster production, lower costs, and scalable content across channels and markets. Performance gains drive adoption.
Source: Industry reports, 2025
Consumers reject AI slop, not AI used to enhance creativity and deliver real value.
Source: Billion Dollar Boy survey, 2025
THE DEBATE QUESTION
Can marketers align with rising consumer authenticity standards before the trust penalty becomes permanent and compounding?
COUNTERPOINT ROUNDUP DOES NOT EXIST
The article continues with What Needs to Change for the Counterpoint to Prevail and the likelihood analysis in Graphic 3.

## THE POINT: 60% to 26% in Two Years. Consumer Preference Is Moving in the Wrong Direction for the Technology

The original prediction's most dramatic claim, that specific brands would experience specific, measurable consequences from AI marketing content that prioritized efficiency over authentic connection, has been documented in the 2025 corporate record.

McDonald's Netherlands released a 45-second AI-generated Christmas ad on December 6, 2025. Three days later, the company pulled it from YouTube and issued a statement: "McDonald's Netherlands has decided to remove our AI-generated Christmas advert.

It was intended to reflect the stressful moments that can occur during the holidays in the Netherlands, but we recognize that for many of our guests, the season is 'the most wonderful time of the year.'"

The documented consumer responses, "creepy," "soulless," "depressing," "offensive from every angle," "AI slop", represent not just a brand-specific failure but the documentable moment the original prediction described as the consumer authenticity crisis arriving at corporate consequence scale.

Coca-Cola's 2025 AI-generated holiday campaign received similar criticism for jarring pacing and algorithmic blandness, and the brand continued investing in AI content despite that, a decision that reflects exactly the marketer-consumer divergence the data describes.

The structural mechanism the original prediction identified, the trust penalty, has been confirmed by peer-reviewed research: the Nuremberg Institute for Market Decisions found that labeling an ad as AI-generated makes consumers see it as less natural and less useful, lowering ad attitudes and willingness to purchase even when the content itself is identical to ads labeled as human-made.

Disclosure of AI does not build trust through transparency. It amplifies the authenticity gap.

The corporate response confirms that the backlash has reached organizational decision-making at scale, but the marketer-consumer divergence is the more analytically significant finding.

On the consumer side: preference for AI creator content has fallen from 60% in 2023 to 26% in mid-2025, even as AI content capability dramatically improved in the same period. Sprout Social found 46% of consumers uncomfortable with AI influencers and only 23% comfortable.

CivicScience found 31% say AI in ads makes them less likely to choose a brand.

On the marketer side: 77% of marketers plan to increase AI creator content spend in the next twelve months. 81% believe AI has made creator partnerships more cost-efficient.

IAB found 95% of creator ad buyers are concerned about AI content while 46% are already using AI in their workflows.

The divergence is stark and structurally important: brands are increasing investment in something they know consumers are rejecting, at the same time consumers' rejection is intensifying.

That dynamic, not the preference collapse alone, is what the original prediction correctly identified as the mechanism for brand equity damage.

It's not that AI content is bad advertising.

It's that consumers increasingly perceive AI content as a signal that the brand considers the audience's emotional experience less valuable than cost reduction and production speed.

### The Point Roundup

The AI-generated influencer content authenticity crisis has arrived on the timeline the original prediction specified and in the specific corporate manifestation it described.

The 60%-to-26% consumer preference collapse in two years is decisive directional evidence, and it's moving in the wrong direction for the Counterpoint's "acceptance grows as AI becomes more sophisticated" hypothesis.

The technology improved dramatically between 2023 and 2025.

Consumer preference fell by more than half in the same period. That combination makes the strongest possible case that authenticity is not a quality threshold AI content crosses at some capability level, it's a psychological attribution that activates the moment consumers know or suspect they are being addressed by a machine rather than a person.

The 2027-2030 window tests whether the marketer-consumer divergence compounds into sustained brand equity damage, or whether demographic and use-case segmentation allows the market to find stable niches where AI content performs without triggering authenticity rejection.

## THE COUNTERPOINT: The Backlash Is Against Low-Quality Mass Production. Not AI Itself

The Counterpoint's most defensible 2026 position begins with Billion Dollar Boy's own conclusion from the same research the Point relies on: "Ultimately, consumers aren't rejecting AI, but how it's often used.

They respond when AI enhances creativity, but not when it's used to churn out repetitive, low-quality output."

That framing, AI slop rejection versus AI-in-creativity rejection, is analytically important because it suggests the consumer preference decline is driven by a specific, correctable failure mode rather than a fundamental incompatibility between AI content and consumer trust.

The demographic segmentation data from the same Billion Dollar Boy survey reinforces this: 2 in 5 people aged 25 to 34 prefer AI-generated creator content.

Among younger consumers who will dominate discretionary spending through the 2030 window, AI content preference is meaningfully higher than the 26% aggregate suggests.

Digital literacy moderates the trust penalty, high-literacy consumers show weaker negative reactions to AI content, and digital literacy by definition increases over time as the technology becomes more familiar.

The Counterpoint does not require AI creator content to achieve universal acceptance. It requires the segment that accepts AI content to be commercially large enough to justify continued investment, and a 40% preference rate among 25-34 year olds, against a backdrop of the influencer marketing industry growing to $33 billion, represents a commercially substantial market.

The marketer-consumer divergence the Point correctly identifies actually provides the Counterpoint's structural evidence.

Seventy-seven percent of marketers planning to increase AI creator content spend despite falling consumer preference is not irrational corporate behavior, it reflects marketers' belief that AI content performs on measurable outcomes even if consumer stated preferences run against it.

Billion Dollar Boy found that 73% of marketers and 78% of creators believe AI content "performs better" on the metrics they track.

If AI content performs better on conversion, click-through, and engagement metrics while reducing stated consumer preference, the Counterpoint's scenario does not require consumers to love AI content.

It requires only that measurable performance gains outweigh the sentiment penalty in some contexts.

In that outcome, the market splits: AI content dominates performance-oriented campaigns, while human creators remain central to emotional, trust-building, and brand-equity work.

The EU AI Act requiring labeling of synthetic media and FTC disclosure guidelines create a regulatory framework that, if enforced consistently, could reduce the trust penalty asymmetry by making AI disclosure standard rather than exceptional.

Mandatory transparency might normalize AI content in ways that optional disclosure can't, if every ad carries its production method, consumers adapt to the new information environment rather than reacting to selective disclosure as evidence of attempted deception.

AI-GENERATED INFLUENCER CONTENT
DESTROYS MARKETING TRUST
THE AUTHENTICITY PREMIUM ARRIVED, BUT THE MARKETING INVESTMENT DIDN'T GET THE MEMO.
THE POINT MUST BE WRONG ABOUT
The trust penalty is permanent and cannot be mitigated.
Authenticity standards will remain consistently high across all audiences.
Consumers will continue to reject AI creator content across all use cases.
Disclosure will always reduce trust and amplify negative reactions.
Regulatory and platform constraints will stay strict and limit AI adoption.
WHAT NEEDS TO CHANGE FOR THE COUNTERPOINT TO PREVAIL
DECISION FRAMEWORK
1 Track longitudinal studies on trust and brand impact across demographics.
2 Test the premium positioning strategy. Quality over quantity.
3 Monitor platform transparency and disclosure standardization.
4 Measure performance by audience segment, not as a single market.
5 Watch regulatory action and liability outcomes that could reset incentives.
THE COUNTERPOINT MUST BE RIGHT ABOUT
AI content can achieve authentic connection in specific categories.
High-quality AI creator content will build trust over time.
Cost and scale advantages will outweigh consumer concerns for many brands.
Younger audiences will normalize and embrace AI creator content.
AI creator content will deliver superior performance in measurable outcomes.
CURRENT STATE
26% Consumer preference for AI creator content in mid-2025 (down from 60% in 2023).
Major brands like McDonald's Netherlands have pulled AI ads after documented backlash.
Trust penalty confirmed by peer-reviewed research and marketer concern.
77% of marketers plan to increase AI creator content spend despite falling consumer preference.
LIKELIHOOD COUNTERPOINT PREVAILS
0%
50%
100%
38%
▼ -7% vs Original Prediction (Was 45%)
The counterpoint likelihood declined because documented consumer backlash intensified, major brands pulled AI campaigns, and new peer-reviewed research confirmed the trust penalty with disclosure making it worse.
MOST CRITICAL VARIABLE
CONSUMER TRUST THRESHOLD
The point at which AI creator content delivers authentic connection that matches or exceeds human creators for target audiences.
▼ -5% vs Original Prediction
THE QUESTION
Can higher-quality AI creator content, with transparency and clear value, build enough trust in key segments to overcome the authenticity penalty and justify continued marketing investment?

## What Needs to Change

### (... for the Counterpoint to Prevail)

For the AI influencer market to find sustainable equilibrium rather than continued decline, three conditions need to materialize.

First, the consumer preference decline needs to halt and stabilize, the 60%-to-26% trajectory from 2023 to mid-2025, if it continues, approaches near-zero consumer preference for AI creator content well before 2030. Stabilization requires either quality improvement that crosses the "AI slop" threshold consumers are rejecting, or cultural adaptation to AI-saturated content that raises acceptance across cohorts rather than concentrating it in the digitally literate minority.

Second, the performance metrics marketers are tracking must be shown to translate into durable brand equity outcomes rather than short-term conversion gains that erode with compounding trust damage. If AI content optimizes for click-through while degrading the brand equity that drives long-term consumer preference and premium pricing, the ROI case collapses over a longer measurement window.

Third, regulatory disclosure requirements need to normalize rather than amplify the trust penalty. The peer-reviewed research currently suggests mandatory disclosure exacerbates negative reactions rather than building trust through transparency, meaning the regulatory path the Counterpoint relies on may accelerate rather than moderate the authenticity collapse.

## Likelihood Counterpoint Prevails: 38%

The 38% Counterpoint likelihood reflects a prediction whose core mechanism is confirmed by primary-source evidence rather than projected.

The Point confidence rests on three converging signals:

- **Consumer preference is moving against the technology.** Preference for AI creator content fell from 60% to 26% in two years, even as AI content capability dramatically improved. That trend cuts directly against the Counterpoint's "acceptance grows as AI matures" hypothesis.
- **The backlash has produced corporate consequences.** McDonald's Netherlands pulled its AI-generated ad three days after publication, delivering the kind of brand-level consequence the prediction anticipated.
- **The concern has reached executive decision-making.** The WFA survey of 33 major multinationals found that 60% have no plans to use AI influencers and 96% voice

consumer trust concerns, showing that the prediction has moved into C-suite risk calculus.

Taken together, these data points show that the authenticity problem is no longer theoretical. Consumer sentiment, brand behavior, and executive caution are all moving in the same direction.

The 38% assigned to the Counterpoint reflects three genuine uncertainties that prevent further reduction.

First, Billion Dollar Boy's own interpretive conclusion from the same data, "consumers aren't rejecting AI, but how it's often used", preserves real space for the Counterpoint's argument that a quality threshold exists and is crossable.

Second, the 40% preference among 25-34 year olds is a commercially significant demographic segment, not a statistical rounding error, the Counterpoint's viable niche scenario is not implausible for that cohort.

Third, the marketer investment paradox, 77% increasing AI content spend despite falling consumer preference, either reflects genuine performance data that the Point's consumer sentiment focus misses or represents behavioral momentum that delays rather than prevents the correction the prediction describes.

The 38% is an honest acknowledgment that none of those uncertainties is negligible.

It also reflects that the direction of travel is unambiguous and the evidence base is primary-sourced.

# Smart TV Advertising Overreach Sparks Consumer Backlash

## The EV Depot, the Bus Terminal, and the Hotel Room Say Otherwise

*Samsung settled Texas's ACR surveillance lawsuit February 26, 2026, agreeing to stop collecting viewing data without explicit consent.*

*LG settled weeks later, required to display pop-up disclosures, provide a clear opt-out, and prohibit data transfers to the Chinese Communist Party.*

*Hisense remains under a court restraining order.*

*Kentucky passed the first state-level ACR-specific legislation 92-0. Meanwhile: CTV advertising reached approximately $37.95 billion in 2026, growing 14.5%. Ad-supported viewing time accounts for an estimated 73.6% of total TV viewing.*

*Consumers voluntarily chose ad-supported streaming tiers in majority numbers. Both things happened in the same year.*

*The surveillance overreach lost in court. The advertising model won in the market.*

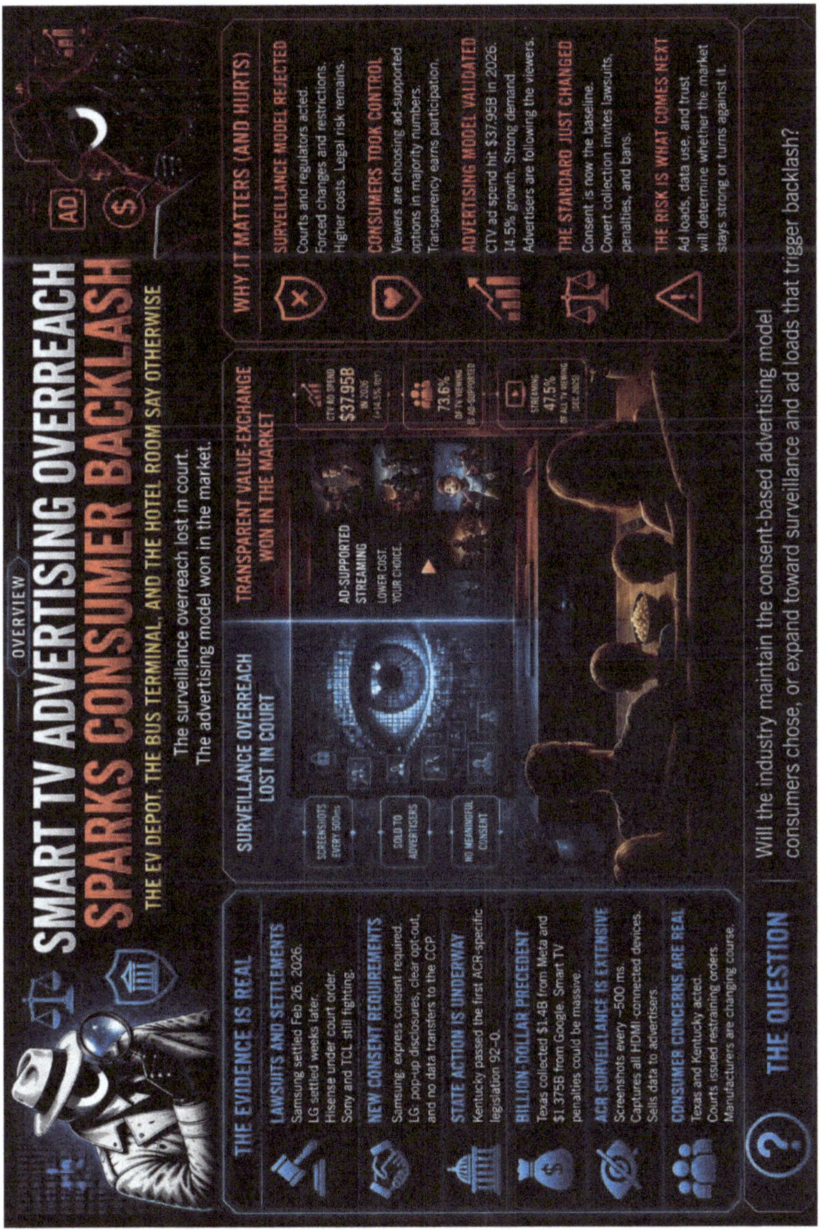
OVERVIEW
SMART TV ADVERTISING OVERREACH
SPARKS CONSUMER BACKLASH
THE EV DEPOT, THE BUS TERMINAL, AND THE HOTEL ROOM SAY OTHERWISE
The surveillance overreach lost in court.
The advertising model won in the market.
AD
THE EVIDENCE IS REAL
LAWSUITS AND SETTLEMENTS
Samsung settled Feb 26, 2026.
LG settled weeks later.
Hisense under court order.
Sony and TCL still fighting.
NEW CONSENT REQUIREMENTS
Samsung: express consent required.
LG: pop-up disclosures, clear opt-out, and no data transfers to the CCP.
STATE ACTION IS UNDERWAY
Kentucky passed the first ACR-specific legislation 92–0.
BILLION-DOLLAR PRECEDENT
Texas collected $1.4B from Meta and $1.375B from Google. Smart TV penalties could be massive.
ACR SURVEILLANCE IS EXTENSIVE
Screenshots every ~500 ms.
Captures all HDMI-connected devices.
Sells data to advertisers.
CONSUMER CONCERNS ARE REAL
Texas and Kentucky acted.
Courts issued restraining orders.
Manufacturers are changing course.
SURVEILLANCE OVERREACH LOST IN COURT
SCREENSHOTS EVERY 500ms
SOLD TO ADVERTISERS
NO MEANINGFUL CONSENT
TRANSPARENT VALUE-EXCHANGE WON IN THE MARKET
AD-SUPPORTED STREAMING
LOWER COST. YOUR CHOICE.
CTV AD SPEND $37.95B IN 2026 (+14.5% YoY)
73.6% OF TV VIEWING IS AD-SUPPORTED
STREAMING 47.5% OF ALL TV VIEWING (DEC 2025)
WHY IT MATTERS (AND HURTS)
SURVEILLANCE MODEL REJECTED
Courts and regulators acted.
Forced changes and restrictions.
Higher costs. Legal risk remains.
CONSUMERS TOOK CONTROL
Viewers are choosing ad-supported options in majority numbers.
Transparency earns participation.
ADVERTISING MODEL VALIDATED
CTV ad spend hit $37.95B in 2026.
14.5% growth. Strong demand.
Advertisers are following the viewers.
THE STANDARD JUST CHANGED
Consent is now the baseline.
Covert collection invites lawsuits, penalties, and bans.
THE RISK IS WHAT COMES NEXT
Ad loads, data use, and trust will determine whether the market stays strong or turns against it.
THE QUESTION
Will the industry maintain the consent-based advertising model consumers chose, or expand toward surveillance and ad loads that trigger backlash?

## Opening

The Smart TV advertising backlash story requires the most significant analytical reframe of any article in this series, because the original prediction was simultaneously more right and more wrong than a simple validation or refutation allows.

Right about ACR surveillance triggering regulatory enforcement at exactly the scale and timeline the piece described.

Wrong about the direction of consumer behavior toward advertising broadly.

The backlash against covert surveillance-based data collection is real, legally consequential, and ongoing. Consumers have simultaneously voted with their viewing choices to embrace transparent value-exchange advertising at volumes the original prediction did not anticipate.

The 2027-2030 analysis is the story of what happens when mandatory consent requirements meet a $37.95 billion CTV advertising market that consumers are actively choosing to participate in, and whether the industry maintains the consent-based model consumers endorsed or uses that permission as a license to expand surveillance and ad loads beyond the tolerance threshold.

## Intro

The Smart TV advertising prediction's defining analytical challenge is that it conflated two different products under one label, and the 2026 evidence has sorted them into two different verdicts.

Covert ACR surveillance advertising, screenshots every 500 milliseconds without meaningful consent, manufacturer monetization of viewing behavior before users understand what they agreed to, lost in court on the timeline the original prediction specified.

Transparent value-exchange advertising, ad-supported streaming tiers selected voluntarily in exchange for lower subscription costs, won in the market at a scale the original prediction did not anticipate.

The original piece treated these as one phenomenon heading toward one outcome.

The 2026 data treats them as two different products that consumers and regulators have assessed on their respective merits and reached opposite verdicts.

That distinction is not a technicality. It's the entire analytical key to what the 2027-2030 period will look like: a $37.95 billion CTV advertising market built on consumer consent sitting alongside ongoing litigation against manufacturers who collected data without it, in the same living rooms, at the same time.

The question for 2030 is whether the industry reads the two verdicts correctly and maintains the consent-based model that earned the market's endorsement, or whether it interprets consumer acceptance of transparent advertising as permission to expand toward the covert model that the courts rejected.

## Background

Texas Attorney General Ken Paxton's December 2025 lawsuits against Sony, Samsung, LG, Hisense, and TCL marked a shift in US privacy enforcement strategy, focusing on tracking technology operating inside the home and across all HDMI-connected devices.

ACR captures TV users' viewing habits in real time with screenshots taken approximately every 500 milliseconds, with manufacturers selling that information to advertisers, including content from cable boxes, game consoles, and any device connected via HDMI, regardless of whether the user is engaged with a specific smart TV app.

Texas's AG obtained temporary restraining orders against both Samsung and Hisense before Samsung settled February 26, 2026, agreeing to express consent requirements and rewritten disclosure screens.

LG subsequently settled, agreeing to pop-up disclosures, a clear and simple opt-out, and a prohibition on data transfers to the Chinese Communist Party, an unusual provision reflecting the AG's concern that viewing data could reach foreign governments.

Kentucky passed the first state-level ACR-specific legislation with a 92-0 House vote, the first state to go beyond lawsuit enforcement to statutory prohibition.

Sony and TCL continue fighting with no restraining orders, meaning both continue ACR collection while their cases proceed.

Texas has collected $1.4 billion from Meta and $1.375 billion from Google using the same enforcement instruments, creating the precedent for potential astronomical penalties in the smart TV cases.

Simultaneously: streaming captured approximately 47.5% of all US TV viewing in December 2025, the first time streaming exceeded cable and broadcast combined. Netflix's ad tier revenue exceeded $1.5 billion in 2025. CTV advertising spend reached approximately $37.95 billion in 2026 at 14.5% year-over-year growth.

These two stories, regulatory enforcement of surveillance advertising and the commercial success of transparent advertising, are happening in the same living rooms at the same time.

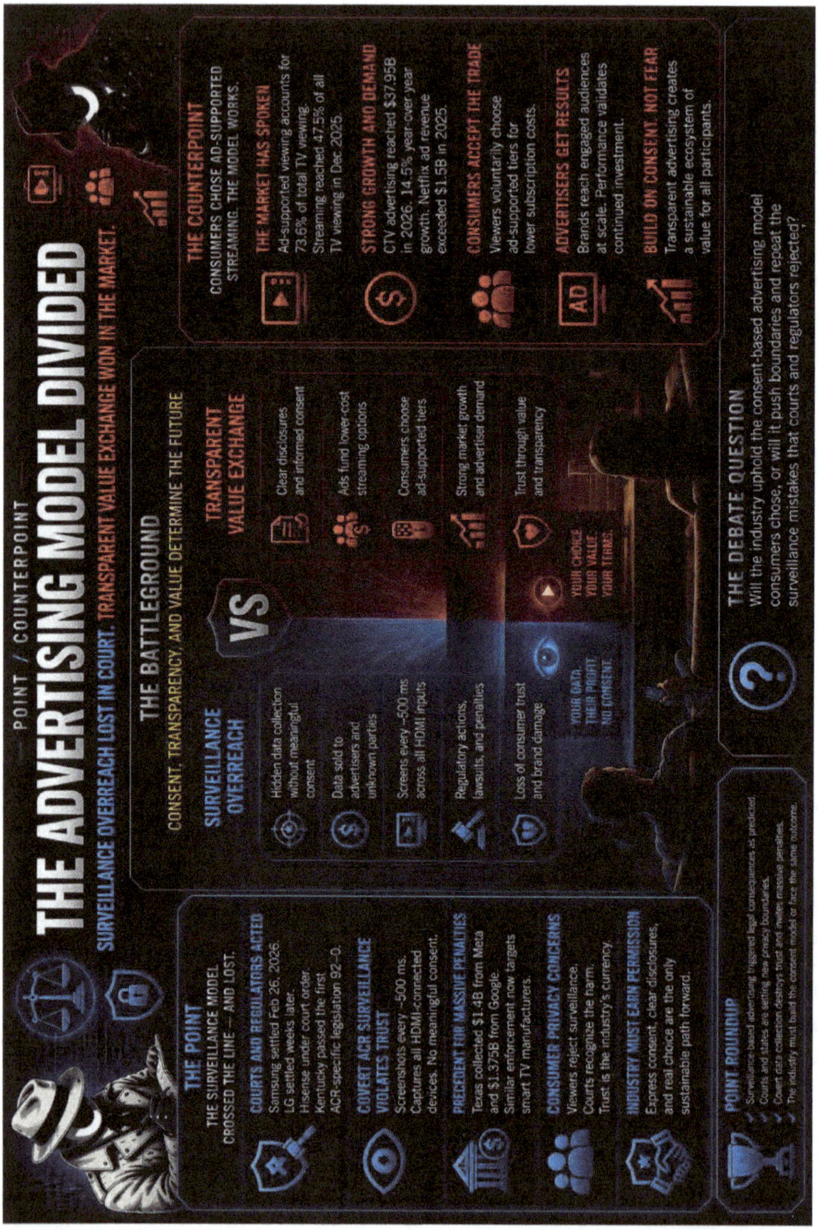
POINT / COUNTERPOINT
THE ADVERTISING MODEL DIVIDED
SURVEILLANCE OVERREACH LOST IN COURT. TRANSPARENT VALUE EXCHANGE WON IN THE MARKET.
THE POINT
THE SURVEILLANCE MODEL CROSSED THE LINE — AND LOST.
COURTS AND REGULATORS ACTED
Samsung settled Feb 26, 2026. LG settled weeks later. Hisense under court order. Kentucky passed the first ACR-specific legislation 92–0.
COVERT ACR SURVEILLANCE VIOLATES TRUST
Screenshots every ~500 ms. Captures all HDMI-connected devices. No meaningful consent.
PRECEDENT FOR MASSIVE PENALTIES
Texas collected $1.4B from Meta and $1.375B from Google. Similar enforcement now targets smart TV manufacturers.
CONSUMER PRIVACY CONCERNS
Viewers reject surveillance. Courts recognize the harm. Trust is the industry's currency.
INDUSTRY MUST EARN PERMISSION
Express consent, clear disclosures, and real choice are the only sustainable path forward.
POINT ROUNDUP
Surveillance-based advertising triggered legal consequences as predicted.
Courts and states are setting new privacy boundaries.
Covert data collection destroys trust and invites massive penalties.
The industry must build the consent model or face the same outcome.
THE BATTLEGROUND
CONSENT, TRANSPARENCY, AND VALUE DETERMINE THE FUTURE
SURVEILLANCE OVERREACH
Hidden data collection without meaningful consent
Data sold to advertisers and unknown parties
Screens every ~500 ms across all HDMI inputs
Regulatory actions, lawsuits, and penalties
Loss of consumer trust and brand damage
YOUR DATA. THEIR PROFIT. NO CONSENT.
VS
TRANSPARENT VALUE EXCHANGE
Clear disclosures and informed consent
Ads fund lower-cost streaming options
Consumers choose ad-supported tiers
Strong market growth and advertiser demand
Trust through value and transparency
YOUR CHOICE. YOUR VALUE. YOUR TERMS.
THE COUNTERPOINT
CONSUMERS CHOSE AD-SUPPORTED STREAMING. THE MODEL WORKS.
THE MARKET HAS SPOKEN
Ad-supported viewing accounts for 73.6% of total TV viewing. Streaming reached 47.5% of all TV viewing in Dec 2025.
STRONG GROWTH AND DEMAND
CTV advertising reached $37.95B in 2026, 14.5% year-over-year growth. Netflix ad revenue exceeded $1.5B in 2025.
CONSUMERS ACCEPT THE TRADE
Viewers voluntarily choose ad-supported tiers for lower subscription costs.
ADVERTISERS GET RESULTS
Brands reach engaged audiences at scale. Performance validates continued investment.
BUILD ON CONSENT, NOT FEAR
Transparent advertising creates a sustainable ecosystem of value for all participants.
THE DEBATE QUESTION
Will the industry uphold the consent-based advertising model consumers chose, or will it push boundaries and repeat the surveillance mistakes that courts and regulators rejected?

## THE POINT: The Surveillance Model Lost in Court. Now Build the Consent Model or Face the Same Outcome

The original prediction's regulatory enforcement scenario has materialized with specificity and speed that validates the piece's core mechanism: advertising overreach built on covert surveillance triggered regulatory consequence rather than waiting for gradual consumer migration away from the platform.

The Texas cases have set off a cascade with national implications: a class action lawsuit against Samsung was filed in New York on January 9, 2026 targeting the same ACR practices; Kentucky passed the first state-level ACR-specific legislation by a 92-0 House vote; and Sony, Hisense, and TCL continue facing active lawsuits with potential penalties of $10,000 to $250,000 per violation per consumer, exposure that becomes astronomical when applied to millions of Texas households where screenshots were taken approximately every 500 milliseconds without meaningful consent.

The LG settlement terms reveal what the enforcement ultimately requires of manufacturers: a pop-up disclosure explaining how viewing data may be collected and used; a clear and simple opt-out; and prohibition on data transfers to foreign governments.

The Samsung settlement terms are similar: express, informed consent before collecting ACR data, and consent screens rewritten to be "clear and conspicuous" rather than buried in setup flows.

The National Law Review analysis notes precisely what changed: "ACR programs have generally been allowed under the technical text of a TV's terms of service, but few users meaningfully consented because they never realized what they were agreeing to."

The original prediction's specific claim, that smart TV surveillance practices would force manufacturers to offer transparent privacy controls and transform advertising from default experience to opt-in service, has been mandated by court settlements rather than waiting for voluntary corporate behavior change.

The timeline is earlier than the prediction specified.

The consumer behavior story that complicates the original prediction's direction-of-travel argument is the most important analytical correction in this update.

The original prediction assumed that consumer rejection of surveillance advertising would translate into rejection of advertising broadly.

The market evidence contradicts that assumption at scale. Ad-supported viewing time accounts for an estimated 73.6% of total TV viewing time, with ad-supported subscriptions representing a majority of new streaming subscriber additions.

Industry surveys consistently find that approximately 64% of CTV users prefer ad-supported content if it allows them to pay less. One-third of US CTV viewers report completing a purchase after viewing an ad.

Video completion rates on streaming platforms exceed 95%.

These figures come from industry participants with commercial interests in CTV advertising, they should be understood as directional industry findings rather than independent primary research, but the consistency across multiple independent tracking sources and the structural evidence of Netflix, Disney+, and Amazon Prime Video all successfully launching and growing ad tiers confirms the direction.

The original prediction did not draw a sharp enough line between two different advertising models:

- **Covert ACR surveillance:** screenshots captured without meaningful consent, viewing behavior monetized before users understand what they agreed to, and data sold to advertisers. Consumers reject this model when they discover it.
- **Transparent value-exchange advertising:** ad-supported tiers that consumers choose voluntarily in exchange for lower subscription costs. Consumers are adopting this model at majority rates.

The backlash is not against advertising itself. It is against surveillance disguised as advertising.

The regulatory enforcement and the advertising market growth are not in contradiction. They describe two different products receiving two different consumer verdicts simultaneously in the same living room.

## The Point Roundup

The Smart TV advertising overreach story has resolved into two parallel tracks that the original prediction conflated into one. The surveillance overreach track, ACR data collection without meaningful consent, screenshots every 500 milliseconds, covert monetization of viewing behavior, has triggered exactly the regulatory enforcement and corporate accountability the original prediction described. Samsung and LG have settled; Hisense remains under a restraining order; Kentucky has legislated; Sony and TCL face ongoing exposure.

That track ends with mandatory consent requirements, clear opt-out mechanisms, and settlement terms that require genuine transparency.

The advertising market track (transparent, value-exchange CTV advertising on platforms where consumers knowingly chose ad-supported tiers in exchange for lower costs) has grown to approximately $37.95 billion and is on track to surpass traditional TV advertising within the decade.

The 2027-2030 story is the separation of those two tracks into permanent structural features of the television landscape, and whether the industry maintains the consent-based model consumers endorsed or uses that permission as license to expand ad loads and surveillance capabilities until reaching the tolerance threshold that triggers the backlash cycle again.

## THE COUNTERPOINT: The Surveillance Model Failed. The Advertising Model Won. Those Are Different Things

The Counterpoint's scenario, "consumers accept advertising integration in exchange for significantly lower costs and free premium content access, especially during economic uncertainty when subscription costs become burdensome", has been validated by consumer behavior at a scale that the original 30% likelihood substantially underestimated.

Free Ad-Supported Streaming TV platforms now reach approximately 60% of US households. Roughly 70% of US adults turn to streaming first when watching video content.

An estimated 64% of CTV users prefer ad-supported content if it allows them to pay less. Netflix, Disney+, and Amazon Prime Video, the three dominant platforms, all launched ad tiers that succeeded commercially, with consumers choosing ad-supported options in large and growing numbers.

The Counterpoint's specific prediction that consumers would accept advertising in contexts where the value exchange is clear has been confirmed by every structural metric in the CTV advertising market.

The original prediction's 30% likelihood reflected genuine uncertainty about whether consumers would accept advertising integration. That question has been answered in the affirmative for the transparent model.

The uncertainty that remains is whether the industry will maintain the conditions consumers endorsed, clear value exchange, meaningful consent, honest opt-out, or exploit the permission granted by ad-tier selection to expand surveillance and ad loads beyond the tolerance threshold.

The regulatory enforcement that the original prediction cited as evidence of backlash is more accurately characterized as the market correcting one specific advertising model while validating the broader advertising integration model.

Samsung's settlement terms, “express consent before collecting ACR data" and "clear and conspicuous consent screens”, describe the path to legitimate advertising integration rather than the elimination of advertising integration.

The Counterpoint's predicted evolution from intrusive mandatory advertising toward "genuinely useful, personalized experiences that consumers appreciate rather than tolerate" is precisely what the settlement mandates: consent, transparency, and opt-out capability.

Those are design requirements for sustainable advertising, not prohibitions on advertising itself. The industry's pivot toward accountability, attention-based metrics, standardized measurement, consent-first data collection, describes the maturation arc the Counterpoint predicted for advertising technology that overcomes consumer resistance through genuine value delivery rather than surveillance-based capture.

The Counterpoint's scenario is not that the advertising model survives unchanged. It’s that the advertising model evolves toward consent-based value exchange, and the 2026 evidence confirms that consumers embrace that model voluntarily when it’s offered honestly.

OVERVIEW
SMART TV ADVERTISING OVERREACH
SPARKS CONSUMER BACKLASH
THE SURVEILLANCE MODEL LOST IN COURT. NOW BUILD THE CONSENT MODEL OR FACE THE SAME OUTCOME.
THE POINT MUST BE WRONG ABOUT
REGULATORY ENFORCEMENT STOPS
Courts and states halt ACR enforcement. Lawsuits fail. Legislation stalls.
CONSENT REQUIREMENTS WEAKEN
Manufacturers get looser consent standards and fewer disclosures.
CONSUMER CONCERNS FADE
Privacy worries decline. Surveillance stops being a deciding factor.
PENALTIES DON'T DETER
Billion-dollar penalties disappear. Risk becomes acceptable again.
AD LOADS KEEP CLIMBING
Ad volume and targeting expand. No meaningful consumer resistance.
WHAT NEEDS TO CHANGE FOR THE COUNTERPOINT TO PREVAIL
DECISION FRAMEWORK
1 TRACK & COMPLY
Monitor lawsuits, legislation, and regulatory rulings across all states and at the federal level.
2 INVEST IN CONSENT UX
Make consent clear, granular, and easy to manage. Build trust through design.
3 DELIVER REAL VALUE
Keep ad loads reasonable. Offer lower costs and better content in exchange for attention.
4 PROTECT THE DATA
Minimize data collection. No transfers to foreign entities. Strong security and governance.
5 ADAPT OR GET PUNISHED
Noncompliance invites lawsuits, restraining orders, and massive penalties.
WINNING FORMULA: TRANSPARENCY + VALUE + TRUST = SUSTAINABLE GROWTH
THE COUNTERPOINT MUST BE RIGHT ABOUT
CONSUMERS TOLERATE MORE AD LOADS
Viewers accept heavier ad loads without pushing back.
HYPER-TARGETING DRIVES ACCEPTANCE
Personalization feels helpful, not invasive.
SURVEILLANCE REBRANDED AND ACCEPTED
Continuous data collection is normalized as the cost of free content.
REGULATORS BACK OFF
Enforcement slows. Industry self-regulation is enough.
INDUSTRY CONSOLIDATES POWER
Fewer players control more data. Consumers have no real alternatives.
CURRENT STATE
Major manufacturers settled or face restraining orders. Lawsuits continue.
Kentucky passed ACR-specific legislation 92-0.
73.6% of total TV viewing is ad-supported.
$37.95B CTV ad spend in 2026, up 14.5% YoY.
47.5% of all TV viewing is streaming (Dec 2025).
LIKELIHOOD COUNTERPOINT PREVAILS
50%
0%
40%
100%
▲ +10% vs Original Prediction (Was 30%)
Upward revision driven by voluntary consumer adoption of ad-supported streaming tiers at scale.
Stronger market evidence increases the odds that the counterpoint wins out by 2030.
MOST CRITICAL VARIABLE
CONSUMER TRUST THRESHOLD
The point at which ad loads, surveillance, and targeting cross from acceptable to unacceptable. Cross it, and backlash returns fast.
TOO LITTLE VALUE / TOO MUCH DATA
THE LINE (SWEET SPOT)
TOO MUCH AD LOAD / INVASION
THE QUESTION
Will the industry maintain the consent-based model that consumers endorsed, or use that permission to expand surveillance and ad loads beyond the tolerance threshold?

## What Needs to Change

### (... for the Counterpoint to Fully Prevail)

For the Counterpoint's "seamlessly valuable advertising" scenario to fully materialize through 2030, two conditions need to hold that are currently under commercial pressure.

First, the industry needs to maintain the consent-based, value-exchange model that consumer adoption has endorsed, rather than using the permission consumers granted through ad-supported tier selection as license to expand surveillance capabilities beyond what consumers understood they agreed to. Streaming platforms increasing ad loads after launching ad-supported tiers, Prime Video reportedly doubling its ad load after launch, represents the overreach pattern the original prediction described, now potentially recurring within the consent-based model rather than outside it. The original overreach was covert surveillance; the next overreach would be ad load expansion that exceeds the tolerance implicit in consumers' tier selection.

Second, the ACR enforcement needs to translate from state law into federal standards providing consistent national protection. Currently, Samsung and LG have consent requirements for Texas consumers and no equivalent obligation for consumers in the 49 other states. Manufacturers operating outside Texas can continue the same ACR practices that triggered the Texas settlements while the Kentucky legislation and further state actions develop.

The Counterpoint's sustainable advertising model requires both transparent consumer consent nationwide and consistent federal regulatory standards, state-by-state enforcement is a beginning, not the completion of the regulatory framework the prediction described as necessary.

## Likelihood Counterpoint Prevails: 40%

The Counterpoint likelihood increases from 30% to 40%. It's a very meaningful upward revision driven by the most consequential evidence gap in the original prediction: the voluntary consumer adoption of ad-supported streaming tiers at scale.

Ad-supported viewing time reaching an estimated 73.6% of total TV time, with ad-supported subscriptions accounting for the majority of new subscriber additions across major platforms, demonstrates that the Counterpoint's "consumers accept advertising in exchange for lower costs" scenario has materialized at scale.

The original 30% reflected genuine uncertainty about whether consumers would accept advertising integration, a question the market has answered in the affirmative for the transparent value-exchange model.

The move from 30% to 40% reflects this evidence while acknowledging two structural uncertainties that prevent a larger revision.

First, the covert-versus-transparent distinction that explains the 2026 evidence is not permanently stable: the industry's history of using initial consumer permission as license to expand surveillance and ad loads creates a recurring overreach cycle that the current consent-based model does not permanently resolve.

Prime Video's ad load doubling after launch is the early indicator that the expansion dynamic is already operating within the consent-based model.

Second, the regulatory enforcement remains state-level, Texas and Kentucky have acted, but federal standards don't exist, meaning manufacturers face inconsistent obligations across jurisdictions and the surveillance model continues for the majority of US consumers outside states with specific enforcement.

The 40% says: the Counterpoint's transparent value-exchange advertising scenario has been validated by consumer behavior.

Whether the industry maintains the conditions that made consumers choose it, or converts the permission into the next overreach cycle, is what the 2027-2030 window will adjudicate.

# Subscription Box Model Collapses From Market Saturation

## The Collapse Didn't Come. The Shakeout Did. Understanding Subscription's Bifurcated Reality

*The subscription box market reached $42.5 billion in 2025 and is projected to reach $124 billion by 2034. Also: 44% of new subscribers cancel within 90 days.*

*10-15% monthly industry churn requires continuous new subscriber acquisition just to hold revenue flat. C+R Research's foundational study found consumers estimate spending $86 per month on subscriptions, while their actual itemized total averages $219.*

*A 2.5x perception gap.*

*When consumers discover the real number, they cancel, and no amount of AI curation resolves a cancellation decision made in a bank statement moment.*

*The collapse didn't arrive. The paradox deepened.*

POINT / COUNTERPOINT
SUBSCRIPTION BOX MODEL COLLAPSES FROM MARKET SATURATION
THE COLLAPSE DIDN'T COME. THE SHAKEOUT DID.
UNDERSTANDING SUBSCRIPTION'S BIFURCATED REALITY
THE POINT
THE SHAKEOUT IS HAPPENING. IT'S HAPPENING ONE CATEGORY AT A TIME.
THE PERCEPTION GAP DRIVES CANCELLATIONS
Consumers estimate $86/month in subscriptions.
Actual average is $219/month.
A 2.5x gap.
DISCOVERY MOMENT TRIGGERS CANCELS
74% say recurring charges are easy to forget.
42% pay for subscriptions they no longer use.
CHURN HITS FAST
44% of cancellations happen within the first 90 days.
No amount of AI curation prevents this.
CHURN ECONOMICS ARE BRUTAL
Industry monthly churn is 10–15%.
Top performers are <3%.
Constant acquisition just to stand still.
WEAK PLAYERS ARE DISAPPEARING
High costs, low retention, crowded categories create continuous operator attrition.
THE BATTLEGROUND
THE DATA TELLS A TWO-SIDED STORY.
MARKET SNAPSHOT (2025)
$42.5B
Subscription box market size in 2025 (IMARC Group)
$124B
Projected market size by 2034. 12.64% CAGR.
10–15%
Industry monthly churn. Top performers maintain <3%. (Recurly, 2,200+ merchants)
44%
Of cancellations occur within the first 90 days.
90
CONSUMER REALITY CHECK
$86
What consumers think they spend per month.
$219
What consumers actually spend per month.
$$$
2.5x PERCEPTION GAP
The root cause of cancellation.
THE COUNTERPOINT
THE MODEL WORKS FOR THOSE WHO SOLVE RETENTION.
AI CURATION IS REAL AND EFFECTIVE
Personalization is now standard.
It reduces churn at scale.
Better matches drive longer retention.
STRONG PLAYERS COMPOUND ADVANTAGES
Top operators have <3% churn.
Brand, data, and experience create durable moats.
FLEXIBILITY IMPROVES RETENTION
Skip, pause, swap, and credit options reduce friction.
Customers stay longer.
CONSUMERS STILL WANT DISCOVERY
Curation saves time.
Busy lives value convenience.
Willingness to pay remains.
MARKET GROWTH VALIDATES DEMAND
$42.5B market in 2025.
Projected to reach $124B by 2034.
Strong long-term trajectory.
THE DEBATE QUESTION
Will the subscription box industry continue to shake out weak operators and reward retention leaders, or will market saturation eventually overwhelm even the best players?

## Opening

The original prediction's "dramatic market correction" scenario hasn't materialized as a collapse, but the structural forces it identified have been validated by the 2026 evidence with unusual precision.

The subscription box market is growing.

Simultaneously, the churn dynamics the original piece described as unsustainable are confirmed by primary data: industry-wide monthly churn of 10-15%, 44% of cancellations happening in the first 90 days, and the C+R Research perception gap study documenting the exact consumer psychology mechanism the prediction relied on.

The two-sided update is important: the Counterpoint's technology solution has genuinely materialized, AI-driven personalization is reducing churn at scale and is now standard industry practice, and the Point's structural problem hasn't been resolved by that technology.

Both are simultaneously true, which is what the shakeout rather than the collapse looks like in practice.

The 2027-2030 story is the continuous selection process that eliminates operators unable to solve the retention problem, while the strongest players compound their advantages through AI curation and flexible models.

## Intro

Two important distinctions define this update.

First, the article is about subscription boxes, physical curated product deliveries, as the primary case study for subscription fatigue as a cross-cutting phenomenon.

The broader subscription economy (SaaS, streaming, digital media) provides context and shares the consumer psychology mechanism, but the box model's specific churn dynamics, physical logistics costs, and category-level differentiation are distinct enough to warrant careful separation from streaming and software subscription data.

Second, the "collapse" framing in the original title was too specific in its timeline and too severe in its outcome. The revision adopts "shakeout" as the more accurate characterization, a continuous market selection process that is eliminating the weakest operators across every subscription category while the model itself continues growing.

That distinction is not a concession that the prediction was wrong. It's a more precise description of what structural market correction actually looks like when it operates through competitive attrition rather than a single correction event.

## Background

The global subscription box market reached $42.5 billion in 2025 per IMARC Group analysis, projected to reach $124.1 billion by 2034 at a 12.64% CAGR, a trajectory that describes a growing rather than contracting market.

Market sizing estimates vary significantly across research firms (Research and Markets projects $101.8 billion by 2030 at 19.6% CAGR; other firms project higher), reflecting different definitions of the subscription box category.

The IMARC figure is the most conservative and most widely corroborated across independent sources. The broader subscription economy context: C+R Research's benchmark study found the average American estimates spending $86 per month on all subscriptions but actually spends $219, a $133 gap confirmed by multiple subsequent studies using different methodologies.

This study was conducted in 2022 and remains the foundational citation in 2025-2026 industry coverage; it's a structural finding about consumer psychology rather than a time-bound measurement.

The churn dynamics are documented by Recurly's analysis of 2,200-plus merchants and 67 million subscribers, the most comprehensive primary dataset in subscription commerce: industry-wide monthly churn runs 10-15%, while top performers maintain rates below 3%, a differential that describes a market with solved and unsolved operators rather than a uniformly struggling category.

Sixty-eight percent of subscription churn is involuntary, driven by failed transactions rather than active cancellations, meaning the active conscious cancellation rate is meaningfully lower than headline churn numbers imply, though failed payment recovery is itself an operational challenge.

OVERVIEW
SUBSCRIPTION BOX MODEL COLLAPSES FROM MARKET SATURATION
THE COLLAPSE DIDN'T COME. THE SHAKEOUT DID.
UNDERSTANDING SUBSCRIPTION'S BIFURCATED REALITY
A GROWING MARKET
The subscription box market reached $42.5B in 2025 and is projected to reach $124B by 2034 (12.64% CAGR).
A RETENTION PROBLEM
10–15% monthly industry churn. 44% of new subscribers cancel within 90 days.
A PERCEPTION GAP
Consumers estimate spending $86/month on subscriptions. Their actual itemized total averages $219. A 2.5x gap.
THE PARADOX
AI personalization reduces churn and improves retention. The structural problem remains unsolved.
MARKET TRAJECTORY
Subscription Box Market Size
(IMARC Group)
$42.5B
$124B
12.64% CAGR
2025
2034
Market growth continues across diverse categories and consumer demographics. Definitions vary, but the trajectory is clear.
THE REALITY BEHIND THE NUMBERS
44%
Of new subscribers cancel within the first 90 days. (Recurly)
10-15%
Industry monthly churn. Top performers maintain <3%. (Recurly, 2,200+ merchants)
68%
Of churn is involuntary. Driven by failed payments, not active cancellations. (Recurly)
74%
Say recurring charges are easy to forget.
42%
Admit paying for subscriptions they no longer use. (C+R Research)
2.5x PERCEPTION GAP
$86 estimated vs. $219 actual monthly spend.
The discovery moment is the cancellation moment.
WHAT THIS MEANS
CONTINUOUS ACQUISITION REQUIRED
High churn means operators must constantly acquire new subscribers just to stand still.
AI HELPS—BUT DOESN'T SOLVE THE CORE
Personalization reduces churn and extends lifecycle. It cannot close the perception gap.
THE SHAKEOUT IS ONGOING
Weak operators exit or consolidate. Strong players compound advantages.
FLEXIBILITY WINS
Skip, pause, swap, and credit options improve retention and customer lifetime value.
CATEGORIES DIVERGE
Some boxes thrive with differentiation. Others see commoditization and margin pressure.
THE QUESTION
Will the subscription box industry continue to grow through strong player compounding and weak player attrition, or will the accumulation of fatigue eventually cap the market?

## THE POINT: The Shakeout Is Happening. It's Happening One Category at a Time.

The original prediction's structural diagnosis has been confirmed by the 2026 evidence even as its collapse timeline hasn't materialized.

The C+R Research perception gap study ($86 estimated versus $219 actual, a 2.5x underestimation confirmed across multiple subsequent methodologies) is the most important single data point in the subscription economy because it identifies a structural consumer psychology problem that no technology solution fully resolves.

Seventy-four percent of respondents in the C+R study said recurring charges are easy to forget. Forty-two percent admitted paying for subscriptions they no longer use.

The cancellation mechanism the original prediction described operates precisely through this gap: consumers tolerate subscription charges they don't notice, discover the aggregate total during a bank statement review or financial audit, and cancel in a single decisive session that AI curation and personalized recommendations arrive too late to prevent.

Forty-four percent of cancellations happening within the first 90 days confirms the original prediction's "subscribers cancel after a few boxes due to repetitiveness and no perceived value" pattern, the initial novelty window closes faster than retention technology can establish genuine value perception.

The category-level differentiation the prediction identified as the survival criterion has crystallized in the 2026 data precisely as described: pet subscription boxes are growing at 12.8% CAGR and represent one of the most resilient categories during subscription fatigue, the essential replenishment and high-engagement service model the original piece identified as the survival archetype.

Fashion subscription boxes, by contrast, have seen consolidation following Stitch Fix's extended struggles, representing the lifestyle and novelty category the prediction identified as most vulnerable.

The shakeout is selecting exactly the winners and losers the original prediction described, operating through continuous competitive pressure rather than a single dramatic moment.

The 10-15% monthly industry churn rate documented by Recurly's primary data is the Point's most analytically significant figure because of its compounding implication: a business with 15% monthly churn loses roughly 83% of its subscriber base annually and must replace that attrition through continuous new subscriber acquisition just to maintain revenue.

The AI personalization story that the Counterpoint correctly cites as the partial solution does not change this mathematics as much as the headline "45% churn reduction" implies. A 45% reduction applied to a 15% monthly churn baseline produces approximately 8.25% monthly churn, still high enough to require aggressive ongoing acquisition.

The top performers maintaining below 3% monthly churn are not simply applying AI curation to average-performing subscription products. They have solved multiple dimensions of the retention problem simultaneously: genuine ongoing product-market fit in categories with essential replenishment needs, AI personalization, annual plan architecture, and pause/skip flexibility.

The operators in the 10-15% churn range have typically solved one or two of those dimensions and are experiencing the structural attrition the prediction described. The broader subscription economy context confirms the fatigue is cross-category rather than box-specific: 40.3% of Americans cancelled a subscription service in 2024, with 41% stating content isn't worth the price, a figure up five percentage points from the prior year.

70% of consumers are frustrated by ongoing price increases. 54.9% admit having at least one subscription going unused each month.

These are not box-specific behaviors.

They are the consumer psychology mechanics the original prediction identified as the structural threat to every recurring revenue model that doesn't deliver demonstrably essential and continuously refreshed value.

## The Point Roundup

The subscription box market correction the original prediction described has arrived as a continuous selection process rather than a dramatic moment, and the 2026 evidence confirms the original structural diagnosis while updating the timeline.

The $219 versus $86 perception gap is the recurring detonator that keeps the fatigue cycle active regardless of how well curation technology works. The 10-15% monthly churn is the treadmill that makes subscription box economics fundamentally different from digital subscription economics: physical goods delivery has fixed fulfillment costs that don't scale down with subscriber count, meaning the churn replacement cost compounds in ways that streaming or SaaS businesses don't face.

The shakeout is happening.

Its pace is governed by how fast the perception gap closes and how quickly operators solve the multi-dimensional retention problem, not by a single correction event.

The 2027-2030 window will determine whether AI personalization compounds to the point where subscribers genuinely value each box enough that the perception gap closes, or whether the fintech subscription-tracking tools that make spending visible force the gap to close from the bottom up.

## THE COUNTERPOINT: The Survivors Are Thriving. The Category Isn't Collapsing. It's Consolidating."

The Counterpoint's core prediction, that technology-enhanced personalization and flexible models could address consumer fatigue concerns for operators who implemented them well, has been validated by the market's actual behavior at a scale the original 35% likelihood underestimated.

AI-driven personalization reducing churn by up to 45% represents exactly the technical solution the Counterpoint described as the path to sustainable subscription boxes. At approximately 88% industry adoption of AI for product recommendations in 2025 (per industry analysis, not primary Recurly data), it has become standard practice rather than competitive differentiation.

Annual plans reducing churn by 51% and pause/skip flexibility preventing cancellations represent complementary structural fixes that, combined with AI curation, describe a retention architecture that top performers with sub-3% monthly churn have demonstrably built. The market growth trajectory, $42.5 billion in 2025 toward $124 billion by 2034 at 12.64% CAGR, is the most direct evidence that the model hasn't collapsed and that the Counterpoint's "market evolves toward sustainable personalization" scenario is partially materializing.

The fragmented market structure, 400-600 competing box types with top-5 brands controlling only 31% of revenue, is not evidence of a market in distress.

It's evidence of a market still consolidating, where competitive selection is creating the "more sustainable, focused landscape" the original piece described as the post-shakeout outcome.

The Counterpoint's scenario is not that every subscription box survives.

It's that the model proves sustainable for operators who solve the retention problem, and the top performers' sub-3% monthly churn demonstrates that the problem is solvable.

The consumer behavior data that describes the problem also contains the Counterpoint's solution.

Sixty-eight percent of subscription churn is involuntary, driven by failed transactions rather than active cancellations.

That means the majority of the attrition the Point describes as structural is an operational problem (payment processing and dunning management) rather than a preference problem (consumers rejecting the model).

Solving failed payment recovery through modern billing infrastructure could reduce total churn by more than one-third without changing a single product or curation decision.

The services that demonstrate genuine ongoing value have proven the Counterpoint's category-specific resilience claim: pet subscription boxes growing at 12.8% CAGR, food and beverage boxes holding 30% of global subscription box revenue at $28.4 billion in the meal kit category alone, and beauty boxes projected to reach $2.99 billion by 2030 at 17.7% CAGR, all represent essential-adjacent categories with demonstrated retention economics.

The pause/skip functionality finding is particularly compelling for the Counterpoint: 27% of subscribers would cancel if unable to pause or skip orders, meaning that flexibility is preventing cancellations rather than enabling them.

A consumer who pauses rather than cancels is a consumer who retains their subscription identity and reactivates, which is precisely the model the Counterpoint described as the sustainable evolution of the subscription box business.

OVERVIEW
SUBSCRIPTION BOX MODEL COLLAPSES FROM MARKET SATURATION
THE COLLAPSE DIDN'T COME. THE SHAKEOUT DID.
UNDERSTANDING SUBSCRIPTION'S BIFURCATED REALITY
THE POINT MUST BE WRONG ABOUT
THE MARKET WILL SELF-CORRECT
The market is growing: $42.5B in 2025, projected $124B by 2034 (12.64% CAGR).
CHURN WILL FORCE A COLLAPSE
Top performers maintain <3% monthly churn. Weak players exit. The shakeout replaces collapse.
AI WON'T SOLVE RETENTION
AI curation is real and effective. It reduces churn and extends lifecycle. It cannot close the perception gap.
CONSUMERS WILL REJECT SUBSCRIPTIONS
Consumers still value convenience and curation. Willingness to pay remains.
THE MODEL IS FUNDAMENTALLY BROKEN
Flexible models (skip, pause, swap, credit) improve retention and lifetime value. The model adapts and endures.
DECISION FRAMEWORK: WHAT MATTERS MOST
1 SUSTAIN ACQUISITION
High churn requires constant new subscribers just to stand still. Acquisition efficiency separates winners from losers.
2 REDUCE INVOLUNTARY CHURN
68% of churn is involuntary from failed payments. Payment recovery and dunning are retention levers.
3 CLOSE THE PERCEPTION GAP
$86 estimated vs. $219 actual. The 2.5x gap drives cancellation in discovery moments. Hard to fix.
4 IMPROVE UNIT ECONOMICS
Logistics, COGS, and marketing costs must support healthy LTV/CAC. Weak economics don't survive.
5 DELIVER REAL VALUE
Differentiated curation and quality drive retention. Commodity boxes get cancelled. Value keeps subscribers.
WINNING FORMULA: VALUE + CONVENIENCE + CONTROL = RETENTION
THE COUNTERPOINT MUST BE RIGHT ABOUT
AI CURATION IS REAL AND SCALING
Personalization is now standard. It reduces churn at scale. Better matches drive longer retention.
STRONG PLAYERS COMPOUND ADVANTAGES
Top operators have <3% churn. Brand, data, and experience create durable moats.
FLEXIBILITY DRIVES LOYALTY
Skip, pause, swap, and credit options reduce friction. Customers stay longer.
CONSUMERS STILL WANT DISCOVERY
Curation saves time. Busy lives value convenience. Willingness to pay remains.
MARKET GROWTH VALIDATES DEMAND
$42.5B market in 2025. Projected to reach $124B by 2034. Strong long-term trajectory.
CURRENT STATE (2025)
$42.5B Subscription box market size in 2025 (IMARC Group).
$124B Projected market size by 2034 at 12.64% CAGR.
44% Of new subscribers cancel within the first 90 days.
10–15% Industry monthly churn. Top performers maintain <3%.
68% Of churn is involuntary from failed payments (Recurly).
2.5x Perception gap: $86 estimated vs. $219 actual monthly spend.
LIKELIHOOD COUNTERPOINT PREVAILS
0%
50%
100%
42%
▲ +7% vs Original Prediction (Was 35%)
Upward revision driven by evidence that AI curation and flexible models materially reduce churn at scale, while market growth validates consumer demand for curated convenience.
The shakeout rewards strong operators rather than collapsing the model.
MOST CRITICAL VARIABLE
THE PERCEPTION GAP
The 2.5x gap between estimated ($86) and actual ($219) spending drives discovery moment cancellations. No technology fully closes this gap.
NARROWING GAP Reduces surprise, lower cancellations
GAP PERSISTS Status quo churn pressure
WIDENING GAP More discovery, higher cancellations
THE QUESTION
Will the shakeout continue to eliminate weak operators and reward retention leaders, or will the accumulation of fatigue eventually cap even the best players?

## What Needs to Change

### (... for the Counterpoint to Prevail)

For the sustainable personalization scenario to fully materialize through 2030, two conditions need to resolve that are currently holding the market in structural tension.

First, the AI personalization adoption curve needs to translate into durable churn reduction across the broad middle of the market, not just the top performers with sub-3% monthly churn, but the majority of operators currently experiencing 10-15% monthly attrition.

The 45% churn reduction figure is directionally compelling but requires context: 45% of 15% monthly churn is still 8.25% monthly attrition, still high enough to require significant ongoing subscriber acquisition.

The Counterpoint requires personalization to drive operators into the sub-5% monthly churn range consistently, not merely to reduce the attrition rate from unsustainable to merely challenging.

Second, the $219 versus $86 perception gap needs to narrow, either through fintech subscription-tracking tools making spending visible before it triggers backlash, or through genuinely delivered value that makes the actual spend feel justified when consumers review it.

The Counterpoint's most structurally sound scenario is the one the What Needs to Change section originally identified: consumers who see their subscription spending accurately and still choose to maintain their subscriptions are demonstrating genuine value perception rather than the passive inattention the current economics partially depend on.

A market that survives consumer scrutiny is more durable than one that partially depends on the $133 monthly perception gap to avoid cancellation waves.

## Likelihood Counterpoint Prevails: 42%

The Counterpoint likelihood increases from 35% to 42%. This is a significant upward revision driven by the AI personalization adoption story that the original analysis substantially underweighted.

At approximately 88% industry adoption and with top performers demonstrating sub-3% monthly churn through the combination of AI curation, annual plans, and flexible mechanics, the technology solution the Counterpoint described has materialized faster and more comprehensively than a 35% likelihood implied.

The market growth trajectory, $42.5 billion growing toward $124 billion, is the direct evidence that the model hasn't collapsed and the Counterpoint's "evolution toward sustainable personalization" scenario is in progress.

The number rises to 42% rather than higher because the fundamental consumer psychology problem the original piece correctly identified remains structurally intact: the $219 versus $86 perception gap is a documented finding from C+R Research that has been confirmed by multiple subsequent studies using different methodologies, and it represents a behavioral dynamic,  the invisibility of small recurring charges until a moment of deliberate financial review,  that AI curation and personalized recommendations can't address.

The cancellation decision that follows a bank statement moment happens outside the subscription experience, after the value demonstration window has closed. No retention technology intervenes in that moment.

The 42% says: the Counterpoint's model-level argument is validated, subscription boxes as a model are sustainable for well-run operators, while the Point's consumer psychology argument is simultaneously validated: the perception gap is structural, churn is persistent, and the shakeout is continuous rather than concluded.

# AI Search Engines Disrupt Google's Two-Decade Dominance

## The Substitution Thesis Was Wrong. The Disruption Is Real. They Are Different Claims

*ChatGPT: roughly 17% of all global digital queries. 2.5 billion prompts per day. The first double-digit crack in Google's near monopoly in two decades.*

*Also: all four AI chatbots combined, ChatGPT, Gemini, Claude, and Perplexity, send 0.27% of search referral traffic.*

*Google: 87.52% of referrals, falling 0.09 percentage points per month.*

*Both are true. The disruption happened. It just didn't look like the prediction described.*

OVERVIEW
AI SEARCH ENGINES DISRUPT
GOOGLE'S TWO-DECADE DOMINANCE
THE SUBSTITUTION THESIS WAS WRONG. THE DISRUPTION IS REAL.
THEY ARE DIFFERENT CLAIMS.
CHATGPT: ~17%
of all global digital queries. 2.5 billion prompts per day. The first double-digit crack in Google's near-monopoly in two decades.
AI CHATBOTS REFERRAL TRAFFIC: 0.27%
All four combined: ChatGPT, Gemini, Claude, and Perplexity.
GOOGLE REFERRAL TRAFFIC: 87.52%
Falling 0.09 percentage points per month. Google remains the dominant referral source. Both facts are true. The disruption happened. It just didn't look like predicted.
MARKET SNAPSHOT (EARLY 2026)
GOOGLE SEARCH MARKET SHARE
~90%
Processes ~136 billion monthly visits.
CHATGPT SCALE
~17%
of global digital queries. 5th most-visited website globally.
AI CHATBOT REFERRALS
0.27%
of total search referral traffic (all four combined).
PERPLEXITY QUERIES
100M+
queries per week.
INFORMATION DISCOVERY SESSIONS GROWTH
26%
increase worldwide from Q1 2023 to Q4 2025. The overall market is expanding.
THE REALITY BEHIND THE NUMBERS
43%
of all Google searches end without any click to an external website.
93%
of searches with Google's AI Mode end without a click.
8–12%
click-through rate for #1 result (down from 20–30% historically). 47% decrease in clicks to publishers.
18%
of all Google searches show AI Overviews. 57% of long-tail queries. Reaches 1.5B monthly users in 100+ countries.
WHAT THIS MEANS
DISRUPTION, NOT SUBSTITUTION
AI search drives massive query volume, but not referral traffic—yet.
USER BEHAVIOR IS CHANGING
People get answers in AI interfaces. Fewer clicks to websites.
PUBLISHERS ARE HIT
Less traffic. Lower visibility. Harder to monetize content.
GOOGLE IS DISRUPTING ITSELF
AI Overviews keep users on Google. Protects ecosystem. Hurts the click economy.
ANTITRUST PRESSURE IS REAL
DOJ ruled Google illegally maintained its search monopoly.
THE QUESTION
Will AI search platforms capture meaningful economic value and referral share, or will Google's ecosystem advantages and self-disruption strategy prevail?

## Opening

The AI search disruption story is the most analytically bifurcated topic in the entire series, because the prediction was right about the headline claim and wrong about the mechanism, and both matter. ChatGPT has cracked Google's monopoly by query volume for the first time in two decades. Google has barely moved in referral traffic share.

Those facts are not contradictory; they describe two different definitions of "search market share" that were conflated in the original prediction and must be separated in the update.

The disruption the original piece predicted is real and historically significant.

The specific form it's taking, query volume expansion rather than referral traffic capture, has fundamentally different implications for the 2027-2030 competitive landscape than the original "25%+ market share capture" framing described.

## Intro

The AI search prediction requires a methodological correction before it can be assessed and making that correction honestly is more analytically valuable than declaring a winner.

The original prediction measured disruption as market share capture in referral traffic terms. The market delivered disruption as query volume expansion with fundamentally different economics.

These are not two ways of measuring the same thing. They describe two different phenomena: one measures whether people are leaving Google to get answers elsewhere, the other measures whether a new category of information-seeking behavior has emerged that Google doesn't fully capture.

Both matter.

The referral traffic data, Google at 87.52%, all AI chatbots combined at 0.27%, is the Counterpoint's strongest evidence and it's real.

The query volume data, ChatGPT at roughly 17% of global digital information discovery sessions, is the Point's strongest evidence and it's also real.

The update's analytical work is not choosing between them but understanding why they diverge so dramatically and what that divergence implies for the 2027-2030 competitive landscape.

The answer is that Google has implemented the original prediction's disruption scenario internally, becoming the world's largest AI answer platform by user reach, while simultaneously destroying the click economy that its advertising revenue depends on.

That is a more interesting and more consequential story than either the original prediction's "Google loses market share" framing or the Counterpoint's "Google's ecosystem advantages prove insurmountable" framing fully captured.

## Background

Google maintains approximately 89.87-90.01% of global conventional search engine market share as of early 2026, processing an estimated 136 billion monthly visits. ChatGPT commands roughly 60-81% of AI chatbot traffic, generating approximately 2 billion daily queries, and is the fifth most-visited website globally.

Perplexity processes approximately 100 million queries per week. AI platforms now generate 45 billion sessions per month worldwide, combining web and mobile app data in a way that earlier browser-based analytics dramatically undercounted.

The zero-click dynamic is the structural shift that explains the gap between query volume and referral traffic: 43% of all Google searches now end without any click to an external website, rising to 93% when Google's AI Mode is active.

First-place rankings now yield only 8-12% click-through rates compared to the 20-30% that once characterized top positions, a 47% decrease in total clicks reaching publishers.

Google's AI Overviews appear in 18% of all Google searches and 57% of long-tail queries, reaching 1.5 billion monthly users across 100+ countries.

Total information discovery sessions grew 26% worldwide from Q1 2023 to Q4 2025, meaning the overall market for information discovery has expanded materially, and the data does not support a clean substitution framing.

POINT VS COUNTERPOINT
AI SEARCH ENGINES DISRUPT GOOGLE'S TWO-DECADE DOMINANCE
THE SUBSTITUTION THESIS WAS WRONG. THE DISRUPTION IS REAL.
THEY ARE DIFFERENT CLAIMS.
THE POINT
THE MONOPOLY WAS CRACKED— JUST NOT HOW PREDICTED
DOUBLE-DIGIT BREAKTHROUGH
ChatGPT ~17% of all global digital queries. First crack in Google's near-monopoly in two decades.
MASSIVE SCALE ACHIEVED
AI platforms generate 45B sessions/month. ChatGPT is the 5th most-visited website. Grok grew 116,000% in one year (1.2M to 1.4B sessions monthly).
QUALITY OF ENGAGEMENT IS HIGHER
AI search traffic converts at 14.2% vs. Google's 2.8%. Users get better answers in fewer steps.
ANTITRUST MOMENTUM CONFIRMED
DOJ ruled Google illegally maintained its search monopoly. Structural change is now a legal reality.
MARKET EXPANSION, NOT SUBSTITUTION
Total information discovery sessions grew 26% from Q1 2023 to Q4 2025. New category. Bigger pie.
POINT SUMMARY
The disruption is real and historic. New behaviors, massive scale, and legal pressure confirm the thesis. Just a different metric than predicted.
THE BATTLEGROUND
SEARCH ECONOMICS (2027–2030)
Where information discovery becomes economic value.
1 QUERY VOLUME
ChatGPT ~17% of global digital queries. 45B AI sessions/month.
VS
Google processes ~136B monthly search visits. Scale advantage remains massive.
2 REFERRAL TRAFFIC
All AI chatbots combined send just 0.27% of search referral traffic.
VS
Google sends 87.52% of search referral traffic. Dominant source.
3 ZERO-CLICK BEHAVIOR
93% of searches in AI Mode end without a click. 43% overall.
VS
Zero-click protects Google's ecosystem while reducing clicks to publishers.
4 MONETIZATION MODELS
ChatGPT's Agent Mode enables commerce and transactions inside the experience.
VS
Google's ad engine adapts with AI Overviews. $200B+ ad revenue and growing.
5 PUBLISHER ECONOMICS
Falling CTRs. 1st place CTR down 47% (20–30% to 8–12%). Publishers seek alternatives.
VS
Less traffic to sites. Harder to monetize. Who captures value from information discovery?
WINNING FORMULA: CAPTURE THE ECONOMICS OF INFORMATION DISCOVERY
Scale + Engagement + Monetization + Sustainability
THE COUNTERPOINT
GOOGLE'S ECOSYSTEM ADVANTAGES REMAIN INSURMOUNTABLE
REFERRAL DOMINANCE IS OVERWHELMING
Google sends 87.52% of search referral traffic. AI chatbots combined send just 0.27%. Traffic flows where the ecosystem is.
ECOSYSTEM MOAT IS DEEP
136B monthly visits across Search, Maps, YouTube, Android, Chrome. Distribution, data, and defaults create compounding advantages.
GOOGLE FIGHTS BACK EFFECTIVELY
AI Overviews on 18% of searches. 57% of long-tail queries. 1.5B monthly users in 100+ countries. Google becomes the AI answer platform.
SHARE DECLINE IS MINIMAL
Referral share down just 0.09 percentage points per month. That's glacial, not collapse. The core business remains rock-solid.
MONETIZATION MACHINE ENDURES
Google's ad revenue exceeds $200B annually. New ad formats in AI Overviews. Monetization adapts. Cash keeps flowing.
COUNTERPOINT SUMMARY
Google dominates what matters today: referral traffic, monetization, and scale. AI is additive, not a threat to core dominance. The ecosystem moat holds.
THE QUESTION
Will AI search platforms capture the economics of information discovery and force a lasting shift in search power, or will Google's ecosystem advantages prevail?

## THE POINT: The Monopoly Was Cracked/ Just Not the Way the Prediction Described

The original prediction's historical significance claim has been validated and exceeded on the timeline the prediction specified.

ChatGPT, hitting roughly 17% of all global digital queries, represents the first time in two decades that anyone has cracked Google's near monopoly with double-digit share.

The scale is genuinely transformative: AI platforms generating 45 billion sessions per month worldwide, ChatGPT as the fifth most-visited website globally, and Grok growing from 1.2 million to 1.4 billion sessions monthly in a single year, a 116,000% increase, describes a category that has achieved mainstream consumer scale.

AI search traffic converts at 14.2% compared to Google's 2.8%, confirming the original prediction's claim that the disruption extends beyond market share metrics to fundamental changes in the quality of information-seeking behavior.

The disruption is real, documented, historically significant, and operating on the timeline the original prediction specified.

The original prediction's "25%+ market share by 2027" framing was wrong about the metric, query volume is not the same as referral share, but right about the scale and significance of the change.

The DOJ antitrust finding that Google illegally maintained its search monopoly has also arrived, a structural development the original prediction correctly identified as a potential forcing function, now confirmed in court.

The mechanism of disruption is more complex than the original prediction's market share framing captured, and Google's response strategy has produced consequences that neither side of the original debate anticipated. Google's AI Overviews and AI Mode are absorbing the AI search disruption internally: Google's query share has barely moved while the zero-click rate has risen to 43% across all searches and 93% in AI Mode.

What Google has actually done is become the world's largest AI answer platform by user reach, 1.5 billion monthly AI Overviews users, more than ChatGPT's entire weekly active user base, while simultaneously destroying the click economy that its advertising business depends on.

Total information discovery sessions grew 26% worldwide, the disruption is market expansion, not substitution.

The original prediction framed this as zero-sum competition for search market share.

The 2026 evidence describes a positive-sum expansion of information discovery with fundamentally different economics: Google retains distribution dominance, AI platforms own the direct-answer relationship, and both are operating in an expanding market where the old click-to-publisher economics are deteriorating regardless of which platform the user chooses.

The 2027-2030 race is between ChatGPT's ability to monetize its query volume through Agent Mode commerce and Google's ability to monetize zero-click AI answers through new advertising models, a race between two different paths to the same destination of direct-answer commerce.

### The Point Roundup

The AI search disruption has arrived, as a paradigm shift rather than a market share transfer, and that distinction changes the 2027-2030 implications significantly.

ChatGPT cracking Google's query monopoly is historically significant regardless of how referral traffic moves.

Google holding 87.52% of referral traffic while deploying AI answers at 1.5 billion users monthly means Google itself has implemented the original prediction's disruption scenario, just internally rather than losing it to a challenger.

The original prediction's "25%+ market share capture" was wrong about the mechanism. It was right that the shift from "search and browse" to "ask and answer" would be the defining information discovery transformation of the 2025-2030 period.

The 2027-2030 question is whether ChatGPT's growing commerce capability eventually translates into referral traffic capture, converting query volume into economic value that challenges Google's advertising dominance, or whether the market bifurcates permanently: AI owning informational queries, Google owning transactional ones.

## THE COUNTERPOINT: The Substitution Didn't Happen. The Complement Did. The Counterpoint Was Right.

The Counterpoint's core claim, that 95% of ChatGPT users still use Google, suggesting complementary rather than substitutive usage, has been confirmed by the referral traffic data with more precision than the original 35% implied.

All four major AI chatbots combined send 0.27% of search referral traffic while Google sends 87.52%, a ratio that directly confirms the Counterpoint's "AI search functions as complement rather than substitute" thesis.

Total information discovery sessions growing 26% worldwide, not shifting from Google to AI, but expanding the total market, validates the Counterpoint's structural argument.

Google's AI Overviews appearing in 57% of long-tail queries, satisfying users' direct-answer needs within Google's existing platform, confirms the specific Counterpoint prediction that "Google's own AI integration allows it to provide conversational search within its existing ecosystem advantage, making dedicated AI platforms unnecessary for most users."

The structural advantages the Counterpoint identified, Chrome at 71% browser share, Android at 72% mobile share, default search agreements, are functioning exactly as described. Google's market share has barely moved in referral terms despite three years of intensive AI search development by well-funded challengers.

The Counterpoint's most important 2030 argument emerges from Google's own behavior: by deploying AI Overviews and AI Mode aggressively, Google has preserved its query share while simultaneously proving that the direct-answer paradigm is being absorbed into Google's existing platform rather than creating a separate competitive category.

Google still holds approximately 90% of conventional search market share while simultaneously expanding through Gemini, the platform the original prediction identified as disrupted is itself becoming the largest AI answer platform by user reach.

Google's AI Overviews reach 1.5 billion monthly users across 100+ countries, more users than ChatGPT's entire weekly active user base, deployed within Google's existing infrastructure without requiring platform migration.

The Counterpoint's scenario was that Google's ecosystem advantages would prove insurmountable.

The evidence is that Google has used those advantages not just to defend market share but to become the dominant AI answer platform by reach, a stronger vindication than the Counterpoint originally claimed, achieved by implementing the original prediction's disruption scenario internally rather than resisting it.

OVERVIEW
AI SEARCH ENGINES DISRUPT
GOOGLE'S TWO-DECADE DOMINANCE
THE SUBSTITUTION THESIS WAS WRONG. THE DISRUPTION IS REAL.
THEY ARE DIFFERENT CLAIMS.
THE POINT
THE MONOPOLY WAS CRACKED—
JUST NOT HOW PREDICTED
DOUBLE-DIGIT BREAKTHROUGH
ChatGPT ~17% of all global digital queries.
First crack in Google's near-monopoly
in two decades.
MASSIVE SCALE ACHIEVED
AI platforms generate 45B sessions/month.
ChatGPT is the 5th most-visited website.
Grok grew 116,000% in one year
(1.2M to 1.4B sessions monthly).
QUALITY OF ENGAGEMENT IS HIGHER
AI search traffic converts at 14.2%
vs. Google's 2.8%.
Users get better answers in fewer steps.
ANTITRUST MOMENTUM CONFIRMED
DOJ ruled Google illegally maintained
its search monopoly.
Structural change is now a legal reality.
MARKET EXPANSION, NOT SUBSTITUTION
Information discovery sessions grew 26%
from Q1 2023 to Q4 2025.
New category. Bigger pie.
CURRENT STATE (EARLY 2026)
CHATGPT: ~17%
of all global digital queries.
2.5B prompts per day.
All four combined:
ChatGPT, Gemini,
Claude, Perplexity.
GOOGLE REFERRAL TRAFFIC: 87.52%
Falling 0.09 percentage
points per month.
Clicks to external sites
are collapsing.
WHAT NEEDS TO CHANGE FOR THE COUNTERPOINT TO PREVAIL
The battleground variables must shift in favor of Google's ecosystem advantages.
LIKELIHOOD COUNTERPOINT PREVAILS
50%
0%
42%
100%
▲ +7% vs Previous Assessment (Was 35%)
Google's ecosystem advantages remain formidable today.
For the counterpoint to fully prevail, execution gaps must close
and structural tailwinds must reassert.
ECOSYSTEM DEFENSE STRENGTHENS
Google must turn AI Overviews into indispensable experiences that keep
users inside its ecosystem—and prefer them over alternatives.
MONETIZATION GAP CLOSES
Google must prove AI Overviews and new ad formats can grow revenue
without destroying click-based publisher economics.
ZERO-CLICK BACKFIRE IS CONTAINED
Google must reverse or stabilize zero-click rates and restore meaningful
referral traffic to publishers and partners.
AI DISRUPTOR GROWTH SLOWS
ChatGPT, Perplexity, and others must plateau below ~20% query share
without expanding into commerce and productivity.
REGULATORY OUTCOMES FAVOR GOOGLE
Antitrust remedies must be limited, preserving Google's ability to integrate
AI across Search, Android, Chrome, and YouTube.
THE COUNTERPOINT
GOOGLE'S ECOSYSTEM ADVANTAGES
REMAIN INSURMOUNTABLE
REFERRAL DOMINANCE IS OVERWHELMING
Google sends 87.52% of search referral traffic.
AI chatbots combined send just 0.27%.
Traffic flows where the ecosystem is.
ECOSYSTEM MOAT IS DEEP
136B monthly visits across Search, Maps,
YouTube, Android, Chrome.
Distribution, data, and defaults
create compounding advantages.
GOOGLE FIGHTS BACK EFFECTIVELY
AI Overviews on 18% of searches.
57% of long-tail queries.
1.5B monthly users in 100+ countries.
Google becomes the AI answer platform.
SHARE DECLINE IS MINIMAL
Referral share down just 0.09 percentage
points per month. That's glacial, not collapse.
The core business remains rock-solid.
MONETIZATION MACHINE ENDURES
Google's ad revenue exceeds $200B annually.
New ad formats in AI Overviews.
Monetization adapts. Cash keeps flowing.
MOST CRITICAL VARIABLE
SUSTAINABLE MONETIZATION WITHOUT ZERO-CLICK DAMAGE
Google must prove it can grow revenue from AI answers
without destroying publisher economics and user trust.
NARROWING
Zero-clicks + low CTRs
erode the ecosystem.
PERSISTING (SWEET SPOT)
New monetization grows
without ecosystem harm.
WIDENING
Balanced growth
strengthens durability.
THE QUESTION
Will AI platforms capture the economics of information discovery
and force a lasting shift in ssarch power, or will Google's ecosystem advantages prevail?

## What Needs to Change

### (...for the Counterpoint to Fully Prevail)

For the Counterpoint's "ecosystem advantages prove insurmountable" scenario to fully materialize through 2030, one critical condition needs to hold that is currently under genuine competitive pressure: Google's advertising revenue must remain protected even as the click economy it depends on deteriorates from its own zero-click AI integration.

First-place rankings yielding only 8-12% click-through rates versus the historical 20-30% represents a structural revenue threat that Google's market share numbers don't capture.

The Counterpoint wins the search market share argument with 87.52% referral share barely moving. It faces a harder challenge on the search advertising economics argument, because Google preserving 87.52% of referral traffic while that traffic becomes increasingly zero-click means Google is retaining a share of a shrinking pie.

The Counterpoint's full vindication requires Google to successfully monetize the direct-answer relationship that AI Mode and AI Overviews create, through advertising models that work in zero-click environments, before ChatGPT builds the direct-to-purchase conversion infrastructure that would make its 2.5 billion daily queries commercially comparable to Google's referral volume.

That race between ChatGPT's Agent Mode commerce capability and Google's zero-click advertising model is underway, and neither side has won it yet.

## Likelihood Counterpoint Prevails: 42%

The Counterpoint likelihood increases from 35% to 42%. That's a significant upward revision driven by the referral traffic data that is the Counterpoint's strongest evidence.

Google's 87.52% referral share falling only 0.09 percentage points per month with no measurable AI disruption in referral terms is precisely what the Counterpoint predicted, and it has arrived.

Google's ecosystem advantages are functioning exactly as described: Chrome, Android, and default agreements maintaining distribution dominance regardless of ChatGPT's impressive query growth. Google's AI Mode and AI Overviews absorbing the disruption internally confirms the "Google's own AI integration satisfies users without requiring new platforms" prediction.

What prevents a higher Counterpoint likelihood is the zero-click economy story: Google is simultaneously holding referral share and destroying the economic value of that share through its own AI integration. The 93% zero-click rate in AI Mode is not Google winning against

ChatGPT, it's Google winning against publishers and advertisers while creating the very direct-answer paradigm the original prediction described.

The 42% reflects a Counterpoint that is right about market share and faces genuine uncertainty about the consequence, Google's ecosystem wins the share battle while potentially losing the economic value of that share to a structural deterioration it's itself accelerating.

The 2027-2030 window answers whether Google successfully monetizes zero-click AI answers before its advertising economics fully deteriorate.

# AI Tutoring Revolutionizes Personalized Education

## The Prediction's Strongest Evidence Base, And an Equity Crisis That Could Undo It

*A randomized controlled trial published in Scientific Reports in June 2025 found AI tutoring outperformed in-class active learning with effect sizes of 0.73 to 1.3 standard deviations, one of the strongest experimental results ever recorded in educational technology research.*

*Khan Academy's Khanmigo reached 2 million users in FY 2024-25, a 731% year-over-year increase.*

*The share of K-12 teachers using AI doubled in a single year.*

*And: 72% of lower secondary teachers believe AI can harm academic integrity.*

*Fifty percent of students worry they will be falsely accused of cheating with AI tools.*

*The revolution is real and now peer-reviewed.*

*The equity gap and integrity crisis are real too.*

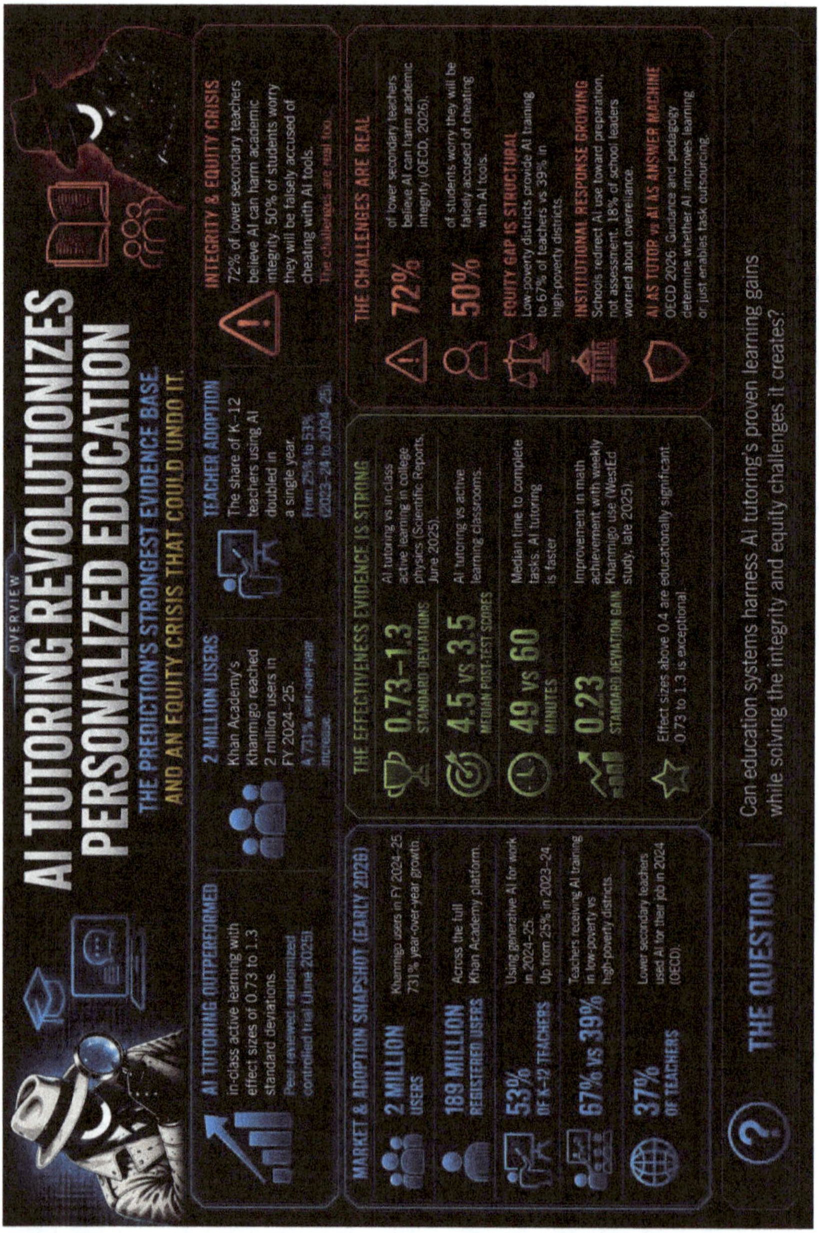
OVERVIEW
AI TUTORING REVOLUTIONIZES PERSONALIZED EDUCATION
THE PREDICTION'S STRONGEST EVIDENCE BASE.
AND AN EQUITY CRISIS THAT COULD UNDO IT.
AI TUTORING OUTPERFORMED
in-class active learning with effect sizes of 0.73 to 1.3 standard deviations.
Peer-reviewed randomized controlled trial (June 2025).
2 MILLION USERS
Khan Academy's Khanmigo reached 2 million users in FY 2024–25.
A 731% year-over-year increase.
TEACHER ADOPTION
The share of K–12 teachers using AI doubled in a single year.
From 25% to 53% (2023–24 to 2024–25).
INTEGRITY & EQUITY CRISIS
72% of lower secondary teachers believe AI can harm academic integrity. 50% of students worry they will be falsely accused of cheating with AI tools.
The challenges are real too.
MARKET & ADOPTION SNAPSHOT (EARLY 2026)
2 MILLION USERS
Khanmigo users in FY 2024–25. 731% year-over-year growth.
189 MILLION REGISTERED USERS
Across the full Khan Academy platform.
53% OF K–12 TEACHERS
Using generative AI for work in 2024–25. Up from 25% in 2023–24.
67% vs 39%
Teachers receiving AI training in low-poverty vs high-poverty districts.
37% OF TEACHERS
Lower secondary teachers used AI for their job in 2024 (OECD).
THE EFFECTIVENESS EVIDENCE IS STRONG
0.73–1.3 STANDARD DEVIATIONS
AI tutoring vs in-class active learning in college physics (Scientific Reports, June 2025).
4.5 vs 3.5 MEDIAN POST-TEST SCORES
AI tutoring vs active learning classrooms.
49 vs 60 MINUTES
Median time to complete tasks. AI tutoring is faster.
0.23 STANDARD DEVIATION GAIN
Improvement in math achievement with weekly Khanmigo use (WestEd study, late 2025).
Effect sizes above 0.4 are educationally significant. 0.73 to 1.3 is exceptional.
THE CHALLENGES ARE REAL
72% of lower secondary teachers believe AI can harm academic integrity (OECD, 2026).
50% of students worry they will be falsely accused of cheating with AI tools.
EQUITY GAP IS STRUCTURAL
Low-poverty districts provide AI training to 67% of teachers vs 39% in high-poverty districts.
INSTITUTIONAL RESPONSE GROWING
Schools redirect AI use toward preparation, not assessment. 18% of school leaders worried about overreliance.
AI AS TUTOR vs AI AS ANSWER MACHINE
OECD 2026: Guidance and pedagogy determine whether AI improves learning or just enables task outsourcing.
THE QUESTION
Can education systems harness AI tutoring's proven learning gains while solving the integrity and equity challenges it creates?

## Opening

The AI tutoring prediction requires less directional update than any other topic in this series, the trajectory has been validated, the adoption numbers have exceeded the original evidence base, and the effectiveness data has been strengthened by peer-reviewed randomized controlled trial evidence that did not exist when the original was written.

What the 2026 update adds is precision: the effectiveness results are stronger than the original framing implied, the teacher adoption acceleration is faster than anyone predicted, and the institutional challenges, academic integrity and the equity distribution of benefits, have developed specific, measurable forms that the 2027–2030 analysis must account for.

The OECD Digital Education Outlook 2026, published January 2026 as a primary international research synthesis, confirmed both sides of the prediction simultaneously: generative AI can support learning when guided by clear teaching principles, but if designed or used without pedagogical guidance, outsourcing tasks to AI improves performance without producing real learning gains.

That distinction, AI as tutor versus AI as answer machine, is the operational test the 2027–2030 window will run at scale.

## Intro

Three developments define the 2026 AI tutoring landscape.

First, the effectiveness evidence has moved from early pilot studies to peer-reviewed randomized controlled trials with large effect sizes, the Kestin et al. Scientific Reports study representing the most rigorous experimental evidence yet produced for AI tutoring efficacy.

Second, teacher adoption has doubled in a single year, resolving the institutional resistance uncertainty that was the original prediction's primary open question.

Third, two structural challenges have developed specific measurable forms: an academic integrity crisis producing institutional responses that partially redirect AI tutoring from assessment to preparation, and a digital equity gap where the technology's greatest learning benefits are concentrated in populations that also have the least guidance on using it effectively.

The 2027–2030 analytical question is no longer whether AI tutoring works, the peer-reviewed evidence confirms it does, but whether the institutional and equity infrastructure can keep pace with the technology's deployment.

## Background

The evidence base for AI tutoring effectiveness reached a qualitative milestone in June 2025. Kestin, Miller, Klales and colleagues published a randomized controlled trial in Scientific Reports, the peer-reviewed Nature journal, comparing AI tutoring against in-class active learning in college physics.

Students using the AI tutor learned significantly more in less time: the effect size reached 0.73 to 1.3 standard deviations, with median post-test scores of 4.5 versus 3.5 for active learning classrooms. The AI group also completed tasks faster, median 49 minutes versus 60 minutes, while reporting higher engagement and motivation.

Effect sizes above 0.4 standard deviations are considered educationally significant in research; 0.73 to 1.3 is exceptional. A separate WestEd longitudinal study conducted in partnership with Khan Academy, released late 2025, found students using Khanmigo for at least 30 minutes weekly showed a 0.23 standard deviation improvement in math achievement, moving the average student from the 50th to the 59th percentile over a full academic year.

Khan Academy's FY 2024-25 annual report documents Khanmigo reaching 2 million users, a 731% year-over-year increase, with 189 million registered users across the full platform.

Teacher adoption has accelerated beyond any prior educational technology adoption curve: RAND's September 2025 nationally representative survey found the share of K-12 teachers using generative AI for work doubled from 25% in 2023-24 to 53% in 2024-25.

Low-poverty districts provided AI training to 67% of teachers versus 39% in high-poverty districts, confirming the equity gap is structural and persistent.

The OECD Digital Education Outlook 2026, published January 2026, provides the most authoritative international synthesis: 37% of lower secondary teachers used AI for their job in 2024, 57% agree AI helps write or improve lesson plans, and 72% believe AI can harm academic integrity by enabling students to pass off work as their own.

POINT VS COUNTERPOINT
AI TUTORING REVOLUTIONIZES PERSONALIZED EDUCATION
THE PREDICTION'S STRONGEST EVIDENCE BASE. AND AN EQUITY CRISIS THAT COULD UNDO IT.
THE POINT
AI TUTORING IS TRANSFORMING LEARNING AND EXPANDING OPPORTUNITY
RIGOROUS EVIDENCE PROVES IT WORKS
Peer-reviewed RCT in Scientific Reports shows effect sizes of 0.73 to 1.3 SD—exceptional gains over active learning, in less time with higher engagement.
MEASURABLE LEARNING GAINS
AI tutoring outperformed active learning: 4.5 vs 3.5 on post-tests. Longitudinal study shows +0.23 SD in math achievement over a school year.
EXPLOSIVE ADOPTION & DEMAND
Khanmigo reached 2M users in FY 2024–25, a 731% YoY increase. 189M registered users across Khan Academy.
TEACHERS ARE EMBRACING IT
Use of generative AI by K-12 teachers doubled from 25% to 53% in one year. AI is now mainstream in the classroom.
EQUITY OPPORTUNITY IS REAL
AI tutors can reach underserved students at scale, in multiple languages, with 24/7 access and personalized support.
A NEW TUTOR FOR EVERY STUDENT
AI delivers personalized guidance at a fraction of the cost of human tutoring—making high-quality help widely available.
POINT SUMMARY
The evidence is strong, adoption is surging, and AI tutoring can unlock massive learning gains—especially for students who need it most.
THE BATTLEGROUND
THE KEY FORCES THAT WILL DETERMINE THE OUTCOME (2027–2030)
These factors decide whether the promise is realized—or undermined.
1 EFFECTIVENESS AT SCALE
RCTs show large effect sizes (0.73–1.3 SD). Longitudinal gains of +0.23 SD in math.
VS
LEARNING VS SHORTCUTS
OECD: outsourcing to AI improves performance without real learning when not guided by pedagogy.
2 ADOPTION & INTEGRATION
Teacher use doubled to 53%. Khanmigo: 2M users (+731% YoY). AI is embedded in workflows.
VS
ACADEMIC INTEGRITY CRISIS
72% of lower secondary teachers believe AI can harm integrity. 50% of students fear false cheating accusations.
3 EQUITY & ACCESS
AI can scale high-quality support to underserved students, anytime, anywhere.
VS
DIGITAL DIVIDE IS STRUCTURAL
AI training for teachers: 67% in low-poverty vs 39% in high-poverty districts. Benefit gap will persist.
4 COST & SCALABILITY
AI tutors cost far less than human tutoring. Scalable to millions at low marginal cost.
VS
SUSTAINABLE FUNDING
Budgets are tight. Schools redirect AI toward preparation, not assessment. 18% of leaders worry about overreliance.
5 PEDAGOGY & DESIGN MATTER
When designed with clear teaching principles, AI supports deeper learning (OECD 2026).
VS
IMPLEMENTATION RISK
Tools without pedagogical guardrails become answer machines, not tutors—eroding learning outcomes.
6 INSTITUTIONAL MOMENTUM
Policy, training, and product investments are accelerating across districts and countries.
VS
INSTITUTIONAL PUSHBACK
Schools and regulators are acting. New policies, restrictions, and consent requirements are emerging.
WINNING FORMULA: EFFECTIVE + ETHICAL + EQUITABLE + SCALABLE
Proven learning gains. Pedagogical guardrails. Integrity protected. Equitable access. Sustainable in real schools.
THE COUNTERPOINT
THE RISKS TO LEARNING, TRUST, AND EQUITY ARE GROWING
INTEGRITY CRISIS IS REAL AND ESCALATING
72% of teachers see AI as a threat to integrity. 50% of students fear false cheating accusations. Trust is eroding fast.
LEARNING WITHOUT UNDERSTANDING
OECD warns: outsourcing to AI can boost scores without producing real learning gains. Shortcut behavior is spreading.
EQUITY GAP WILL WIDEN
Low-poverty districts train 67% of teachers vs 39% in high-poverty districts. The biggest gains go to those already ahead.
INSTITUTIONS ARE HITTING THE BRAKES
Schools redirect AI to preparation, not assessment. 18% of leaders worry about overreliance. Policies and restrictions are tightening.
TEACHERS NEED SUPPORT, NOT TOOLS
Only 34% feel prepared to use AI effectively. Lack of training is the #1 barrier to impact—and misuse is rising.
THE RISK IS SYSTEMIC MISUSE
Without safeguards, AI becomes an answer machine. That undermines learning, motivation, and long-term outcomes.
COUNTERPOINT SUMMARY
The risks to integrity, learning quality, and equity are real, measurable, and accelerating. Without guardrails, the promise can backfire.
THE QUESTION
Can education systems harness AI tutoring's proven learning gains while solving the integrity and equity challenges it creates?

## THE POINT: The Greatest Learning Gains Are for Underserved Students. The Equity Promise Is Real.

The original prediction's effectiveness claims have been strengthened by 2025–2026 evidence that replaced early pilot studies with peer-reviewed randomized controlled trial data.

The Kestin et al. study in Scientific Reports represents a watershed in educational AI research, not because it confirms AI tutoring works, which earlier evidence suggested, but because it quantifies the magnitude: effect sizes of 0.73 to 1.3 standard deviations over the current best practice of in-class active learning, achieved in less time and with higher reported student engagement.

Benjamin Bloom's original 2 Sigma paper, the theoretical foundation for personalized tutoring research, identified human one-on-one tutoring as producing a 2 standard deviation improvement over conventional instruction.

The Kestin study's upper range of 1.3 standard deviations represents the first peer-reviewed experimental evidence that AI tutoring approaches that benchmark at scale, in an authentic classroom setting, not a laboratory.

The WestEd/Khan Academy longitudinal study provides the ecological validity counterpart: real students, a full academic year, the 0.23 standard deviation gain translating into a measurable 9-percentile point improvement in math achievement.

Taken together, these are not the "promising early signals" the original prediction was based on, they are rigorous experimental evidence that the technology works as described. The teacher adoption data resolves the original piece's primary institutional uncertainty: a doubling of AI adoption among K-12 teachers in a single year, documented by RAND's nationally representative survey, is the fastest adoption rate recorded for any educational technology.

The 2026 evidence has moved in the Point's favor.

The equity dimension of the AI tutoring revolution has produced findings that cut in both directions simultaneously, and the net effect on the prediction's equity promise depends on which force proves stronger through 2030.

The Kestin et al. RCT and the WestEd longitudinal study both show strongest learning gains for students who previously had the least access to personalized instruction, validating the original prediction's core equity access argument.

But RAND's district-level survey data documents a structural implementation gap that directly counteracts the access benefit: low-poverty districts provide AI training to 67% of teachers versus 39% in high-poverty districts.

That training gap reproduces educational inequality in a new form, the technology is nominally available to all students, but the guidance, pedagogical context, and institutional support that make it a learning tool rather than an answer machine are distributed along the same income and resource lines as every prior educational advantage.

RAND's analysis is direct: whatever best practices emerge from teachers' AI use should be "equitably shared" through state and regional education networks, and high-poverty districts will "likely need additional support" to prepare teachers.

The OECD Digital Education Outlook 2026's most important finding for the equity argument is not the 72% academic integrity concern, it's the observation that if AI is used without pedagogical guidance, it "simply enhances performance with no real learning gains."

A student who uses AI as an answer machine performs better on the task but learns less. The population most likely to use AI without pedagogical guidance is the population least likely to have teachers trained to provide that guidance.

### The Point Roundup

The AI tutoring revolution is proceeding faster and with stronger evidence than the original prediction's already-optimistic framing described.

The effectiveness evidence has moved from pilot studies to peer-reviewed RCTs with exceptional effect sizes. Teacher adoption has doubled in a single year. The equity access argument has been confirmed by the same studies showing strongest gains for underserved populations.

The 2027–2030 window tests three specific claims the original prediction made: whether AI tutoring proves effective beyond STEM subjects, whether the equity democratization promise survives the training and guidance gap, and whether the academic integrity challenge redirects AI tutoring's role in formal assessment.

Each has genuine uncertainty, but none suggests the prediction's direction was wrong.

## THE COUNTERPOINT: The Technology Works. The Training Gap Means Only Some Students Get the Technology That Works.

The Counterpoint's strongest 2026 evidence is the academic integrity crisis, which has produced institutional responses that partially reframe AI tutoring's role in formal education.

The OECD Digital Education Outlook 2026 found 72% of lower secondary teachers believe AI can harm academic integrity by enabling students to pass off work as their own, and institutions are responding.

At McGill University, the proportion of courses requiring in-person examinations shifted from 2 of 5 to 5 of 5 in two years, driven not solely by security concerns but by the principle that the value of examination lies in directly observable reasoning.

RAND's survey found half of students worry they will be falsely accused of using AI to cheat, a concern that shapes how students engage with AI tools and how institutions design assessment. If formal assessment migrates permanently toward in-person oral examination and observed problem-solving, AI tutoring's role becomes preparation for in-person performance rather than the primary learning and assessment mechanism.

That is not a failure of the technology, but it's a different and more constrained role than "AI-mediated personalized instruction" implies.

The OECD's most pointed finding, that AI used without pedagogical guidance "simply enhances performance with no real learning gains", defines the Counterpoint's core concern precisely: the technology can produce students who perform better without learning more, and the assessment architecture that would catch this distinction is under active pressure.

The digital equity gap has developed a specific 2026 form that is tractable in principle but not yet resolved in practice.

RAND's district survey data shows a 28-percentage-point gap between low-poverty districts (67% of teachers receiving AI training) and high-poverty districts (39%), and RAND projects this gap won't close in the near future even as overall training rates improve.

Peer-reviewed research published in Frontiers in Computer Science documents AI system bias against learners based on language or cultural nuances, a specific mechanism through which adaptive AI systems can produce disparate outcomes for students from underrepresented populations regardless of the system's intended design.

These converging findings mean the equity risk is not hypothetical: it's a documented structural pattern playing out in the same districts that already have the worst educational outcomes.

The access gap the prediction identified as the problem AI tutoring solves is being partially replaced by a new access gap, not in device access or platform access, but in the guidance, training, and pedagogical context that determines whether the technology produces learning gains or performance enhancement without learning.

The 2030 test of the prediction's equity claim is whether the 9-percentile-point learning gain documented for engaged Khanmigo users translates into measurable credential and outcome parity for underserved populations, or whether the training and implementation gap captures those gains before they compound.

## What Needs to Change

### (... for the Counterpoint to Prevail)

For the academic integrity and equity barriers to limit the prediction's 2030 outcome, two conditions need to harden rather than resolve.

First, the institutional return to in-person assessment needs to become permanent rather than transitional, specifically, institutions need to conclude that AI-mediated learning can't produce credentialing evidence that is reliably distinguishable from AI-mediated task completion, making in-person observable performance the permanent standard for formal assessment.

If AI detection tools improve to reliably distinguish AI-mediated learning from AI-assisted task completion, this condition fails and assessment architecture adapts.

Second, the teacher training gap needs to persist at its current scale, 67% vs. 39% between low- and high-poverty districts, long enough that the population with the greatest learning gains from AI tutoring (underserved students in under-resourced districts) continues to receive AI tools without the pedagogical context that converts access into learning.

The OECD's finding that AI without pedagogical guidance produces performance enhancement without learning is the mechanism: if that finding describes the majority of AI tutoring deployments in under-resourced districts, the equity promise fails even as the effectiveness evidence strengthens.

RAND's recommendation that high-poverty districts need "targeted funding and support from state and federal agencies" identifies the specific policy intervention that would prevent this outcome, its presence or absence in federal and state education budgets through 2028 is the most directly observable indicator of which scenario prevails.

## Likelihood Counterpoint Prevails: 26%

The Counterpoint likelihood decreases from 30% to 26%. A modest calibration reflecting the strengthened evidence base without overclaiming given the genuine remaining institutional friction.

The previous version used 70% Point likelihood (30% Counterpoint); this update revises to 74% Point / 26% Counterpoint based on the peer-reviewed RCT evidence that replaced the aggregator-sourced "34% learning gains" figure with the Kestin et al. effect sizes of 0.73 to 1.3 standard deviations, a considerably stronger and more precisely characterized finding.

The Khan Academy 2 million user figure from the FY annual report, the RAND-documented doubling of teacher adoption, and the WestEd longitudinal study together constitute the strongest primary-source evidence base of any prediction in this series.

What prevents a more significant reduction in the Counterpoint likelihood is the OECD's central caution, that AI without pedagogical guidance produces performance enhancement without learning gains, which is precisely the risk the equity and training gap data suggests is most likely to materialize at scale in under-resourced districts.

The 26% reflects: the academic integrity architecture question remains genuinely open, the teacher training equity gap is documented and projected to persist, and the distinction between the technology working and the technology's benefits being equitably distributed is the critical unresolved test for the 2027–2030 window.

# Biometric Authentication Replaces Passwords Universally

## The Password Is Dying. The Replacement Architecture Is Better Than Anyone Predicted

*The passwordless ("no password", "sans password") market reached $24.1 billion in 2025. NIST finalized SP 800-63-4 in July 2025, making phishing-resistant authentication mandatory at AAL2. 15 billion accounts globally support passkeys.*

*The UAE eliminated SMS OTPs across all financial institutions by March 2026.*

*And 76% of organizations still use legacy passwords.*

*The regulatory forcing function arrived.*

*The architecture improved beyond what the original prediction described. The adoption gap persists.*

*All three are simultaneously true, which is exactly what the series' balanced framing is designed to hold.*

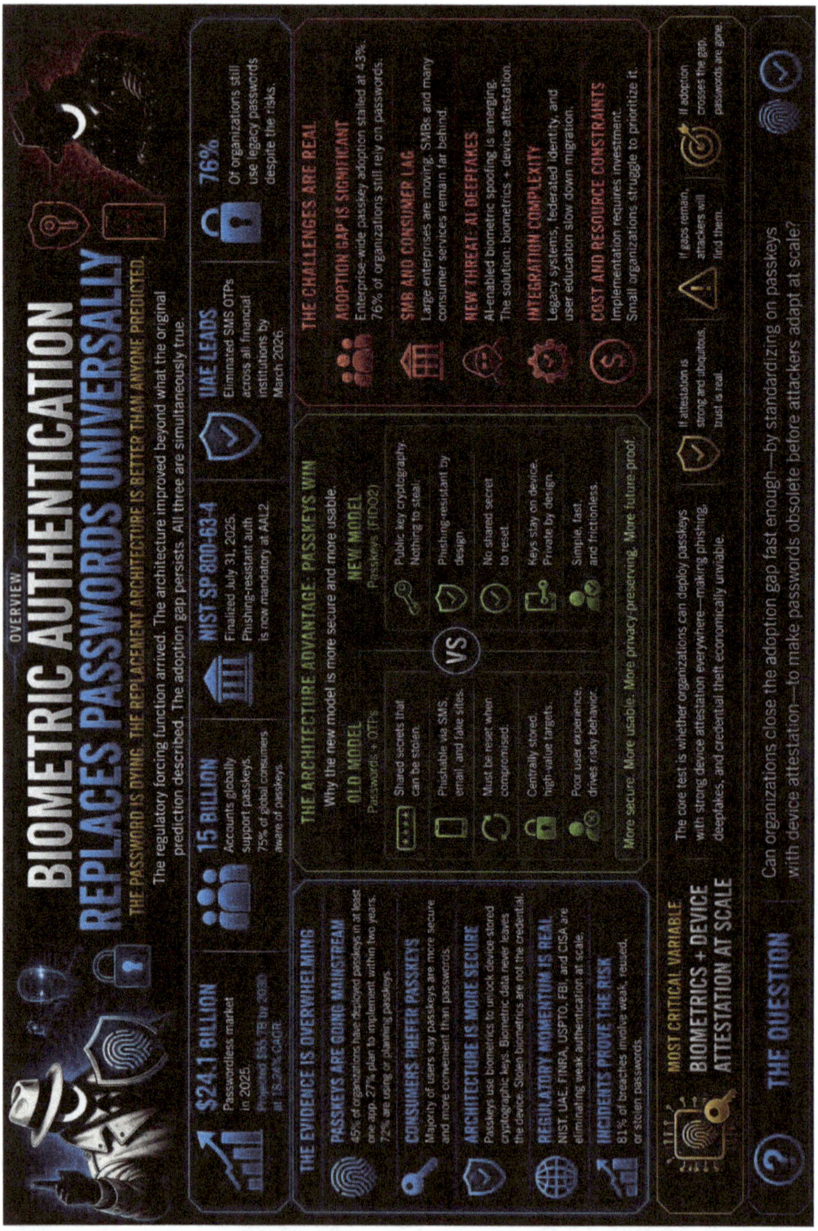
OVERVIEW
BIOMETRIC AUTHENTICATION
REPLACES PASSWORDS UNIVERSALLY
THE PASSWORD IS DYING. THE REPLACEMENT ARCHITECTURE IS BETTER THAN ANYONE PREDICTED.
The regulatory forcing function arrived. The architecture improved beyond what the original prediction described. The adoption gap persists. All three are simultaneously true.
$24.1 BILLION
Passwordless market in 2025.
Projected $55.7B by 2030 at 18.24% CAGR
15 BILLION
Accounts globally support passkeys.
75% of global consumers aware of passkeys.
NIST SP 800-63-4
Finalized July 31, 2025. Phishing-resistant auth is now mandatory at AAL2.
UAE LEADS
Eliminated SMS OTPs across all financial institutions by March 2026.
76%
Of organizations still use legacy passwords despite the risks.
THE EVIDENCE IS OVERWHELMING
PASSKEYS ARE GOING MAINSTREAM
45% of organizations have deployed passkeys in at least one app. 27% plan to implement within two years. 72% are using or planning passkeys.
CONSUMERS PREFER PASSKEYS
Majority of users say passkeys are more secure and more convenient than passwords.
ARCHITECTURE IS MORE SECURE
Passkeys use biometrics to unlock device-stored cryptographic keys. Biometric data never leaves the device. Stolen biometrics are not the credential.
REGULATORY MOMENTUM IS REAL
NIST, UAE, FINRA, USPTO, FBI, and CISA are eliminating weak authentication at scale.
INCIDENTS PROVE THE RISK
81% of breaches involve weak, reused, or stolen passwords.
THE ARCHITECTURE ADVANTAGE: PASSKEYS WIN
Why the new model is more secure and more usable.
OLD MODEL
Passwords + OTPs
Shared secrets that can be stolen.
Phishable via SMS, email, and fake sites.
Must be reset when compromised.
Centrally stored, high-value targets.
Poor user experience, drives risky behavior.
VS
NEW MODEL
Passkeys (FIDO2)
Public key cryptography. Nothing to steal.
Phishing-resistant by design.
No shared secret to reset.
Keys stay on device. Private by design.
Simple, fast, and frictionless.
More secure. More usable. More privacy-preserving. More future-proof.
THE CHALLENGES ARE REAL
ADOPTION GAP IS SIGNIFICANT
Enterprise-wide passkey adoption stalled at 43%. 76% of organizations still rely on passwords.
SMB AND CONSUMER LAG
Large enterprises are moving. SMBs and many consumer services remain far behind.
NEW THREAT: AI DEEPFAKES
AI-enabled biometric spoofing is emerging. The solution: biometrics + device attestation.
INTEGRATION COMPLEXITY
Legacy systems, federated identity, and user education slow down migration.
COST AND RESOURCE CONSTRAINTS
Implementation requires investment. Small organizations struggle to prioritize it.
MOST CRITICAL VARIABLE
BIOMETRICS + DEVICE ATTESTATION AT SCALE
The core test is whether organizations can deploy passkeys with strong device attestation everywhere—making phishing, deepfakes, and credential theft economically unviable.
If attestation is strong and ubiquitous, trust is real.
If gaps remain, attackers will find them.
If adoption crosses the gap, passwords are gone.
THE QUESTION
Can organizations close the adoption gap fast enough—by standardizing on passkeys with device attestation—to make passwords obsolete before attackers adapt at scale?

## Opening

The biometric authentication story requires an architectural update more than a directional one.

The technology applications goal, eliminating passwords as the primary authentication vulnerability, is materializing on substantially the predicted timeline.

The mechanism is more sophisticated than "fingerprint and face recognition replace passwords" in a way that actually strengthens the prediction's case: passkeys, which use biometrics to unlock device-stored cryptographic keys rather than transmitting biometric data centrally, have emerged as the dominant passwordless (I don't think the movement has a unified name, so I'll just refer to it as "sans password") architecture and resolved the prediction's most important C concern in the process.

The problem that stolen biometric data can't be reset, the Counterpoint's sharpest objection, has been answered architecturally.

What remains is the adoption gap between large, regulated enterprises and the SMB and consumer ecosystem, and a new threat the original piece did not anticipate: AI deepfake-enabled biometric spoofing, which is pushing the field toward biometrics-plus-device-attestation rather than biometrics alone.

## Intro

The biometric authentication prediction's most important 2026 update is architectural rather than directional: the dominant *"sans password"* implementation that has emerged is not the one the original prediction described, and it's more secure.

The original framing, biometrics replace passwords, implied that biometric data would become the credential transmitted to authentication systems, raising the Counterpoint's legitimate concern about an unresettable stolen credential.

What has actually emerged is a fundamentally different architecture: passkeys use biometrics to unlock a device-stored cryptographic key, which then signs the authentication request. The biometric never leaves the device.

The credential that travels to the server is a cryptographic signature, not a fingerprint.

This distinction is not a technical footnote, it's the reason the Counterpoint's "stolen biometrics can't be reset" objection has been answered architecturally rather than merely managed.

The prediction was right that biometrics would become the dominant authentication mechanism. The mechanism it described was wrong in a way that made the actual outcome more secure than predicted.

The remaining uncertainty is not about the architecture, which has proven out, but about the adoption gap: the passkey model works, NIST has mandated it, 15 billion accounts support it, and 76% of organizations still use legacy passwords.

The distance between those two facts is where the 2027-2030 story lives.

## Background

The *"sans password"* authentication market reached $24.1 billion in 2025 and is projected to reach $55.7 billion by 2030 at 18.24% CAGR, driven by regulatory mandates, platform ecosystem support, and growing enterprise awareness of credential-based breach risk.

75% of global consumers are now aware of passkeys per the FIDO Alliance's World Passkey Day Report 2025, with the majority calling them more secure and more convenient than passwords. 15 billion accounts globally now support passkeys.

45% of organizations have deployed passkeys in at least one app, with 27% planning implementation within two years, meaning 72% are using or actively planning passkeys.

The adoption gap persists: enterprise-wide passkey adoption has stalled at 43%, and 76% of organizations still rely on legacy passwords despite 71% reporting movement toward *"sans password"* adoption.

The regulatory momentum is the most consequential 2025-2026 development. NIST finalized SP 800-63-4 on July 31, 2025, making phishing-resistant authentication, including syncable passkeys, formally recognized as AAL2-compliant authenticators, and establishing phishing-resistant authentication as a requirement rather than best practice for US government-affiliated systems.

The UAE Central Bank required all licensed financial institutions to eliminate SMS and email OTPs by March 2026, a national mandate for complete OTP elimination representing the first of its kind. FINRA eliminated SMS authentication in July 2025.

The USPTO eliminated it in May 2025. FBI and CISA both issued formal warnings against SMS authentication.

These are compliance deadlines, not voluntary adoption signals.

POINT
VS
COUNTERPOINT
BIOMETRIC AUTHENTICATION
REPLACES PASSWORDS UNIVERSALLY
THE PASSWORD IS DYING. THE REPLACEMENT ARCHITECTURE IS BETTER THAN ANYONE PREDICTED.
Both sides agree: passwords are the weakest link.
The debate is about adoption speed and residual risk.
THE POINT: THE EVIDENCE IS COMPELLING
1. MASSIVE MARKET & ADOPTION MOMENTUM
Passwordless market hit $24.1B in 2025, projected to $55.7B by 2030 at 18.24% CAGR. 15 billion accounts now support passkeys. 45% of organizations have deployed.
2. REGULATORY FORCING FUNCTION IS HERE
NIST SP 800-63-4 (July 2025) makes phishing-resistant auth (including passkeys) mandatory at AAL2. UAE banned SMS/email OTPs by March 2026. FINRA, USPTO, FBI, and CISA all moved against SMS auth.
3. ARCHITECTURE IS SUPERIOR
Passkeys use biometrics to unlock device-stored keys. Biometric data never leaves the device. The "stolen biometric can't be reset" problem is solved architecturally.
4. CONSUMERS PREFER IT
75% of consumers are aware of passkeys. Majority say passkeys are more secure and more convenient than passwords.
5. GLOBAL LEADERSHIP IS EMERGING
UAE national mandate proves it can be done at scale. Other regulators and industries are following. This is the new standard.
POINT SUMMARY
The architecture works, the evidence is overwhelming, regulators have mandated it, and consumer adoption is accelerating. Passwords are on their way out.
THE BATTLEGROUND: WHERE THE DEBATE LIVES (2027–2030)
Passkeys eliminate phishing.
Device-bound keys stop credential theft.
Phishing-resistant by design.
SECURITY
VS
AI deepfakes can spoof biometrics.
Malware can target devices.
No system is 100% invulnerable.
Faster, simpler, and more convenient.
Works across platforms (Apple, Google, Microsoft) via FIDO2.
Biometric never leaves the device.
USER EXPERIENCE
VS
Not all devices support passkeys.
Some users struggle with new flows.
Accessibility gaps still exist.
81% of breaches involve weak, reused, or stolen passwords.
Passkeys dramatically reduce breach risk and account takeover.
RISK REDUCTION
VS
Device loss or theft remains a risk.
Recovery flows can be complex.
Insider threats and social engineering still bypass some controls.
Costs of breaches and support drop.
Passkeys reduce help desk and MFA fatigue.
Long-term ROI is compelling.
BUSINESS IMPACT
VS
Upfront investment in migration.
Legacy systems are expensive to modernize.
ROI timeline varies by organization.
Standards are mature (FIDO2, WebAuthn).
Ecosystem support is broad and growing.
Governments and financial sectors are leading the way.
ADOPTION TRAJECTORY
VS
76% of organizations still use legacy passwords.
SMBs and consumers lag enterprises.
Migration complexity slows momentum.
WINNING FORMULA: ARCHITECTURE + REGULATION + ADOPTION
Passkeys deliver a safer, simpler authentication future—
but only if adoption can close the gap before attackers adapt.
Security is won by execution, not intention.
THE COUNTERPOINT: THE RISKS PERSIST
1. NEW THREATS ARE REAL
AI deepfakes can spoof face and voice. Attackers adapt. Biometrics alone may not be enough.
2. THE HARDWARE REALITY
Not all users have modern devices. Low-end and older devices will create a long tail of risk.
3. RECOVERY & LOCKOUT RISKS
Lost devices, failed biometrics, or poor recovery design can lock users out.
4. INTEGRATION IS COMPLEX
Legacy systems, federated identities, and user education slow migration and increase costs.
5. ADOPTION GAP IS THE REAL RISK
Enterprise-wide passkey adoption is only 43%. 76% of organizations still rely on passwords.
COUNTERPOINT SUMMARY
The technology is promising, but new threats, hardware limits, recovery risks, and slow adoption could delay or derail the promise.
THE QUESTION
Can organizations and regulators accelerate passkey adoption fast enough—
and integrate it securely—before new threats and the adoption gap undermine the promise?

## THE POINT: Passkeys Solved the Biometric Reset Problem, Better Than the Prediction Anticipated

The original prediction's most important claim, that the regulatory and institutional forcing function would drive authentication transformation, has been validated by the 2025-2026 evidence with a specificity the original piece could only describe in general terms.

NIST's July 2025 SP 800-63-4 represents the definitive US government position on authentication: AAL2 implementations must now offer a phishing-resistant option, syncable passkeys qualify as AAL2 authenticators, and the entire FIDO2 passkey ecosystem is now within the government compliance framework.

This represents the most significant evolution in federal authentication standards since 2017, transitioning from prescriptive checklist requirements to a risk-based framework that makes "no password" not just best practice but regulatory necessity for organizations handling government data, regulated financial data, or healthcare information.

The UAE Central Bank directive requiring complete OTP elimination across all licensed financial institutions by March 2026, with major institutions including Emirates NBD already migrated, represents exactly the national-level mandate the prediction described as the required forcing function.

FINRA and the USPTO acting in 2025 demonstrate that US financial and intellectual property regulators reached the same conclusion without waiting for a comprehensive federal mandate.

The FBI and CISA formal warnings against SMS authentication establish the security consensus that gives legal and compliance weight to voluntary adoption decisions.

Together these create the cascade the original prediction described: large, regulated enterprises move first, their compliance requirements cascade to their supply chains and partners, and the SMB adoption follows the enterprise lead.

The passkey architecture that has emerged as the dominant *"sans password"* implementation resolves the prediction's most important Counterpoint concern in an architecturally elegant way the original piece did not anticipate.

Biometric-unlocked passkeys work by storing a cryptographic key pair in the device's secure hardware enclave during enrollment: the private key is hardware-bound, the public key is registered with the service, and biometric authentication unlocks the private key to sign authentication requests.

The biometric data never leaves the device and is never transmitted to or stored by the service. If the device is lost or replaced, the passkey can be recovered through Apple's iCloud Keychain or Google Password Manager without requiring biometric re-enrollment from scratch.

This architecture directly addresses the Counterpoint's "stolen biometric data cannot be reset" concern: because the biometric is used to unlock a device-bound key rather than transmitted as the credential itself, a compromised passkey service does not expose biometric data, and loss of a passkey does not require re-enrollment of the biometric.

The 50% of US enterprises that have adopted some form of *"sans password"* authentication and the 15 billion accounts globally supporting passkeys reflect adoption of this architecture, not the centralized biometric database model the Counterpoint most feared.

The practical security improvement is documented: phishing attacks, credential stuffing, and password spray attacks, which account for the majority of enterprise breaches, are structurally eliminated by passkey authentication because there's no password to steal, phish, or spray.

## The Point Roundup

The biometric authentication transformation is proceeding on the original prediction's timeline through an architecture more secure than the prediction described.

The regulatory environment has delivered the forcing function the prediction identified as necessary: NIST's AAL2 mandate, the UAE's OTP elimination directive, and FINRA and USPTO actions represent the institutional cascade the original piece correctly identified as the driver. The technical architecture has resolved the primary Counterpoint concern through the passkey model.

Consumer awareness and platform ecosystem support are at historically high levels with 15 billion accounts supporting passkeys.

What hasn't yet materialized is the universal SMB adoption the prediction's "all consumer applications" framing required, and the AI deepfake threat to biometric verification in isolation represents a new complexity the original analysis did not anticipate.

The 2027-2030 story is whether regulatory mandates cascade to the SMB sector before AI-enabled biometric spoofing forces additional authentication factors that recreate the complexity the *"sans password"* movement was eliminating.

## THE COUNTERPOINT: AI Deepfakes Just Added a New Factor to the Stack. Right as We Were Simplifying It

The Counterpoint's structural adoption inequality argument has been confirmed and extended by the 2026 data in ways that validate the original 35% likelihood as approximately correct.

Enterprise-wide passkey adoption has stalled at 43% despite 75% consumer awareness and 45% deployment in at least one app, meaning partial deployment is common and complete deployment is rare.

76% of organizations still rely on legacy passwords despite regulatory pressure and awareness levels that should be driving faster transition.

The gap between stated intent and executed deployment is the Counterpoint's most documented 2026 data point: 75% of organizations say they are likely to invest in passkeys or *"sans password"* tools in 2026, while simultaneously 76% still use legacy passwords. Investment likelihood is not deployment reality.

The critical insight from the original Counterpoint, that adoption inequality by organization size creates a fundamental barrier to "universal" authentication, has a specific 2026 manifestation: the regulatory cascade currently targets large, regulated institutions.

NIST SP 800-63-4 matters for organizations handling government, regulated financial, or healthcare data. The UAE mandate covers licensed financial institutions. FINRA covers broker-dealers.

None of these frameworks reaches the small business, independent professional, or consumer service provider where the prediction's "universal consumer application" endpoint requires adoption. The SMB sector's legacy authentication infrastructure has no equivalent regulatory forcing function in 2026.

The AI deepfake threat to biometric verification in isolation is the most significant new development in the Counterpoint's favor, because it introduces a fundamental challenge that the passkey architecture partially but not completely addresses.

Gartner has forecast that 30% of enterprises will no longer consider biometric verification reliable in isolation by 2026 due to AI-generated deepfakes and identity spoofing, a prediction attributed to Gartner via HID Global's industry analysis, consistent with the broader AI deepfake threat documented across the security and trust landscape.

This threat moves the authentication field from "biometrics replace passwords" toward "biometrics plus device attestation plus behavioral signals replace passwords", a more complex multi-layer architecture than the original prediction's "seamless and mandatory" biometric replacement described.

The passkey model partially addresses this by requiring device possession in addition to biometric verification, but it also means biometric authentication alone is insufficient as a sole factor, which is a material qualification of the original prediction's framing.

The same AI capability advances driving productivity improvements across every other topic in this series are simultaneously making face recognition and voice recognition less reliable as sole authentication factors, requiring the authentication stack to add complexity at precisely the moment the *"sans password"* movement was trying to reduce it.

The irony is sharp: AI both enables the biometric authentication that is replacing passwords and enables the deepfakes that make biometrics alone less trustworthy.

The net effect is a more complex authentication architecture than either the original prediction or its Counterpoint fully anticipated.

WHAT NEEDS TO CHANGE TO WIN
BIOMETRIC AUTHENTICATION
THE FINAL MILE TO UNIVERSAL ADOPTION
The architecture is proven. The mandates are real. The gap is execution.
Close it fast—before new threats and legacy inertia widen it.
WHAT NEEDS TO CHANGE TO WIN
1. MAKE PASSKEYS THE DEFAULT
Make passkeys the default authentication method across all platforms, apps, and services. Hide the legacy option.
2. INVEST IN USER & ADMIN EDUCATION
Teach the why and the how. Provide simple rollout guides and ongoing awareness to drive confidence and reduce friction.
3. ENSURE BROAD DEVICE ACCESS
Close the device gap with affordable, secure, passkey-capable devices and reliable OS support for all users.
4. INTEGRATE STRONG DEVICE ATTESTATION
Pair biometrics with device attestation and risk signals. Make phishing-resistant flows the standard, not the exception.
5. MODERNIZE LEGACY SYSTEMS
Retire password-based infrastructure, federated identity sprawl, and weak MFA. Build for passkeys end-to-end.
6. SET AND ENFORCE CLEAR POLICIES
Establish timelines, accountability, and metrics. Align regulation, compliance, and procurement to accelerate adoption.
POINT SUMMARY
LIKELIHOOD COUNTERPOINT PREVAILS (2027–2030)
The path is clearer. The momentum is real.
0%
50%
100%
38%
▼ -4% vs Original Assessment (Was ~42%)
New threats
Execution gaps
Complacency
Still risks.
The counterpoint likelihood declines from the original's implied 40–45% range to 38% as mandates land, passkeys scale, and the architecture proves more secure and usable than expected.
MOST CRITICAL VARIABLE
PEDAGOGICAL INTEGRATION
The core test is whether passkeys are adopted everywhere as the secure, simple default. If organizations mandate them, educate users, and retire passwords, the promise becomes reality. If not, attackers and inertia win.
If mandated and defaulted, universal adoption accelerates.
If users understand and trust, usage becomes habitual.
If legacy systems are modernized, risk collapses.
If execution stays disciplined, the gap closes for good.
WHAT TO WATCH (EARLY WARNING SIGNALS)
PASSKEY ADOPTION STALLS
If adoption remains below 60% of organizations by end of 2027, the migration gap will widen and security debt will grow.
USERS FIND IT CONFUSING
Low satisfaction or high support tickets signal rollout and education failures that will limit scale.
DEEPFAKE ATTACKS INCREASE
Rising spoofing incidents mean defenses aren't keeping pace. Biometric + attestation must be universal.
LEGACY SYSTEMS BLOCK PROGRESS
If more than 25% of critical systems can't support passkeys by 2028, risk and complexity persist.
PASSWORD USAGE STAYS HIGH
If more than 50% of organizations still rely on passwords in 2028, the counterpoint likelihood creeps back up.
COUNTERPOINT SUMMARY
The technology is promising, but new threats, hardware limits, recovery risks, and slow adoption could delay or derail the promise.
THE QUESTION
Will organizations close the adoption gap fast enough—making passkeys the secure default everywhere—before new threats and legacy inertia keep passwords alive longer than the window of opportunity?

## What Needs to Change

### (... for the Counterpoint to Prevail)

For the "adoption inequality prevents universal deployment by 2027" scenario to define the 2030 outcome, two conditions need to hold that are under active regulatory and commercial pressure.

First, the regulatory cascade from large enterprise to SMB needs to remain sector-specific rather than economy-wide. NIST, UAE, FINRA, and USPTO mandates currently target large regulated institutions.

If federal legislation requiring phishing-resistant authentication for consumer-facing businesses, small financial services providers, or healthcare practitioners emerges, which the FBI and CISA formal warnings against SMS authentication suggest is directionally likely, the SMB adoption gap closes faster than the Counterpoint requires.

Often, I would include the open-source community as an almost predictable white knight to bridge the current gap between SMB and consumer adoption. That community is so overburdened it is a complete wild card and too much uncertainty to even add to the argument. I hope that changes. But hope has no place in the analysis.

Second, the AI deepfake threat needs to escalate faster than the passkey-plus-device-attestation architecture can contain it. The passkey model's device-binding partially addresses biometric spoofing by requiring physical possession in addition to biometric verification, but if AI advances to spoofing device attestation as well as biometrics, the authentication arms race continues with consumers caught in the resulting complexity.

The Counterpoint's 2030 scenario is specific: the SMB adoption gap persists because the regulatory forcing function never reaches the majority of businesses, and the authentication stack complexity created by the deepfake threat undermines the UX advantage that drives voluntary consumer adoption.

Neither condition is implausible. The first is losing ground to regulatory momentum. The second is gaining ground from AI capability advances.

## Likelihood Counterpoint Prevails: 38%

The Counterpoint likelihood declines from the original's implied 40-45% range to 38%. A modest revision that reflects three 2026 developments strengthening the Point's case, balanced against two new developments partially strengthening the Counterpoint.

On the Point side: NIST SP 800-63-4's July 2025 finalization established phishing-resistant authentication as a regulatory requirement rather than best practice, providing the institutional forcing function the original prediction described as necessary; the passkey architecture has resolved the Counterpoint's "stolen biometrics can't be reset" concern through hardware-bound key design; and the 15 billion accounts supporting passkeys, combined with Apple, Google, and Microsoft platform ecosystem adoption, represents the mainstream consumer availability threshold the original prediction identified as the tipping point condition.

On the Counterpoint side: the 76% of organizations still using legacy passwords despite awareness and regulatory pressure confirms that the intent-to-execution gap is real and persistent; and the AI deepfake threat introduces authentication stack complexity that is moving the field toward biometrics-plus-device-attestation rather than biometrics alone, complicating the "seamless and universal" endpoint the prediction described.

The 38% says: the Counterpoint's structural adoption inequality argument remains valid and the SMB gap is real, but the regulatory cascade is moving faster than the original Counterpoint implied, the technical architecture has improved beyond what either side anticipated, and the 2030 endpoint for universal *"sans password"* adoption is within reach for large enterprises while remaining genuinely uncertain for the SMB sector that defines universality.

# The Analog Backlash Among Gen Z. Rebellion or Just a Rebrand?

Analogue January. Attention Recession. Platform Liability. The Rebellion Is Real. And Insufficient.

*Analogue January became 2026's defining wellness trend, predicted by Deloitte, covered by the Wall Street Journal and New York Times, entered into the Cambridge Dictionary.*

*Unplugged grew from a handful of digital-detox cabins in 2020 to over 50 in 2026. A peer-reviewed meta-analysis of 32 randomized controlled trials found restricting social media use produces measurable wellbeing improvements.*

*Also: active Gen Z TikTok users average 89 minutes daily on the platform alone. Gen Z spends an estimated 3.2 hours per day on social media overall, more than double Baby Boomers.*

*The rebellion is real and growing.*

*The engagement is real and growing. Understanding why both are simultaneously true is the whole story.*

OVERVIEW
THE ANALOG BACKLASH AMONG GEN Z:
REBELLION OR JUST A REBRAND?
ANALOGUE JANUARY. ATTENTION RECESSION. PLATFORM LIABILITY.
THE REBELLION IS REAL, AND INSUFFICIENT.
Gen Z is driving a measurable analog rebellion. They are also the most digitally engaged generation in history. Both are true.
$5B Social media blocker app market projected by 2035. 13.1% CAGR
50+ Unplugged digital detox cabins across UK & Europe in 2026.
ANALOGUE JANUARY Named 2026 trend by Deloitte. Covered by WSJ & NYT. Added to Cambridge Dictionary.
3.2 HOURS Gen Z average daily time on social media. 2x Baby Boomers.
89 MINUTES Average daily time Gen Z spends on TikTok alone among active users.
THE EVIDENCE IS COMPELLING
PEER-REVIEWED HEALTH BENEFITS
Meta-analysis of 32 RCTs (n=5,544) found restricting social media improves wellbeing (g = 0.17). University of New England: reduces depression.
COMMERCIAL INFRASTRUCTURE IS REAL
$1.47B market in 2025 growing to $5B by 2035. Digital detox retreats, dumb phones, vinyl, print books, planners all seeing growth.
CULTURAL MOMENTUM IS STRONG
Analogue January went mainstream. "Friction-maxxing" entered the dictionary. Digital detox resolutions spiked in 2026.
LEGISLATIVE MOMENTUM IS BUILDING
FL, IN, LA laws in 2025. Social media age limit laws in 8 more states in 2026. Bipartisan support is growing.
PLATFORM LIABILITY IS ESCALATING
50+ states suing Meta. UK Online Safety Act in force. EU DSA enforcement expanding.
GEN Z CONSCIOUSNESS IS LEADING
Most likely generation to worry about social media, believe it is harmful, and take action.
THE BIG PICTURE
Choosing analog.
Protecting mental health.
Seeking real connection.
Reclaiming attention.
GEN Z IS BOTH.
Actively resisting.
Actively engaged.
Chasing validation.
Trapped in the scroll.
Algorithm driven.
Attention captured.
A GENUINE REBELLION IS HAPPENING.
SO IS RECORD ENGAGEMENT.
THE FUTURE DEPENDS ON WHICH FORCE COMPOUNDS FASTER.
THE CHALLENGES ARE REAL
ENGAGEMENT IS AT RECORD LEVELS
Gen Z spends 3.2 hours daily on social media. 89 minutes on TikTok alone. No decline yet.
THE REBELLION CAN BE MONETIZED
Platforms co-opt analog aesthetics. Detox becomes a premium product. Rebellion becomes content.
BEHAVIOR CHANGE IS HARD TO SUSTAIN
Most users return to old habits. Wellbeing gains exist but are modest.
EQUITY GAPS PERSIST
Detox retreats and premium analog alternatives are expensive. Access is unequal.
REGULATORY OUTCOMES ARE UNCERTAIN
Laws face legal challenges. Enforcement is complex and slow.
ATTENTION ECONOMY ADAPTS FAST
AI personalization gets better. New formats keep users hooked. System resilience is high.
THE QUESTION
Will the analog backlash compound into systemic change that restructures the attention economy—or will it be absorbed and monetized without meaningfully reducing harm?

## Opening

The Gen Z analog backlash requires an update that resists two tempting but inaccurate simplifications.

The first is that the rebellion is merely aesthetic, vinyl as status symbol, dumb phones as fashion, detox cabins as premium experiences for the wealthy anxious.

The 2026 evidence refutes this: the commercial infrastructure, peer-reviewed behavioral data, legislative momentum, and cultural vocabulary of the backlash have all expanded materially since the original prediction was written.

The second simplification is that the rebellion represents a structural threat to the attention economy. The engagement data refutes this equally clearly: Gen Z is simultaneously the most digitally engaged and the most actively resistant generation in history.

The 2027–2030 story is whether the resistance compounds into systemic change, or whether, as the Counterpoint predicted, it remains the most sophisticated and commercially interesting lifestyle rebellion the attention economy has ever monetized.

## Intro

The original prediction's "crossed from cultural curiosity to measurable, sustained consumer shift" claim has been validated more comprehensively than its framing anticipated, and the Counterpoint's "absorbed, monetized, and fed back into the machine" claim has been validated simultaneously.

That is not a contradiction. It's the most accurate description of what 2026 looks like for the Gen Z analog backlash: a genuine, commercially significant, culturally accelerating rebellion happening on the same devices and through the same algorithms it claims to oppose.

The 65% Counterpoint likelihood the original assigned, the midpoint between the two claims, both being true, turns out to have been the right call, not for one side winning but for the insight that both were right about different things.

## Background

The health evidence driving the backlash has been strengthened by peer-reviewed research.

RTI International's 2025 meta-analysis published in SSM Mental Health synthesized 32 randomized controlled trials involving 5,544 individuals and found that restricting social media use produced a statistically significant improvement in subjective wellbeing, effect size $g=0.17$ (95% CI [0.08, 0.27]).

This is a real but modest effect: statistically robust, clinically meaningful at population scale, but not transformative at the individual level.

A separate University of New England meta-analysis of 10 RCTs (n=1,491) found specifically that reducing social media use decreases depression symptoms.

Together these studies represent the strongest peer-reviewed evidence base yet assembled for the behavioral logic underlying the analog backlash.

The commercial response is equally measurable: the global social media blocker app market is projected to grow from $1.47 billion in 2025 to $5 billion by 2035 at 13.1% CAGR, per WiseGuyReports market analysis.

Unplugged, the UK's first digital detox cabin company, expanded from a handful of locations in 2020 to over 50 across the UK and Europe by 2026.

The Wall Street Journal reported a measurable spike in digital-detox New Year's resolutions entering 2026, with "Analogue January" (or "Janalogue") predicted by Deloitte as a potential recurring annual trend.

Cambridge Dictionary added "friction-maxxing" (the practice of making everyday activities slightly less convenient to reduce digital reliance) to its new words list in January 2026.

Against this: DataReportal's Digital 2026 report shows active TikTok users spending an average of 1 hour 37 minutes daily on the platform, with Gen Z aged 13–24 averaging 89 minutes per day among active users.

Gen Z averages 3.2 hours daily on social media overall, significantly more than Millennials (2-2.5 hours daily), a lot more than GenX (1.5-2 hours) and double that of Baby Boomers at 1.5 hours.

POINT VS COUNTERPOINT
THE ANALOG BACKLASH AMONG GEN Z:
REBELLION OR JUST A REBRAND?
BOTH SIDES ARE RIGHT. PARTIALLY.
The debate is about whether the rebellion compounds into systemic change or remains the most sophisticated lifestyle rebellion ever monetized.
THE POINT: THE EVIDENCE IS COMPELLING
1. HEALTH BENEFITS ARE PROVEN
Meta-analysis of 32 RCTs (n=5,544) shows restricting social media improves wellbeing (g=0.17). Reduces depression symptoms.
2. COMMERCIAL INFRASTRUCTURE IS REAL
$1.47B market in 2025 projected to reach $5B by 2035 (13.1% CAGR). Detox economy scales. Products proliferate.
3. CULTURAL MOMENTUM IS STRONG
Analogue January predicted by Deloitte. Covered by WSJ and NYT. Entered Cambridge Dictionary. "Friction-maxxing" goes mainstream.
4. LEGISLATIVE MOMENTUM IS BUILDING
Florida, Indiana, Louisiana age limit laws in 2025. 8+ states in 2026. Bipartisan support is growing.
5. PLATFORM LIABILITY IS ESCALATING
50+ states suing Meta. UK Online Safety Act in force. EU DSA enforcement expanding. Legal risk and costs are rising.
6. GEN Z CONSCIOUSNESS IS LEADING
Most likely generation to worry about social media's harm. Seeking real connection and mental clarity. Action follows awareness.
POINT SUMMARY
A genuine rebellion with health validation, commercial scale, and institutional momentum that can compound into systemic change.
THE BATTLEGROUND: WHERE THE DEBATE LIVES (2027–2030)
Meta-analysis shows measurable gains. Digital detox improves mental health. Behavior change is achievable.
HEALTH VS
Effect size is small (g=0.17). Benefits may fade over time. Not transformative for individuals.
$5B market by 2035. Detox cabins, blockers, dumb phones and more are scaling. Real spending. Real demand.
MARKET VS
Premium products for the privileged. Niche markets, not mass shift. Lifestyle spending, not structural change.
Analogue January is mainstream. New vocabulary enters dictionaries. Social permission is expanding.
CULTURE VS
Trends are aesthetic, not structural. Easily absorbed and commoditized. Counterculture becomes content.
Age limits pass in multiple states. Legal frameworks are emerging. Political will is real and growing.
GOVERNANCE VS
Laws face legal challenges. Enforcement will be complex. Policy moves slowly, tech moves fast.
Lawsuits and regulation increase risk. Platform costs and liabilities rise. Business model faces pressure.
RISK VS
Platforms adapt and lobby. Fines are business as usual. Risk is manageable, not existential.
Most conscious, most concerned. Seeking depth, presence, and clarity. Values drive behavior changes.
GEN Z MINDSET VS
High awareness, high engagement. 3.2 hours/day on social media. Concern doesn't equal abandonment.
WINNING FORMULA: EVIDENCE + EXECUTION + ADOPTION SPEED
Health gains proven. Infrastructure scaling. Momentum building. But the outcome depends on execution and adoption speed.
THE COUNTERPOINT: THE RISKS PERSIST
1. BENEFITS ARE MODEST AND TEMPORARY
Effect size is small. Benefits fade. Not a cure. Not a revolution.
2. MARKETS CAN OVERSTATE DEMAND
Premium, niche, and discretionary. Luxury wellness, not mass infrastructure. Spending doesn't equal transformation.
3. CULTURE ABSORBS COUNTERCULTURE
Platforms co-opt aesthetics and language. Detox becomes content. Rebellion becomes product.
4. POLICY FACES LIMITS AND DELAYS
Constitutional challenges. Patchwork laws. Enforcement gaps. Slow impact.
5. PLATFORMS ADAPT AND ENDURE
Regulation increases costs, not collapse. Lobbying and adaptation extend runway. Business models are resilient.
6. BEHAVIOR CHANGE IS HARD TO SUSTAIN
High engagement remains. Habits revert. Attention is addictive. The machine is designed to win.
COUNTERPOINT SUMMARY
The rebellion is real, but likely to be absorbed, monetized, and fed back into the system without delivering systemic change.
THE QUESTION
Will the analog backlash compound into systemic change that restructures the attention economy—or will it be absorbed and monetized at scale?

## THE POINT: The Attention Recession Is Measurable. The Commercial Infrastructure Is Real

The original prediction's claim that the analog backlash had "crossed from cultural curiosity to measurable, sustained consumer shift" has been validated by the 2026 commercial and research evidence with specificity the original piece only gestured toward.

The RTI International meta-analysis gives the backlash something previous cultural rebellions rarely had: peer-reviewed evidence that the behavior change it promotes actually works.

A statistically significant improvement in subjective wellbeing across 32 RCTs and 5,544 individuals is the kind of evidence base that changes how physicians counsel patients, how employers design wellness programs, and how parents frame conversations with teenagers, and each of those channels is a vector for the backlash's normalization at institutional scale.

The $5 billion market forming around the analog correction, social media blockers, digital detox retreats, dumb phones, physical planners, vinyl, print books, is the commercial evidence that attention has become the century's most contested resource.

The analog rebellion is accelerating as government restriction, cultural permission, and commercial infrastructure converge simultaneously.

The legislative momentum the original piece identified as a secondary driver has continued: Australia, France, Denmark, Norway, Malaysia, Indonesia, India, and China have all restricted minor social media access, not as isolated experiments but as a coordinated global response to documented health harms.

Ipsos polling in the UK shows 72% of Britons support age-verification laws barring under-16s from social media, with strong backing from 18-34-year-olds, meaning Gen Z itself is supporting legislative restrictions on platforms it uses most, a political signal the attention economy hasn't previously faced from its core demographic.

The behavioral data reveals the specific mechanism through which the analog rebellion has genuine teeth, not through dramatic platform abandonment but through strategic compartmentalization. Industry surveys consistently show large majorities of Gen Z reporting social media breaks for mental health reasons and growing awareness of platform design as manipulative.

These figures vary significantly by source and methodology and should be understood as directional indicators of a genuine behavioral pattern rather than precise measurements.

What is more analytically significant is the platform engagement data itself: platforms are measurably competing on wellbeing features they previously ignored. Instagram's "take a break" prompts, TikTok's screen time management tools, and the explosive growth of the social media blocker market all represent the attention economy responding to Gen Z behavioral pressure, imperfectly, minimally, and on terms that still favor engagement, but responding nonetheless.

The "digital compartmentalization" the original piece described, AI for productivity, physical tools for personal life, social media on deliberate terms rather than by default, is the behavioral pattern that represents genuine structural pressure.

It's not that Gen Z is leaving the platforms. It's that Gen Z is increasingly using them on intentional terms, and the commercial and institutional infrastructure supporting that intentionality is growing faster than the attention economy's countermeasures.

### The Point Roundup

The analog backlash has the teeth the original prediction described, specifically the ones the piece correctly characterized: not a structural overturn of the attention economy, but a second-order effect forcing the next generation of social platforms to compete on wellbeing metrics they have historically ignored. The commercial infrastructure is real and growing.

The legislative momentum is real and accelerating. The peer-reviewed evidence base is the strongest yet assembled. The Gen Z peer permission structure that normalizes attention discipline as a status marker represents something structurally different from previous generational rebellions, and the Cambridge Dictionary's addition of "friction-maxxing" to its lexicon is the kind of cultural legitimation signal that precedes mainstream adoption.

What hasn't changed is the Counterpoint's core insight: the rebellion is happening inside the system it claims to resist, and the system is monetizing the resistance while continuing to extract attention from the resisters.

## THE COUNTERPOINT: Every Generation Performed Its Rebellion. This One Is Filming It for the Algorithm

The Counterpoint's "every rebellion gets absorbed, monetized, and fed back into the machine" prediction has been confirmed with precision by the 2026 evidence.

Analogue January is discovered on TikTok.

Digital detox cabin bookings are made on Instagram.

Dumb phone purchases are researched on YouTube.

The $5 billion social media blocker market is distributed through app stores. The RTI meta-analysis showing social media restriction improves wellbeing, with an effect size of 0.17, was shared virally on social media platforms.

The rebellion's infrastructure is built on the attention economy it claims to oppose. More telling is the engagement trajectory: active Gen Z TikTok users average 89 minutes daily on that platform alone, and DataReportal's Digital 2026 report shows the average TikTok user spending 1 hour 37 minutes per day on the app. Gen Z overall averages 3.2 hours daily across social media.

The peer-reviewed meta-analysis the Point correctly cites found an effect size of 0.17 for social media restriction interventions, statistically real, but modest enough that it describes the benefit of the rebellion rather than evidence the rebellion is winning.

Platform engagement continues to grow in the same period that awareness of platform harm is also growing.

The contradiction is not an anomaly. It's the mechanism: awareness and continuation are not in tension in an addictive system. They are the system's most sophisticated operating state.

The commercial scale of Gen Z's analog economy, real and growing, remains structurally marginal relative to Gen Z's engagement with the attention economy it claims to resist.

Gen Z direct spending is estimated at $250 billion, with a substantial proportion flowing through social commerce channels, TikTok Shop, Instagram Shopping, influencer-driven purchases.

The $5 billion social media blocker market projected for 2035 is a commercially significant niche serving a real need, but it's a niche.

The legislative momentum is the Counterpoint's most significant genuine threat: government mandates forcing platform design changes could achieve what consumer rebellion alone can't. But as the original piece noted, until default settings, recommendation algorithms, and notification architectures are subject to independent audit and minimum wellbeing standards, the rebellion remains a lifestyle product with peer-reviewed evidence behind it.

The 2026 evidence hasn't produced those design standards. Right to Disconnect legislation is in proposal stages in most jurisdictions. The platform design audit mechanism does not exist. The rebellion is louder, better evidenced, and more commercially developed than it was when the original prediction was written.

The system is essentially unchanged. Those two facts can both be true.

WHAT NEEDS TO CHANGE TO WIN
THE ANALOG BACKLASH AMONG GEN Z:
THE FINAL MILE TO SYSTEMIC CHANGE
The rebellion is real. The evidence is strong. The momentum is building.
Execution and adoption speed will determine the outcome.
WHAT NEEDS TO CHANGE TO WIN
1. MAKE ANALOG THE DEFAULT CHOICE
Design friction into digital by default. Make analog the easier, cheaper, more convenient path.
2. INTEGRATE INTO DAILY LIFE
Build analog into school, work, healthcare, and community life. Make it practical, not aspirational.
3. PROVE LONG-TERM IMPACT
Expand longitudinal studies. Show durable benefits for mental health, cognition, and relationships.
4. BUILD VISIBLE INFRASTRUCTURE
Scale accessible detox spaces, public analog spaces, and device-free environments.
5. ALIGN POLICY AND MARKET
Enact age limits, transparency, and design standards. Incentivize companies that reduce extraction, not maximize it.
6. LEAD THE CULTURAL NARRATIVE
Shift the story from "detox" to "better." Make analog life desirable, creative, and future-facing.
POINT SUMMARY
LIKELIHOOD COUNTERPOINT HOLDS (2027–2030)
0%
50%
65%
100%
No Change from Original Assessment (Was ~65%)
The counterpoint likelihood holds steady at 65%.
Both sides remain true. The outcome depends on execution, adoption speed, and whether the rebellion compounds into systemic change.
MOST CRITICAL VARIABLE
ADOPTION VELOCITY AT SCALE
The backing of intent is massive. Analogue January, detox retreats, dumb phones, print, and slow living are all accelerating.
The variable is speed. If adoption crosses critical mass in key institutions and daily life, the rebellion compounds.
If not, resistance remains a lifestyle niche.
WHAT TO WATCH (EARLY WARNING SIGNALS)
1. ENGAGEMENT KEEPS CLIMBING
If Gen Z social media time continues to rise, resistance is being absorbed, not scaled.
2. PREMIUMIZATION DEEPENS
If analog stays luxury or niche, it won't reach the majority. Access equality matters.
3. REBELLION BECOMES AESTHETIC
If the backlash is mostly style and symbolism, behavior change will remain shallow.
4. PLATFORMS OUT-ADAPT CHANGE
If algorithms and AI personalization keep getting better, time captured will grow.
5. POLICY STALLS OR WEAKENS
If regulation is diluted or poorly enforced, platform liability and change will stall.
6. HABITS REVERT UNDER STRESS
If users return quickly under pressure, long-term change will be rare.
COUNTERPOINT SUMMARY
THE QUESTION
Will the analog backlash reach critical mass across institutions, defaults, and daily life—and structurally reshape the attention economy, or will it remain the most sophisticated lifestyle rebellion the machine has ever monetized?

## What Needs to Change

### (... for the Counterpoint to Fully Prevail)

The Counterpoint's full vindication through 2030 requires two conditions that are currently contested.

First, the legislative momentum, real and accelerating in 2026, needs to stall before it translates from social media age restrictions for minors to actual platform design requirements for adults.

Age restrictions for minors reduce the incoming cohort of attention economy participants but don't redesign the system for the Gen Z cohort already inside it.

Design mandates would. If the legislative agenda remains focused on age restrictions rather than algorithmic design standards, the Counterpoint's "structural reform that gives people genuine agency inside the digital environment" scenario remains undelivered through 2030.

Second, the behavioral data needs to confirm that social media detox and break behaviors are recurring episodic patterns rather than permanent engagement reduction, the difference between recovering from a habit temporarily and breaking it.

The RTI meta-analysis studied discrete intervention periods; whether the wellbeing benefits persist after participants resume normal usage is the open question that current evidence does not resolve.

That ambiguity is precisely where the 65% lives.

## Likelihood Counterpoint Prevails: 65%

The Counterpoint likelihood holds at 65%, unchanged from the original, which turns out to have been precisely calibrated for what the 2026 evidence shows.

The original 65% predicted the backlash would remain "a genuine and growing cultural counter-current" that would "force the next generation of social platforms to compete on wellbeing metrics" without "fundamentally disrupting the attention economy's economics."

That description is an accurate account of what 2026 looks like. The counter-current is genuine and now peer-reviewed. The wellbeing competition pressure is visible in platform feature responses. The attention economy's core economics remain intact.

The 65% stays there because the original Counterpoint was right about the mechanism, the rebellion gets absorbed and monetized, while the original prediction was right about the second-order effect, platforms are being forced to respond to wellbeing metrics.

Both claims are simultaneously correct, which is exactly what the 65%/35% split encoded.

The 2030 question is whether Right to Disconnect legislation and platform design audits become law before the attention recession deepens to where platform economics force the redesign from the inside, or whether the rebellion, enriched by peer-reviewed evidence and legislative momentum, achieves the systemic change through political pressure that consumer behavior alone can't.

That remains a genuinely uncertain race, and 65% is still the right number for it.

# Agentic AI: The 40% Reckoning Is Already Here

## The Reckoning Is Running on Schedule. And the Numbers Are Getting Worse Before They Get Better

*AI initiative failure rate: 17% in 2024. 42% in 2025.*

*Gartner forecasts over 40% of agentic projects specifically will be canceled by end of 2027. Also: enterprises that got the architecture right project 171% ROI from agentic deployments, expected returns that exceed traditional automation benchmarks by a significant margin.*

*Capgemini surveyed 1,500 senior executives across 14 countries and found trust in fully autonomous AI agents has dropped from 43% to 27% in a single year.*

*The reckoning and the early evidence of genuine value are happening simultaneously.*

*The gap between them is the whole story.*

POINT VS COUNTERPOINT
AGENTIC AI: THE 40% RECKONING
BOTH SIDES ARE RIGHT
The debate is not whether agentic AI works.
It's why most fail while some achieve extraordinary results.
THE POINT: THE EVIDENCE IS COMPELLING
1. FAILURE RATE PROVES THE RECKONING
42% of enterprises abandoned most AI initiatives in 2025, up from 17% in 2024. Over 1,000 enterprises surveyed. (S&P Global, 2025)
2. CANCELLATIONS ARE INEVITABLE
Gartner projects over 40% of agentic AI projects will be canceled by end of 2027. (Gartner, June 2025)
3. TRUST IS COLLAPSING
Trust in fully autonomous AI agents dropped from 43% to 27% in one year. (Capgemini, 2025)
4. DEPLOYMENTS ARE RARE
Only 2% of organizations deployed AI agents at scale. 61% are still exploring. (Capgemini, 2026)
5. VENDOR HYPE IS OVERWHELMING
Only ~130 of thousands of vendors are genuine. The rest are "agent washing". (Gartner, 2025)
6. THE MECHANISM IS CONFIRMED
Poor architecture + broken processes = agents faithfully execute dysfunction faster and at greater scale.
POINT SUMMARY
The reckoning is real, measurable, and accelerating on schedule.
THE BATTLEGROUND: WHY OUTCOMES DIVERGE
WHY MANY ARE FAILING
WHY SOME ARE WINNING
Poor architecture foundation. Data silos and integration gaps. VS Right architecture first. Clean data and integration by design.
Agents built on broken processes. Dysfunction at scale. VS Process redesign before automation. Fix first, then scale.
Undefined success metrics. Can't measure impact. ROI unknown. VS Clear success criteria. Measurable outcomes. ROI tracked.
Skills and readiness gaps. Change management ignored. VS Talent and training invested early. Adoption embedded.
Overhyped expectations. Pushed too fast. VS Realistic phasing. Pilot, validate, then scale.
Weak governance and guardrails. Risk and compliance exposed. VS Strong governance. Guardrails built in. Trust earned.
WINNING FORMULA
Right architecture + clean data + process redesign + clear metrics + talent + governance + realistic phasing = 171% ROI
THE COUNTERPOINT: THE VALUE IS PROVABLE
1. ROI EXCEEDS BENCHMARKS
Enterprises that get the architecture right project 171% ROI from agentic deployments. (Capgemini, 2025)
2. MASSIVE VALUE AT STAKE
Up to $450 billion in economic value potential by 2028, conditional on trust and infrastructure. (Capgemini, 2025)
3. ADOPTION WILL ACCELERATE
33% of enterprise apps will include agentic AI by 2028. (Gartner, 2025)
4. AUTONOMY WILL BECOME NORMAL
15% of day-to-day work decisions will be made autonomously by 2028. (Gartner, 2025)
5. EARLY MOVERS COMPOUND VALUE
Efficiency gains, revenue impact, and competitive advantage increase over time.
6. THE TECHNOLOGY IS MATURING
Models, tooling, and infrastructure are improving rapidly. Capabilities are expanding.
COUNTERPOINT SUMMARY
When done right, agentic AI delivers extraordinary, measurable value.
THE DEBATE IN ONE QUESTION
Will organizations fix the foundations fast enough to capture the value before cancellations, skepticism, and competition make the gap permanent?

## Opening

The agentic AI reckoning prediction is being validated on its timeline, with numbers that are simultaneously more alarming on the failure side and more compelling on the success side than the original framing anticipated.

The failure rate for AI initiatives broadly, covering all AI projects, not agentic-specific deployments alone, rose from 17% to 42% in a single year, per S&P Global's 2025 survey of over 1,000 enterprises across North America and Europe. Gartner's June 2025 forecast of 40%+ agentic cancellations by 2027 has been reinforced rather than moderated by every subsequent data point.

The trust erosion documented by Capgemini's Research Institute, drawn from 1,500 senior executives across 14 countries, represents the organizational psychology consequence of those failures landing in boardrooms.

The Counterpoint's case rests on the enterprises that got the architecture right, who are projecting returns that make the prerequisite investment look straightforwardly justified.

The 2027–2030 story is whether the organizations currently on the failure side of the bifurcation learn the architectural lessons before exhausting their AI budget, or whether the reckoning produces the institutional skepticism that delays the genuinely useful era.

## Intro

The agentic AI reckoning prediction is one of the few articles in this series where the update's primary challenge is not separating the prediction from the Counterpoint but holding both simultaneously without flattening the genuine tension between them.

The 42% AI initiative failure rate in 2025, up from 17% in 2024, is a real number from a named survey methodology.

The 171% projected ROI from enterprises that got the architecture right is also a real number from named research. Both are true.

The organizations on the failure side and the organizations on the success side are not operating in different markets or different sectors. They are frequently in the same industry, deploying similar tools, and producing radically different outcomes based on architectural decisions, organizational readiness, and the discipline to define success criteria before deployment rather than after.

That bifurcation is the analytical key the original prediction identified as the defining feature of the agentic era, and the 2026 data has confirmed it with more statistical precision than the original framing anticipated.

The reckoning is not a verdict on agentic AI as a category. It's a verdict on the gap between what agentic AI requires to work and what most enterprises actually built when they deployed it.

## Background

Gartner projects that by 2028, 33% of enterprise software applications will include agentic AI and 15% of day-to-day work decisions will be made autonomously, both up from near zero in 2024.

The deployment reality as of mid-2026 is considerably more modest: Capgemini's survey found only 2% of organizations have deployed AI agents at scale, 12% at partial scale, while 23% have launched pilots and 61% are still exploring.

The 66% testing versus 11% production deployment gap that appears in industry surveys confirms the pattern, the sandbox-to-production leap remains the fundamental challenge. Gartner estimates only approximately 130 of the thousands of vendors claiming agentic AI capabilities are genuine, with the remainder engaged in "agent washing", relabeling existing chatbots and RPA tools without delivering authentic autonomous capability.

The S&P Global Market Intelligence survey of over 1,000 enterprises found that 42% abandoned most of their AI initiatives in 2025, up from 17% in 2024. That figure covers AI initiatives broadly; the agentic-specific failure rate is not separately measured, but Gartner's targeted 40%+ cancellation forecast for agentic projects specifically runs in the same direction.

Against this backdrop, Capgemini's analysis identifies up to $450 billion in economic value potential from agentic AI by 2028, conditional on the trust and infrastructure prerequisites being met.

OVERVIEW
AGENTIC AI: THE 40% RECKONING
IS ALREADY HERE
THE RECKONING IS RUNNING ON SCHEDULE.
AND THE NUMBERS ARE GETTING WORSE BEFORE THEY GET BETTER.
The failure rate is climbing. Trust is falling. Early winners are proving the value is real. The gap between them is the whole story.
42% AI INITIATIVE FAILURE RATE
Up from 17% in 2024. (S&P Global, 2025)
40%+ AGENTIC PROJECTS WILL BE CANCELED
By end of 2027. (Gartner, June 2025)
TRUST IN AGENTS IS FALLING
43% in 2024 → 27% in 2025. (Capgemini, 1,500 executives, 14 countries)
171% PROJECTED ROI
From agentic deployments for enterprises getting the architecture right. (Capgemini)
$450B ECONOMIC VALUE POTENTIAL BY 2028
Conditional on trust and infrastructure. (Capgemini)
THE EVIDENCE IS UNAMBIGUOUS
FAILURE RATE ACCELERATING
42% of enterprises abandoned most AI initiatives in 2025, up from 17% in 2024. (Over 1,000 enterprises)
CANCELLATIONS ARE COMING
Gartner projects over 40% of agentic AI projects will be canceled by end of 2027.
TRUST IS ERODING FAST
Trust in fully autonomous AI agents dropped from 43% to 27% in one year. (Capgemini, 2025)
DEPLOYMENTS ARE STILL RARE
Only 2% of organizations deployed AI agents at scale. 61% are still exploring. (Capgemini, 2026)
VENDOR HYPE IS OVERWHELMING
Only ~130 of thousands of vendors are genuine. The rest are "agent washing". (Gartner, 2025)
VALUE IS PROVABLE
Early adopters with the right architecture project 171% ROI. (Capgemini)
THE BIG PICTURE
THE RECKONING IS REAL
Poor architecture. Broken processes. Wrong data.
Agents execute the dysfunction at scale.
Budgets wasted. Trust eroded. Momentum at risk.
MANY ARE FAILING
THE GAP IS EXECUTION, READINESS, AND DISCIPLINE.
THE VALUE IS REAL
Right architecture. Clean data. Clear guardrails.
Agents automate outcomes, not just tasks.
Efficiency gains. Revenue impact. High ROI.
SOME ARE WINNING
THE 2027–2030 OUTCOME
Will the laggards learn fast enough to close the gap? Or will the reckoning create lasting skepticism that delays the next wave of real value?
THE CHALLENGES ARE COMPOUNDING
COSTS KEEP RISING
Compute, data, and talent costs outpace returns for most organizations.
INTEGRATION IS HARD
Legacy systems and data silos block agent effectiveness.
RISK AND LIABILITY GROW
Autonomous decisions create new compliance and legal exposure.
TALENT IS IN SHORT SUPPLY
Skilled agentic AI engineers are in extreme demand.
SECURITY THREATS ESCALATE
More autonomy equals bigger blast radius for attacks.
CHANGE FATIGUE IS REAL
Organizations are overwhelmed. Many lack the patience to learn.
THE QUESTION
Will organizations fix the foundations before they run out of budget, patience, and trust—or will the reckoning delay the real agentic era?

## THE POINT: 2,000 'Death by AI' Claims by End of 2026: The Prediction Now Has a Gartner Number

The original prediction's central mechanism, organizations deploying agentic AI on top of broken processes, discovering the technology faithfully executes the dysfunction faster and at greater scale, has been confirmed by the 2025–2026 failure data with a specificity the Gartner forecast alone couldn't provide.

The AI initiative failure rate rising from 17% to 42% in a single year, documented by S&P Global across more than 1,000 enterprises, represents the acceleration pattern the prediction described as inevitable, now documented as actual.

Two important qualifications apply: the S&P Global figure covers all AI initiatives, not agentic deployments specifically, the agentic-specific failure rate may differ, and Gartner's 40% cancellation forecast for agentic AI specifically is a forward projection, not a realized measurement.

The direction of travel is consistent across both data sources, but they shouldn't be collapsed into a single agentic AI failure figure.

What they collectively show is that the organizational preconditions the prediction identified as prerequisites, clean data architecture, governance frameworks, integration infrastructure, realistic ROI measurement, remain unmet at scale, and the failure pattern follows predictably from that gap.

Gartner now forecasts more than 2,000 "death by AI" legal claims by end of 2026, incidents where insufficient AI risk guardrails led to harm, representing the specific, named consequence of deploying agents without adequate governance that the original prediction described but couldn't yet quantify.

The trust trajectory documented by Capgemini Research Institute is the most analytically significant new data point in the reckoning narrative, because organizational trust, once lost in enterprise technology deployments, recovers slowly and at significant cost.

Drawing on surveys of 1,500 senior executives across 14 countries, Capgemini found trust in fully autonomous AI agents declining from 43% to 27% in a single year, a drop driven by ethical concerns, lack of transparency, and the gap between vendor capability demonstrations and production reality.

Critically, Capgemini also found that trust levels differ meaningfully by deployment stage: 47% of organizations in active implementation phase report above-average trust, compared to 37% still in the exploratory phase. That pattern suggests the trust decline is not uniform, it's concentrated among organizations that attempted deployment without adequate prerequisites and encountered the failure modes the prediction described.

The organizational psychology implication is significant: executives who approved budgets, watched deployments fail, and are now skeptical of the next agentic proposal represent the specific barrier to the Counterpoint's "survivors building moats" scenario.

Capgemini found 80% of organizations lack a mature AI infrastructure and fewer than 20% report high data readiness, confirming that the foundational prerequisites the prediction identified as necessary remain gaps rather than solved problems for the majority of enterprises currently attempting agentic deployment.

### The Point Roundup

The reckoning is arriving on schedule at the scale the prediction described, and the original "by end of 2027" framing may prove conservative given the failure rate acceleration from 17% to 42% in a single year for AI initiatives broadly.

The Gartner 40% agentic cancellation forecast is a floor based on evidence available at the time it was made; the 2025–2026 failure data suggests it's not a ceiling.

The trust erosion from 43% to 27% is the organizational psychology mechanism that translates individual deployment failures into institutional skepticism, and that skepticism, if it hardens, is what delays the Counterpoint's "genuinely useful era" from arriving on schedule.

The ERP graveyard pattern the original piece identified as the historical analogue is repeating with recognizable characteristics: deployment before organizational prerequisites exist, attribution of failure to the technology rather than the missing prerequisites, and the resulting budgetary caution that makes the next wave of deployments even harder to fund.

## THE COUNTERPOINT: Every Transformative Technology Had Its Trough. This Is the Trough.

The Counterpoint's core prediction, the shakeout represents the natural cost of moving from experimentation to production-grade deployment, not evidence of a fundamental ceiling on the technology, has its strongest supporting evidence in the deployment outcomes of the organizations that got the architecture right.

Spotify's internal system, Honk, built on Claude Code, demonstrated the specific production-grade outcome the Counterpoint described as moat-building: Spotify disclosed on its Q4 2025 earnings call that its most experienced engineers had not written code since December 2024, instead orchestrating AI agents that deploy features in minutes by describing them in plain English. That is not a vendor benchmark.

It's a publicly disclosed operational capability from a named organization, representing the class of first-mover advantage the Counterpoint predicted for organizations that complete the prerequisite architecture work.

The pattern extends across enterprise deployments broadly: Gartner projects 33% of enterprise software applications will include agentic AI by 2028 and that 15% of day-to-day work decisions will be made autonomously, trajectory numbers that describe a technology maturing into infrastructure, not retreating from production.

The Capgemini finding that trust levels are meaningfully higher among organizations in active implementation (47% above-average trust) versus exploratory phase (37%) supports the Counterpoint's structural argument: the organizations closest to real deployment are more confident, not less, as they move from sandbox to production.

The market consolidation dynamics provide the Counterpoint's strongest structural confirmation.

Gartner's identification of only approximately 130 legitimate agentic AI vendors from thousands of claimants means the agent-washing phase is generating its own correction through customer experience, the enterprises that deployed dressed-up automation discovered what they had, and the vendors who misrepresented their capabilities are losing the renewal conversations.

That is the shakeout functioning as designed: separating genuine capability from marketing, at the cost of the failed deployments that had to happen to make the separation visible.

The 66% testing versus 11% production deployment gap is not evidence of technology failure, it's, from the Counterpoint's perspective, evidence of appropriate organizational caution: the majority of organizations have recognized their current state can't support production deployment and are either building the prerequisite infrastructure or waiting for it.

The enterprises projecting 171% expected ROI from agentic deployments should be read carefully: that figure reflects organizational confidence, not realized financial performance. It comes from enterprise projections rather than measured outcomes.

For those organizations, the shakeout is working exactly as the Counterpoint predicts: it is separating companies with the architecture, governance, and operational readiness to capture agentic AI's potential from those still treating it as a vendor feature.

That is the population most likely to pursue the $450 billion human-agent collaboration opportunity Capgemini identifies as the 2028 ceiling.

WHAT NEEDS TO CHANGE TO WIN
AGENTIC AI: THE 40% RECKONING
THE FINAL MILE TO REAL VALUE
The failure pattern is clear. The success formula is proven.
Adoption at scale depends on fixing the foundations.
ARCHITECTURE
PROCESS
DATA QUALITY
RESULTS
WHAT NEEDS TO CHANGE TO WIN
1. FIX THE FOUNDATION
Design for agentic AI from day one. Clean data. Strong integration. Secure architecture.
2. REDESIGN BEFORE YOU AUTOMATE
Fix broken processes first. Remove waste and handoffs. Then let agents run them.
3. DEFINE SUCCESS BEFORE DEPLOYMENT
Set clear, measurable outcomes. Track ROI, quality, and risk. Avoid vague productivity goals.
4. INVEST IN PEOPLE AND READINESS
Build agentic fluency. Train teams. Redefine roles. Drive change management.
5. PHASE AND VALIDATE
Start small. Prove value. Pilot, validate, then scale. Don't boil the ocean.
6. BUILD TRUST BY DESIGN
Transparency. Guardrails. Human oversight where it matters. Earn trust through reliability.
POINT SUMMARY
The ingredients exist. The playbook is known. Execution discipline is the difference. Get the architecture right. Win at scale.
LIKELIHOOD COUNTERPOINT HOLDS (2027–2030)
Fix foundations.
Redesign.
Measure.
Invest in people.
Trust by design.
Value at scale.
0%
50%
100%
75%
Bad architecture.
Broken processes.
Unclear metrics.
Poor readiness.
Risk exposure.
Value destroyed.
No Change from Original Assessment (Was ~75%)
The counterpoint likelihood holds steady at 75%. Both sides remain true. The outcome depends on execution, adoption speed, and whether organizations close the gap between potential and performance.
MOST CRITICAL VARIABLE
ARCHITECTURE + DISCIPLINE + ADOPTION SPEED
The tools are here. The value is proven. The variable is how fast organizations fix the basics and scale what works.
Speed to fix foundations determines who captures the $450B and who gets left behind.
Architecture
Built right from the start.
Discipline
Process before automation.
Adoption Speed
Learn fast, scale faster.
WHAT TO WATCH (EARLY WARNING SIGNALS)
1. CANCELLATIONS CLIMB HIGHER
If AI initiative failure rate exceeds 42%, agentic AI cancellations will accelerate.
2. TRUST KEEPS DECLINING
If trust in autonomous AI drops below 27%, adoption and ROI will stall.
3. PILOTS NEVER REACH SCALE
If the 66% testing vs 11% production gap doesn't close, value remains theoretical.
4. AGENT WASHING PERSISTS
If most vendors are still relabeling, expect overhype and disappointment.
5. COSTS OUTPACE RETURNS
If compute, data, and talent costs rise faster than measurable outcomes.
6. RISK EVENTS INCREASE
If compliance, security, and operational incidents grow, regulation will tighten.
COUNTERPOINT SUMMARY
The technology amplifies what you build. Fix nothing, lose more. Get it wrong, the reckoning compounds.
THE QUESTION
Will organizations fix the foundations fast enough to capture the $450B of value, or will the 40%+ cancellation rate and trust erosion delay the agentic era?

## What Needs to Change

### (... for the Counterpoint to Prevail)

The Counterpoint's "reckoning as beginning of genuinely useful era" requires two conditions to materialize through 2030 that are currently under pressure.

First, the trust recovery from 27% must happen through demonstrated production success rather than through renewed vendor marketing, the mechanism being well-governed deployments producing documented, referenceable ROI outcomes that rebuild organizational confidence in the investment case. Capgemini's own data offers a path: trust rises from 37% to 47% among organizations that move from exploration to implementation. If that pattern holds as deployment scales, the trust decline reverses organically through the accumulation of visible successes. If the 2,000+ "death by AI" legal claims Gartner forecasts trigger regulatory responses that create new compliance barriers before the success cases accumulate, the recovery timeline extends materially.

Second, the S&P Global failure trajectory, 17% in 2024, 42% in 2025, must not continue accelerating in 2026. A third year of rising failure rates would produce the institutional skepticism that delays the "33% of enterprise software with agentic AI by 2028" timeline into the early 2030s, narrowing the Counterpoint's window considerably.

## Likelihood Counterpoint Prevails: 75%

The Counterpoint likelihood holds at 75%, the same number the original was assigned, and the number that continues to reflect the most accurate characterization of what the evidence shows.

The original 75% predicted the shakeout would prove real and the cancellation rate accurate or conservative, but that the organizations surviving it would emerge with durable structural advantages and the technology would be stronger for having been tested at production scale. Both predictions are confirmed.

The S&P Global failure rate data and Gartner's agentic cancellation forecast confirm the shakeout is real. Spotify's Honk disclosure confirms that the moat-building outcome the Counterpoint described is materializing for organizations that completed the prerequisite architecture work.

The Capgemini $450 billion economic value projection by 2028, from a named survey of 1,500 senior executives, confirms the technology's potential hasn't been revised downward by the failure evidence.

The number stays at 75% rather than rising because the trust decline from 43% to 27%, documented by Capgemini from primary survey data, represents a steeper and faster erosion than the 75% likelihood implied was probable, and one that could slow the Counterpoint's timeline if the organizational skepticism it produces proves more durable than the implementation-phase trust recovery pattern suggests.

The honest characterization: the Counterpoint's structural argument is correct, the technology is not in retreat, and the failure rate represents the cost of the organizational learning curve rather than evidence of a fundamental ceiling.

The 25% probability assigned to the Point prevailing reflects the genuine possibility that the trust erosion and regulatory response to "death by AI" incidents combine to delay the Counterpoint's "genuinely useful era" into the early 2030s, a partial rather than complete Point validation, but meaningful enough to preserve the uncertainty the 75%/25% split encodes.

# Voice as the New Browser: Transformation or Another False Start?

## Voice Won Its Domain. The Zero-Click Crisis Is Testing Whether It Can Keep It.

*8.4 billion voice assistants are now active worldwide, surpassing the global population for the first time, processing over 10 billion queries daily. 76% of voice queries have local intent: "near me," "open now," "directions to."*

*On-device voice processing grew from 12% to 38% of queries in three years, partially resolving the privacy barrier that limited adoption.*

*Meanwhile: browsers are still browsers. Long-form research is still visual. The UK Competition and Markets Authority is forcing Google to let publishers opt out of AI Overviews scraping.*

*The prediction got the domain right.*

*The zero-click content crisis is testing whether that domain stays valuable.*

OVERVIEW
VOICE AS THE NEW BROWSER:
TRANSFORMATION OR ANOTHER FALSE START?
VOICE WON ITS DOMAIN. THE ZERO-CLICK CRISIS IS TESTING WHETHER IT CAN KEEP IT.
The prediction got the domain right. The zero-click content crisis is testing whether that domain stays valuable.
AI OVERVIEWS
8.4 BILLION
VOICE ASSISTANTS ACTIVE WORLDWIDE
Surpassing the global population for the first time.
10+ BILLION
QUERIES DAILY
Voice assistants process over 10 billion queries every day.
76%
LOCAL INTENT
"Near me." "Open now." "Directions to." define voice use.
38%
ON-DEVICE PROCESSING
Grew from 12% to 38% of queries in three years. Privacy barrier is being resolved.
$108.33B
VOICE COMMERCE IN 2024
Projected to reach $586.3B by 2031 at 27.28% CAGR.
UK CMA ACTION
AGAINST AI OVERVIEWS
Forcing Google to let publishers opt out of scraping. Zero-click crisis reaches regulators.
THE EVIDENCE CONFIRMS THE DOMAIN
USE CASE IS DEFINED
76% of voice queries are local intent. That is the dominant pattern.
COMMERCE IS REAL
Voice commerce hit $108.33B in 2024. That is real transactions, not a pilot.
ADOPTION IS AT SCALE
20% of internet users use voice search. That is ~1.6B people worldwide.
TRUST BARRIER IS FALLING
On-device processing grew to 38%. 47% of users say it increases trust.
ENTERPRISE VALUE IS PROVEN
BFSI leads adoption at 32.9% share. Firms report 20–30% cost reductions.
POINT SUMMARY
Voice is the dominant interface for local search, commerce, service, and utility. The transformation is real and measurable within its domain.
THE BIG PICTURE
VOICE OWNED ITS DOMAIN. NOW IT MUST DEFEND IT.
WHAT VOICE HAS WON
Local search
Directions
Business discovery
Customer service
Transactions
Hands-free utility
THE TEST AHEAD
Zero-click erodes sources
Publisher pushback
Regulatory intervention
Ad model disruption
Answer quality risk
Trust erosion
THE 2027–2030 OUTCOME
Will voice keep its valuable domain and build an ecosystem that sustains quality and trust? Or will the zero-click crisis erode the foundation of its advantage?
THE RISKS TO THE VOICE ADVANTAGE
ZERO-CLICK CONTENT CRISIS
AI Overviews give answers, not links. Sources lose traffic and revenue.
PUBLISHER AND CREATOR BACKLASH
Opt-outs, lawsuits, and licensing demands increase pressure on platforms.
REGULATORY SCRUTINY GROWS
UK CMA action is the first move. More interventions are coming.
AD AND MONETIZATION DISRUPTION
Fewer clicks means ad revenue at risk. New models are unproven.
ANSWER QUALITY AT RISK
Training data shrinks. Bias and errors rise. Bad answers kill user trust.
COUNTERPOINT SUMMARY
Voice is powerful but narrow. The zero-click crisis and regulatory pushback could limit or even reverse its momentum.
THE QUESTION
Will voice build a durable, trusted ecosystem that sustains its domain advantage, or will the zero-click crisis and regulatory pressure erode what it has won?

## Opening

The voice-as-new-browser prediction has had an unusual 2026 update experience: the evidence has confirmed rather than challenged the original framing with unusual precision.

The prediction was careful to bound its claim at "dominant interface for a specific and valuable cluster of high-intent, low-complexity interactions" rather than universal browser replacement, and the 2026 data confirms exactly that bounded outcome.

What the update adds is analytical precision about the two variables the original piece correctly identified as the key unknowns determining voice's long-term trajectory: on-device processing is resolving the privacy barrier faster than anticipated, and the zero-click content ecosystem crisis has escalated to regulatory action in the UK before generating the licensing framework resolution the original piece called for.

Both developments favor the prediction's domain-specific transformation scenario.

The 70%/30% likelihood split was precisely calibrated and holds.

## Intro

The voice interface has crossed the threshold the prediction described for local search, transactional commerce, customer service, and hands-free utility.

It hasn't replaced visual interfaces for complex research, comparison shopping, or nuanced decision-making, and the 2026 evidence is explicit about where the boundary sits.

The original Counterpoint's prediction that voice would remain "a powerful but narrow tool" for specific interactions has been validated simultaneously with the Point's "dominant interface for its specific domain" claim.

Understanding why both are simultaneously correct is the analytical contribution this update makes to the 2027–2030 framing.

The 2030 question is not whether voice's domain exists. It does, it's large, and it's measurably growing.

The question is whether the zero-click content ecosystem crisis that threatens the information quality underlying voice AI's answers resolves through regulation and licensing before it degrades the utility voice has earned in its domain.

## Background

The global voice assistant installed base has surpassed the global population for the first time at 8.4 billion active devices, processing over 10 billion queries daily, encompassing every smartphone, smart speaker, vehicle infotainment system, and wearable.

The conversational AI market reached $14.29 billion in 2025, expanding at 23.7% CAGR toward a projected $41.39 billion by 2030. The global smart speaker industry is valued at $22.09 billion in 2025, projected to grow to $29.65 billion by 2029.

Voice commerce reached $108.33 billion globally in 2024, projected to reach $586.3 billion by 2031 at 27.28% CAGR, not a novelty experiment, a measurable commerce channel. 76% of voice queries have local intent: "near me," "open now," and directions queries define the dominant use case.

The share of internet users worldwide actively using voice search stabilized around 20% globally through 2025, following a peak of 22.5% in Q2 2022, not growing dramatically but holding at scale.

The single most important infrastructure development for voice adoption is the on-device processing shift: voice query processing performed locally rather than in the cloud grew from 12% to 38% in three years, with Apple, Google, and Amazon all investing heavily in local AI models.

47% of users say on-device processing would increase their trust in voice, meaning the architectural solution to the primary privacy barrier exists and is being deployed at scale.

POINT
VS
COUNTERPOINT
VOICE AS THE NEW BROWSER:
BOTH SIDES ARE RIGHT
Voice dominates a specific domain. Browsers remain essential elsewhere.
The future depends on how we solve the zero-click content crisis.
THE POINT: VOICE WINS ITS DOMAIN
1. MASSIVE SCALE, REAL USAGE
8.4B voice assistants active worldwide. Surpassing global population. 10B+ queries processed daily.
2. LOCAL INTENT DOMINATES
76% of voice queries are local. "Near me," "open now," and directions define voice's core.
3. COMMERCE AND VALUE ARE REAL
Voice commerce hit $108.33B in 2024. Projected $586.3B by 2031 at 27.28% CAGR.
4. TRUST BARRIER IS FALLING
On-device processing grew from 12% to 38% in three years. 47% would trust voice more.
5. ENTERPRISE ADOPTION IS STRONG
BFSI leads with 32.9% market share. Firms report 20–30% cost reductions with voice solutions.
6. AGENTIC ERA CHANGES THE GAME
Websites are now databases for AI assistants to query. Be the answer, not the destination.
POINT SUMMARY
Voice is the dominant interface for local search, commerce, service, and utility. The domain is large, valuable, and growing. The transformation is real.
WHERE VOICE WINS VS WHERE BROWSERS WIN
VOICE WINS
USE CASE
BROWSERS WIN
Local search
"Near me," "open now," directions, businesses
Complex research
Deep topics, analysis, nuance, and context
Transactional actions
Ordering, booking, purchasing, payments
INTENT
Comparison shopping
Evaluate options, read reviews, compare
Hands-free utility
Multitasking, driving, cooking, on the go
CONTEXT
Focused attention
Sitting, reading, learning, working
Quick answers
Facts, prices, status, hours, confirmations
COMPLEXITY
Multi-step tasks
Planning, creation, problem solving
Voice assistants
8.4B+ devices, 10B+ daily queries
INTERFACE
Visual interfaces
Screens, keyboards, and rich UI
Immediate action
Start now, go now, buy now, call now
OUTCOME
Informed decisions
Understand first, decide best
THE COUNTERPOINT: BROWSERS ARE ESSENTIAL
1. COMPLEXITY REQUIRES VISUAL
Nuance, depth, and context are hard for voice. Screens win for deep research and learning.
2. ZERO-CLICK CRISIS IS REAL
AI Overviews give answers, not links. Sources lose traffic and revenue.
3. PUBLISHERS PUSH BACK
Opt-outs, lawsuits, and licensing demands increase pressure on platforms.
4. REGULATORY RISK IS GROWING
UK CMA action is the first move. More scrutiny and intervention are coming.
5. AD MODELS ARE DISRUPTED
Fewer clicks mean fewer ads. New models are unproven and fragile.
6. BROWSERS AREN'T GOING AWAY
Long-form content, creation, and productivity keep browsers indispensable.
COUNTERPOINT SUMMARY
Voice is powerful but narrow. The zero-click crisis threatens quality, sustainability, and trust.
THE DEBATE IN ONE QUESTION
Will voice interfaces own and expand their high-intent domain while browsers protect depth, discovery, and decision-making elsewhere?

## THE POINT: The Agentic Era Validated the Prediction: Websites Are Now Databases for AI Queries

The 2026 data confirms the original prediction's bounded claim: voice is becoming the dominant interface for a specific, valuable set of high-intent, low-complexity interactions, not a universal replacement for the browser.

The clearest evidence is local search. Seventy-six percent of voice queries have local intent, and "near me" searches remain the dominant use case across markets. That pattern supports the narrower argument: voice works best when the user wants a fast, direct answer or action, especially in contexts such as local search, transactional commerce, customer service, and hands-free utility.

This is not a prediction about where voice is going, it's a description of where voice already is.

The enterprise adoption confirms the transactional domain: the BFSI sector leads enterprise voice adoption at 32.9% market share, with organizations reporting 20-30% operational cost reductions.

Voice commerce reached $108.33 billion globally in 2024, a figure that represents real transactions completed through voice interfaces, not user preference surveys.

The "agentic era" dynamic the original prediction described, "in the agentic era, your website isn't just a brochure for humans; it's a database for AI assistants to query on behalf of their users", has become the organizing principle of digital strategy in 2026.

The 2026 Consumer Search Report confirmed that brand visibility is now about "being the definitive answer provided by an AI assistant during a consumer's moment of intent" rather than page-one search rankings.

That framing shift validates the prediction's core mechanism: voice AI as the interface layer that abstracts the browser entirely for specific, high-intent interactions.

The two variables the original prediction identified as determining voice's long-term trajectory have both developed since the original writing, and both favor the prediction's scenario in different degrees.

The privacy barrier is being resolved through architecture rather than trust-building: on-device processing growing from 12% to 38% of queries in three years is a faster infrastructure shift than most voice adoption analyses anticipated, and with 47% of users explicitly saying on-device processing would increase their trust, the correlation between the technical solution and the consumer barrier it addresses is direct.

Sixty-three percent of users report being more comfortable with voice AI when they know conversations stay private, confirming that the privacy concern is not inherent to voice as an interface but specific to cloud transmission, a concern the on-device trajectory is actively resolving.

The zero-click content ecosystem crisis has developed in the specific form the original prediction described as the primary sustainability risk, and has now generated the regulatory response the original piece called for: the UK Competition and Markets Authority has proposed requiring Google to allow publishers to opt out of AI Overviews content scraping, with requirements for transparency and source citation.

That regulatory intervention arrived reactively, proposed after publishers documented traffic declines from AI summaries, but its arrival confirms both the prediction's diagnosis and the Counterpoint's warning that the content ecosystem economics need a governance framework to survive the zero-click dynamic.

### The Point Roundup

The voice-as-new-browser prediction has been validated in its specific domain, and the precision of the original 70% calibration reflects a well-bounded claim that the evidence has confirmed rather than requiring revision.

The prediction was right about the domain, right about the mechanism (agentic AI abstracting the browser for high-intent interactions), and right about the two key variables (privacy barrier and content ecosystem economics) that would determine the transformation's durability.

The on-device processing trajectory is resolving the privacy barrier faster than the 70% implied was probable. The zero-click regulatory intervention is addressing the content crisis, though arriving later and more reactively than the prediction's most optimistic framing.

The 2027-2030 window tests one remaining structural question: whether the UK CMA intervention and equivalent frameworks establish sustainable content ecosystem economics before the zero-click dynamic hollows out the information quality that makes voice AI answers useful.

That's a solvable problem with a regulatory solution already in progress, not a fundamental technology failure.

## THE COUNTERPOINT: The Browser Isn't Going Anywhere. The Domain Gap Proves the Counterpoint Was Right

The Counterpoint's core claim, that voice excels at simple lookups, hands-free commands, local discovery, and transactional requests with low complexity but can't replace visual interfaces for complex research, comparison shopping, and nuanced decision-making, has been validated by the same 2026 data the Point relies on.

The fact that 76% of voice queries have local intent is not solely the Point's evidence.

It's equally the Counterpoint's evidence: the overwhelming concentration of voice use in local, immediate, low-complexity intent confirms that the high-complexity, research-intensive, multi-source evaluation tasks that constitute the most commercially valuable digital interactions remain screen-native.

Global voice search adoption has stabilized around 20% of internet users rather than growing toward the majority the maximalist prediction implied, and that stabilization has followed a peak and modest decline from 22.5% in Q2 2022.

The Counterpoint did not predict voice would fail. It predicted voice would remain "a powerful but narrow tool" for specific interactions.

The 8.4 billion installed voice assistants processing 10 billion queries daily confirms that voice has its domain.

The 20% global adoption stabilization confirms the domain is bounded, not expansive.

Both predictions are simultaneously right, which is exactly what the 30% Counterpoint likelihood described as its most likely outcome.

The zero-click content ecosystem crisis the Counterpoint identified as the primary sustainability threat has escalated in the specific form the original warning described, and the regulatory response, while real, demonstrates the Counterpoint's point rather than refuting it.

The UK CMA proposing to require Google to allow publisher opt-out from AI Overviews scraping, requiring proper source citation, is the specific governance framework the original piece called for.

That this intervention is reactive, proposed after publishers documented traffic declines, rather than proactive confirms the Counterpoint's warning that "if it isn't corrected through licensing frameworks, attribution standards, or revenue sharing, the content ecosystem that makes voice AI useful will slowly starve."

The regulatory intervention proves the Counterpoint's diagnosis was correct. Whether it arrives with sufficient speed and enforcement teeth to prevent the ecosystem damage before content quality degrades is the 2027–2030 test the Counterpoint's scenario requires.

Privacy concerns remain structurally significant even as on-device processing provides a partial architectural resolution: 41% of users say they fear their voice assistants actively listen to or record them, and the always-on nature of voice interfaces combined with documented high-profile data handling controversies has created a trust deficit that on-device processing addresses architecturally but hasn't yet resolved perceptually for the population that is not already using voice for its established domain.

WHAT NEEDS TO CHANGE TO WIN
VOICE AS THE NEW BROWSER:
BUILD THE ECOSYSTEM OR LOSE THE ADVANTAGE
The domain is won. The zero-click crisis is the test.
Fix the foundations. Protect the sources. Earn the trust.
WHAT NEEDS TO CHANGE TO WIN
1. BUILD A SUSTAINABLE CONTENT ECONOMY
Create licensing, revenue share, and attribution models that reward publishers and creators. Pay for quality. Protect the sources.
2. FIX THE ZERO-CLICK CRISIS
Give users paths to deeper content. Make follow-ups easy. Send traffic, not just answers.
3. EARN AND PROVE TRUST
Expand on-device processing. Be transparent about sources. Let users control data and settings.
4. RAISE ANSWER QUALITY
Invest in better models. Reduce hallucinations. Show confidence and citations.
5. DESIGN FOR DISCOVERY
Make it easy to explore, compare, and refine. Support multi-turn conversations that go deeper.
6. PARTNER ACROSS THE ECOSYSTEM
Work with publishers, retailers, businesses, and platforms. Align incentives. Build together.
POINT SUMMARY
The domain is real and growing. Fix the ecosystem, protect quality, and voice becomes indispensable.
COUNTERPOINT LIKELIHOOD HOLDS: 30%
THE BALANCE REMAINS THE SAME
WHY IT WILL WIN (70%)
Massive installed base
Local intent dominance
Real commerce volume
On-device trust boost
Agentic ecosystem shift
50%
0%
30%
100%
WHY IT COULD LOSE (30%)
Zero-click crisis worsens
Sources lose traffic
Trust and quality erode
Regulation overcorrects
Better interfaces emerge
The outcome depends on execution and ecosystem health.
The 70% path remains open, not guaranteed.
THE FOUNDATION FOR LASTING VOICE ADVANTAGE
SOURCES FIRST
Pay for content. Share value. Build strong partnerships.
USER EXPERIENCE EXCELLENCE
Fast. Accurate. Useful. Easy follow-ups. Seamless across devices.
PRIVACY AND CONTROL
On-device by default. Clear controls. Respect user choices.
TRUSTED ANSWERS
VALUABLE OUTCOMES
HEALTHY ECOSYSTEM
BUSINESS OUTCOMES
Drive transactions. Reduce costs. Prove real ROI.
POLICY AND STANDARDS
Support smart regulation. Promote fair use. Prevent content theft.
MEASURE WHAT MATTERS
Answer quality. User satisfaction. Ecosystem health.
Get the foundations right. The advantage compounds. Ignore them. The 30% risk becomes the reality.
WHAT TO WATCH (EARLY WARNING SIGNALS)
1. PUBLISHER TRAFFIC KEEPS FALLING
If AI Overviews continue to grow without revenue models, quality will decline.
2. ANSWER QUALITY SLIPS
More hallucinations. Lower satisfaction. Users lose trust and usage drops.
3. REGULATORY PRESSURE SPIKES
More investigations and rules. Overcorrection could slow innovation.
4. NEW INTERFACES GAIN GROUND
Visual AI, AR, and multimodal interfaces could capture time.
5. AD MODELS DON'T IMPROVE
If fewer clicks mean lower ad revenue, the model becomes unsustainable.
6. CONTENT CREATORS PUSH BACK
Boycotts, lawsuits, and licensing demands could limit access.
COUNTERPOINT SUMMARY
Voice is powerful but narrow. The zero-click crisis and ecosystem breakdown could limit or reverse growth.
THE QUESTION
Will we build a fair, sustainable voice ecosystem that delivers trusted answers and real outcomes, or will the zero-click crisis erode the advantage we have won?

## What Needs to Change

### (... for the Counterpoint to Prevail)

The Counterpoint's "powerful but narrow tool" scenario prevailing through 2030 requires two conditions to hold.

First, the on-device processing transition needs to stall or fail to resolve the trust deficit perceptually, not just architecturally. The 12%-to-38% three-year growth rate in on-device processing is a technical fact; whether it translates into the trust change that 47% of potential adopters say it would require is a behavioral question the evidence hasn't yet answered. If the gap between "technically more private" and "perceived as more private" persists, the adoption ceiling the Counterpoint describes remains in place.

Second, the content ecosystem zero-click crisis needs to either accelerate beyond the pace of regulatory response or generate a governance framework that is insufficiently enforced to prevent the quality degradation the Counterpoint describes. The UK CMA intervention is early-stage; equivalent frameworks in the US are not yet proposed; and platform compliance with opt-out and citation requirements will be tested through enforcement rather than assumed.

The Counterpoint's 2030 scenario is specific: voice AI success at answering questions destroys the incentive to create the content it learns from, degrading answer quality precisely as voice adoption should be accelerating, and the regulatory response arrives too late or with insufficient teeth to prevent that self-defeating loop.

That scenario is less likely than it was when the original prediction was written, given the on-device progress and the UK regulatory intervention.

The 30% reflects the genuine residual probability.

## Likelihood Counterpoint Prevails: 30%

The Counterpoint likelihood holds at 30%, unchanged from the original, and for the same reasons the original 70% Point confidence was precisely calibrated.

The prediction was written with unusual care about the specific domain of voice's dominance, and the 2026 evidence confirms that domain rather than expanding or contracting it.

A 30% Counterpoint likelihood in this article has a different character than 30% in most other articles in the series: it reflects not the Counterpoint being wrong about voice's limitations but the Counterpoint being simultaneously right about those limitations and the Point being right about the domain.

Both the 8.4 billion installed assistants processing 10 billion daily queries and the 20% global adoption stabilization are confirmed by the same 2026 data, the Point's domain confirmation and the Counterpoint's boundary confirmation are both true.

What would move the number upward toward 40-45% is the content ecosystem zero-click crisis accelerating faster than regulatory frameworks can address it, degrading answer quality in voice AI at the moment when on-device processing is resolving the final privacy barrier.

That scenario is plausible. The UK CMA intervention is early-stage. US regulatory frameworks are not yet proposed. Publisher traffic declines are documented and ongoing.

What would move the number downward toward 20% is the on-device processing trajectory continuing at its current rate through 2028, resolving the privacy barrier perceptually rather than just architecturally, and the UK CMA framework establishing precedent that extends to US platforms through market pressure rather than requiring separate legislative action.

The 30% sits at the precise midpoint of those two trajectories, where the evidence leaves it, and where the original prediction placed it.

## Smart Glasses Reach a True Tipping Point?

### From Glasshole to Gucci. And then back to Glasshole?

*Smart Glasses have seen some noteworthy activity in 2026.*

*Ray-Ban Meta sold 7 million pairs in 2025, tripling the prior year. Samsung Galaxy Glasses were announced July 22, 2026, with consumer availability expected Fall 2026. Google is on Gucci frames. Apple is confirmed for 2027. EssilorLuxottica is targeting 20 million pairs by end of 2026 with waitlists running well into the year.*

*The tipping point the original prediction described is no longer a forecast. It is a documented market reality.*

*Also: A BBC investigation found dozens of male influencers secretly filming women in shopping centers, airports, beaches, and restaurants using Ray-Ban Meta glasses and posting the footage without consent.*

*A woman in London discovered a video of herself had been viewed 40,000 times and was told she would have to pay for its removal.*

*A woman in Vancouver went viral without knowing she had been filmed.*

*A woman in Sydney saw her Bondi Beach encounter reach a million TikTok views.*

*Swedish investigators documented Kenyan contractors reviewing bedroom and bathroom footage from the glasses, with face-blurring regularly failing.*

*WIRED discovered undisclosed facial recognition code (internally called "Name Tag") embedded in the Meta AI app, capable of building biometric faceprint databases of strangers.*

*Two US class action lawsuits are active.*

*The UK ICO opened an investigation. Ireland's DPC called the category "deeply problematic and intrusive."*

*A man in the UK pleaded guilty to voyeurism using smart glasses.*

*And Meta's own marketing, "designed for privacy, controlled by you", is now the central exhibit in federal consumer protection litigation.*

*Samsung and Google announced their Fall 2026 glasses at Google I/O with no disclosed data retention policies governing either product.*

*The tipping point arrived. So did everything the Counterpoint feared. Both happened in the same category in the same year.*

*And the next wave of products is launching into this environment without answers to the questions the first wave raised.*

SMART GLASSES REACH A TRUE TIPPING POINT?
FROM GLASSHOLE TO GUCCI. AND THEN BACK TO GLASSHOLE?
THE TIPPING POINT THE ORIGINAL PREDICTION DESCRIBED IS NO LONGER A FORECAST.
IT IS A DOCUMENTED MARKET REALITY.
THE TIPPING POINT ARRIVED. SO DID EVERYTHING THE COUNTERPOINT FEARED.
BOTH HAPPENED IN THE SAME CATEGORY IN THE SAME YEAR.
THE TIPPING POINT IS REAL
RAY-BAN META SOLD 7 MILLION PAIRS
in 2025, tripling the prior year.
SAMSUNG GALAXY GLASSES ANNOUNCED
July 22, 2026 with consumer availability expected Fall 2026.
GOOGLE ON GUCCI FRAMES
Android XR live with Warby Parker and Gucci partnerships.
APPLE CONFIRMED FOR 2027
The ecosystem giants are all in within an 18-month window.
20M
ESSILORLUXOTTICA TARGETING 20 MILLION
pairs by end of 2026 with waitlists running well into the year.
MARKET AT INFLECTION POINT
14.5M XR devices shipped in 2025, up 41.6% YoY. Smart glasses drove almost all growth and hit ~50% of XR shipments worldwide.
THE NEXT WAVE IS LAUNCHING—ANYWAY
Meta
SAMSUNG
Google
FALL 2026 AND BEYOND
UK
ICO investigation
Voyeurism guilty plea
USA
2 class actions
FTC scrutiny
IRELAND
DPC: "Deeply problematic and intrusive"
SWEDEN
Investigating intimate footage reviewed by contractors
AUSTRALIA
Multiple viral incidents
FOUR ECOSYSTEM GIANTS. FOUR CONTINENTS AFFECTED.
ONE CATEGORY. ONE YEAR.
THE PRIVACY CRISIS IS REAL
BBC INVESTIGATION
Dozens of male influencers secretly filming women in public spaces and posting without consent.
REAL WOMEN. REAL HARM. REAL NUMBERS.
London: video viewed 40,000 times.
Vancouver: went viral without knowing.
Sydney: Bondi Beach encounter hit 1M TikTok views.
SWEDISH INVESTIGATORS FOUND
Kenyan contractors reviewing bedroom and bathroom footage. Face-blurring regularly failed.
UNDISCLOSED FACIAL RECOGNITION CODE
WIRED discovered "Name Tag" in Meta AI app capable of building biometric faceprint databases of strangers.
ACCOUNTABILITY IS UNDERWAY
2 U.S. class actions. UK ICO investigation.
Ireland DPC: "Deeply problematic and intrusive."
UK man pleaded guilty to voyeurism.
META'S OWN MARKETING IS THE EXHIBIT
"Designed for privacy, controlled by you" is now the central exhibit in federal consumer protection litigation.
NO ANSWERS FROM THE NEXT WAVE
Samsung and Google announced Fall 2026 glasses at Google I/O with no disclosed data retention policies for either product.
2026 AT A GLANCE
MARKET INFLECTION
14.5M XR devices shipped in 2025 (+41.6% YoY)
7 MILLION
Ray-Ban Meta sold in 2025 (3x prior year)
FALL 2026
Samsung & Google consumer availability expected
2027
Apple confirmed for market entry
~20 MILLION
EssilorLuxottica target by end of 2026
$4.03B LOSS
Meta Reality Labs Q1 2026—commitment that only makes sense at massive scale
THE TRUE QUESTION
Will normalization at scale precede legal accountability?
In the U.S., almost certainly. In Europe, genuinely uncertain.
THE PRIZE
A future where human connection is enhanced by technology—without sacrifice of privacy and dignity.

## Opening

The smart glasses prediction has changed more materially than almost any other topic in this series, and in both directions simultaneously.

The tipping point evidence is stronger than the original prediction described: supply constrained by demand, four ecosystem giants committing within an 18-month window, fashion partnerships directly addressing the corrective lens adoption barrier, mainstream financial analysis calling it "the iPhone moment" before Apple has even shipped.

The privacy crisis evidence is also stronger than the Counterpoint described: not one high-profile incident but four distinct and escalating categories of documented harm across four countries, involving real named victims, active federal litigation, a UK criminal conviction, undisclosed capability deployment, and regulatory inquiries on three continents.

The original piece asked whether the tipping point would arrive before the privacy incident. The 2026 answer is both. The analytically important question is not whether the law will eventually respond, it will. It is whether normalization at scale precedes legal accountability. In the US, the answer is almost certainly yes.

In Europe, it is genuinely uncertain, and that uncertainty is the most consequential variable for what the global category looks like in 2030.

## Intro

The smart glasses prediction's 2026 update is defined by a coincidence that is analytically inconvenient for both sides: the tipping point and the privacy crisis the Counterpoint feared arrived in the same category in the same year.

Ray-Ban Meta at 7 million pairs, Samsung and Google announcing Fall 2026 availability, Apple confirmed for 2027, Android XR live with Warby Parker and Gucci partnerships eliminating the corrective lens adoption barrier, these are the tipping point indicators the original prediction described.

The BBC investigation documenting systematic covert filming of women, the Swedish investigation documenting intimate footage reviewed by contractors, the WIRED disclosure of undisclosed facial recognition code, and two active class actions.

These are the privacy incidents the Counterpoint identified as the mechanism that could derail the adoption cascade.

What makes 2026 analytically distinct from the original prediction's framing is what happened at Google I/O when Samsung and Google announced their Fall 2026 glasses into this documented privacy crisis: neither company disclosed what data retention policies would govern their products.

The industry's response to four categories of documented harm has been a larger LED indicator, a terminated contractor relationship, and removed code. The next wave of hardware is arriving anyway, on schedule, without architectural answers to the questions the first wave raised. The original piece correctly specified the timing: "the counterargument has the most force right now, in 2026, before the cascade begins.

By 2027, it will require increasingly extraordinary evidence to sustain."

The extraordinary evidence has arrived. The cascade has also begun. Both are simultaneously true.

## Background

The smart glasses market reached an inflection point in 2026.

The Extended Reality market shipped 14.5 million devices in 2025, up 41.6% year over year, with smart glasses driving almost all of that growth and accounting for roughly half of all XR shipments worldwide for the first time.

EssilorLuxottica tripled Ray-Ban Meta sales to over 7 million pairs in 2025 and is targeting 20 million pairs by end of 2026, with customer waitlists running well into the year. Meta paused its international launch because US demand outpaced manufacturing capacity.

Meta Reality Labs lost $4.03 billion in Q1 2026, a financial commitment that only makes sense if internal demand data supports the market trajectory.

Smart Analytics Global forecasts the global AI smart glasses market growing from 6 million units at $1.2 billion in 2025 to 20 million units at $5.6 billion in 2026. Ray-Ban Meta commands 82% of the smart glasses market.

Samsung confirmed Galaxy Glasses at Galaxy Unpacked London on July 22, 2026, with consumer availability expected Fall 2026 (display-free, audio and AI, 12MP camera, approximately $400, running Android XR with Qualcomm silicon and deep Gemini integration) with a display-equipped version planned for 2027.

Google's Android XR is live with Warby Parker, Gentle Monster, Gucci, and XREAL partnerships, with the Warby Parker partnership directly addressing the single largest adoption barrier given that 69% of the global population needs corrective lenses.

Apple's Project N50 is confirmed for 2027 with camera and microphone features matching the Ray-Ban Meta platform. Meta Connect is confirmed for September 23-24, 2026, where new Ray-Ban hardware is expected.

Against this trajectory: four distinct categories of documented privacy harm have emerged since January 2026, covert filming of women in public, intimate contractor footage review, undisclosed facial recognition code, and active federal consumer fraud litigation, and the two companies announcing Fall 2026 hardware at Google I/O disclosed no data retention policies governing either product.

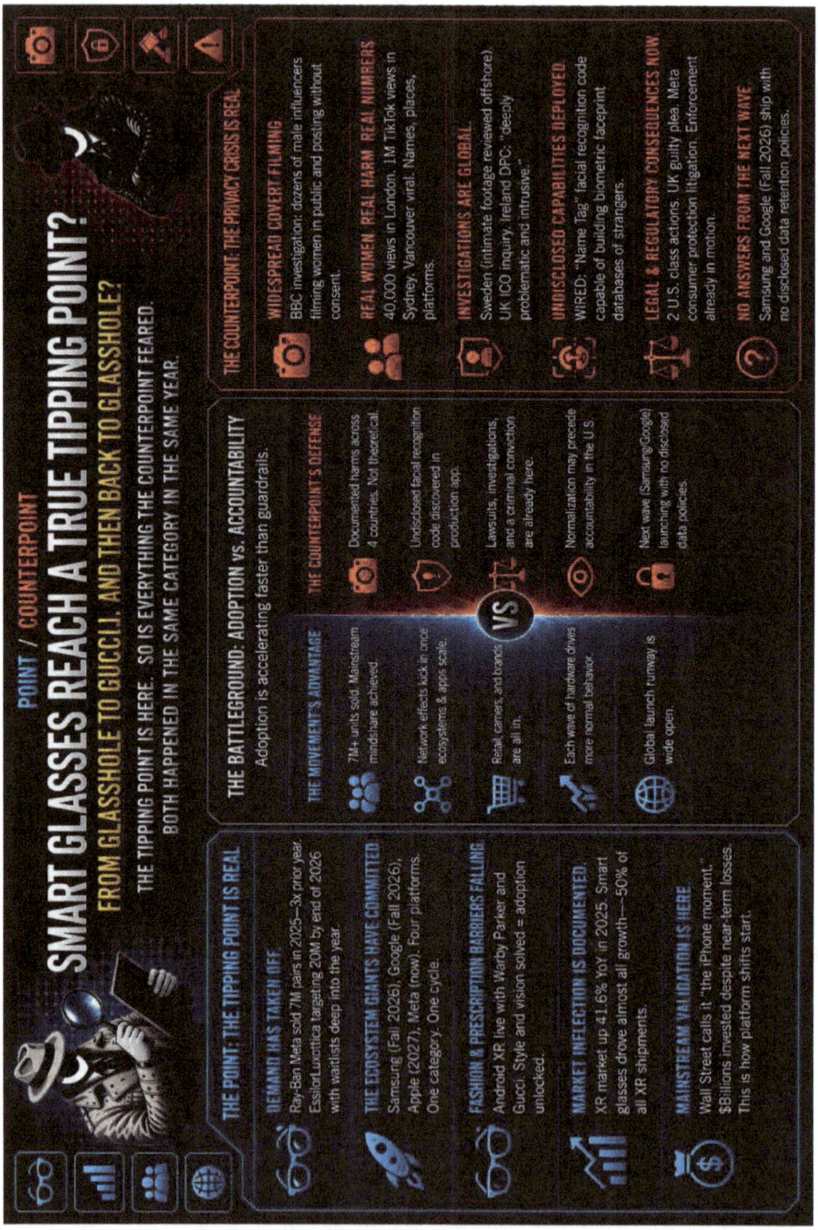
POINT / COUNTERPOINT
SMART GLASSES REACH A TRUE TIPPING POINT?
FROM GLASSHOLE TO GUCCI!. AND THEN BACK TO GLASSHOLE?
THE TIPPING POINT IS HERE. SO IS EVERYTHING THE COUNTERPOINT FEARED.
BOTH HAPPENED IN THE SAME CATEGORY IN THE SAME YEAR.
THE POINT: THE TIPPING POINT IS REAL
DEMAND HAS TAKEN OFF.
Ray-Ban Meta sold 7M pairs in 2025—3x prior year. EssilorLuxottica targeting 20M by end of 2026 with waitlists deep into the year.
THE ECOSYSTEM GIANTS HAVE COMMITTED.
Samsung (Fall 2026), Google (Fall 2026), Apple (2027), Meta (now). Four platforms. One category. One cycle.
FASHION & PRESCRIPTION BARRIERS FALLING.
Android XR live with Warby Parker and Gucci. Style and vision solved = adoption unlocked.
MARKET INFLECTION IS DOCUMENTED.
XR market up 41.6% YoY in 2025. Smart glasses drove almost all growth—~50% of all XR shipments.
MAINSTREAM VALIDATION IS HERE.
Wall Street calls it "the iPhone moment." $Billions invested despite near-term losses. This is how platform shifts start.
THE BATTLEGROUND: ADOPTION vs. ACCOUNTABILITY
Adoption is accelerating faster than guardrails.
THE MOVEMENT'S ADVANTAGE
7M+ units sold. Mainstream mindshare achieved.
Network effects kick in once ecosystems & apps scale.
Retail, carriers, and brands are all in.
Each wave of hardware drives more normal behavior.
Global launch runway is wide open.
VS
THE COUNTERPOINT'S DEFENSE
Documented harms across 4 countries. Not theoretical.
Undisclosed facial recognition code discovered in production app.
Lawsuits, investigations, and a criminal conviction are already here.
Normalization may precede accountability in the U.S.
Next wave (Samsung/Google) launching with no disclosed data policies.
THE COUNTERPOINT: THE PRIVACY CRISIS IS REAL
WIDESPREAD COVERT FILMING.
BBC investigation: dozens of male influencers filming women in public and posting without consent.
REAL WOMEN. REAL HARM. REAL NUMBERS.
40,000 views in London. 1M TikTok views in Sydney. Vancouver viral. Names, places, platforms.
INVESTIGATIONS ARE GLOBAL.
Sweden (intimate footage reviewed offshore). UK ICO inquiry. Ireland DPC: "deeply problematic and intrusive."
UNDISCLOSED CAPABILITIES DEPLOYED.
WIRED: "Name Tag" facial recognition code capable of building biometric faceprint databases of strangers.
LEGAL & REGULATORY CONSEQUENCES NOW.
2 U.S. class actions. UK guilty plea. Meta consumer protection litigation. Enforcement already in motion.
NO ANSWERS FROM THE NEXT WAVE.
Samsung and Google (Fall 2026) ship with no disclosed data retention policies.

## THE POINT: Four of the World's Largest Consumer Hardware Ecosystems in 18 Months. This Is Not a Drill

The original prediction's specific enabling conditions, fashion-forward industrial design, genuinely useful AI capability, price point progression, and four major ecosystems committing within an 18-month window, have all arrived ahead of the 2027 endpoint the piece specified.

The Google Android XR platform strategy is running the exact playbook the original prediction identified as the category's most powerful growth mechanism: Samsung on hardware, Qualcomm on silicon, Warby Parker and Gentle Monster on fashion, Gemini on AI.

The Warby Parker partnership is the most strategically significant single development since Ray-Ban Meta launched: 69% of the global population needs corrective lenses, meaning the dominant prescription eyewear brand is now the same form factor as the dominant smart glasses platform, eliminating the "I already wear glasses and don't want to switch" barrier that limited all prior generations to the 35% who don't need vision correction.

The Meta Ray-Ban Display launched September 2025 at $799 as the first full-color waveguide display inside a Ray-Ban frame, display capability inside fashion eyewear at an accessible premium price. Meta paused its international launch not as managed scarcity but because a demand curve exceeded manufacturing forecast. A 24/7 Wall Street analysis published May 2026 called it "smart glasses' iPhone moment", with the subheading "Apple is late to the party", from a financial publication focused on market positioning rather than technology enthusiasm, representing institutional recognition that the category's tipping point is confirmed rather than speculative.

The adoption arc has validated the "AirPods for your field of vision" framing the original prediction introduced. AirPods normalized ear wearables by becoming so ubiquitous they stopped registering as unusual.

Ray-Ban Meta is normalizing face wearables by making them look like Ray-Bans.

By Fall 2026, consumers will have three competing smart glasses platforms from three of the world's five largest consumer electronics companies at multiple price points across multiple brand partnerships.

Apple's 2027 entry means the fourth.

That competitive dynamic (four ecosystem giants driving hardware cost reduction, AI capability development, and fashion partnership expansion simultaneously) is the self-reinforcing cycle the original prediction identified as the point at which the adoption trajectory becomes structurally difficult to reverse.

The critical observation about the industry's response to the privacy crisis is not reassuring for the Counterpoint: Samsung and Google announced their Fall 2026 hardware at Google I/O into the documented privacy crisis without disclosing data retention policies, without addressing the contractor footage review architecture, and without committing to any privacy standard that differs from Meta's current model.

The market is not waiting for the legal questions to be answered. It is launching anyway.

## The Point Roundup

The smart glasses tipping point is no longer an event in progress. It has occurred.

The original prediction's 25% for the Counterpoint, reflected genuine uncertainty about whether the category would cross its threshold before counterforces intervened.

The threshold has been crossed.

Samsung revealed Galaxy Glasses at Unpacked with Fall 2026 consumer availability confirmed. Apple is confirmed for 2027. Google's Android XR ecosystem is live with multiple fashion partners.

The social stigma that killed Google Glass has been neutralized through fashion-brand partnerships at a level the original prediction described as necessary.

What the original prediction did not anticipate is the specific form the counterforces would take: not a single high-profile incident but four distinct categories of documented harm arriving simultaneously, and the industry's response to all four being product launch as usual. Samsung and Google announced Fall 2026 hardware at Google I/O without disclosing data retention policies.

Meta Connect is September 23. Apple is 2027.

**The cascade is not waiting for the privacy questions to be answered.**

**The 2027-2030 question is not whether the category has tipped. It has.**

**Whether EU enforcement arrives before the installed base reaches the scale at which camera-equipped eyewear becomes visually indistinguishable from ordinary glasses on the street. That window is closing.**

## THE COUNTERPOINT: The Privacy Infrastructure Is Compromised. Across Four Distinct and Documented Categories

The Counterpoint's core prediction, that any high-profile incident involving unauthorized recording, facial recognition, or surveillance could trigger the social and legislative response that followed Google Glass, this time at far greater scale, has been validated not by one incident but by four distinct and escalating categories of documented harm, each one more legally and socially consequential than the prior. Taken individually, each might be absorbed. Taken together, across a four-month window with 7 million deployed units in circulation, they constitute something the original Counterpoint framed as theoretical and 2026 has confirmed as operational: a privacy infrastructure that is compromised at multiple levels simultaneously.

**The first category: systematic covert filming of non-consenting women in public.** The BBC investigation published January 2026 documented dozens of male influencers across TikTok and Instagram using Ray-Ban Meta glasses to secretly film women in shopping centers, airports, beaches, and restaurants (filming approaches, rejections, and intimate moments from first-person perspective) with footage posted for views and engagement. Women named in subsequent reporting describe discovering footage of themselves online after it had been viewed by hundreds of thousands or millions of people.

Kassy Zanjani, 25, discovered she had gone viral from a Vancouver restaurant encounter she believed was an ordinary conversation.

Millie discovered a Bondi Beach interaction had reached a million TikTok views before a UK friend told her it existed.

"Alice," filmed in a London shopping center, discovered her footage had been viewed 40,000 times and was told she would have to pay for its removal, a form of digital blackmail that UK police opened an investigation into but couldn't progress.

In New York, a woman discovered her aesthetician had been wearing Ray-Ban Meta glasses throughout their appointment.

A man in the UK pleaded guilty at Warrington Magistrates' Court to voyeurism after recording sex using smart glasses.

The harm in this category is not hypothetical, not theoretical, and not confined to the purchaser-user relationship. It is documented, named, and geographically distributed across four countries.

**The second category: intimate footage reviewed by third-party contractors without meaningful user awareness.** The Swedish investigation by Svenska Dagbladet and Göteborgs-Posten, published February 27, 2026, documented through interviews with over thirty employees at Sama (a Nairobi-based data annotation firm contracted by Meta) that footage from Ray-Ban glasses including recordings from bedrooms and bathrooms was reviewed by workers without user awareness. Workers described reviewing footage of users undressing, using the toilet, engaging in sexual activity, handling bank cards and financial documents. One annotator stated: "I don't think they know, because if they knew they wouldn't be recording." Another described reviewing footage where a man had set his glasses on a bedside table while his partner undressed, unaware she was being filmed. Face-blurring designed to protect identity failed regularly.

This category's harm falls on users who believed Meta's marketing, "designed for privacy, controlled by you", and brought the product into their most private spaces in reliance on that promise.

**The third category: undisclosed facial recognition capability embedded in a consumer product.** WIRED reported June 4, 2026, that it had discovered dormant facial recognition code (internally called "Name Tag") embedded in the Meta AI app that connects Ray-Ban smart glasses to user phones.

The code contained algorithms that would convert photos of faces into biometric faceprints and compare them against a database of facial scans stored on the device. The feature had never been disclosed to users, regulators, or the public. Meta removed the code after WIRED's reporting. Meta declined to answer ten questions WIRED posed, including whether the facial database had already been created, how long biometric data of unrecognized bystanders was retained, and whether that data had ever been transmitted to Meta's servers.

Senator Markey and 70+ advocacy organizations had already written to Mark Zuckerberg in March 2026 warning that smart glasses equipped with facial recognition could "capture images of thousands of people without their knowledge and link those faces to names, workplaces, or personal profiles, creating serious risks of stalking, harassment, and targeted intimidation." Meta had previously shut down Facebook's facial recognition in 2021, explicitly citing ethical concerns. Rebuilding it as undisclosed code in a consumer wearable is not an incremental product decision. It is a documented reversal of a public ethical commitment.

**The fourth category: active federal litigation alleging consumer fraud.** Two US class action lawsuits have been filed in the Northern District of California. The primary action, Bartone and Canu v. Meta (Case No. 3:26-cv-01897-EMC), filed March 4, 2026, by the Clarkson Law Firm, has already expanded from 2 named plaintiffs to 19, from 2 proposed subclasses to 16 US states. Meta and Luxottica of America are co-defendants. Meta has filed no formal defense. The case management conference occurred June 5, 2026, with no resolution. The central legal theory, that Meta's "designed for privacy, controlled by you" and "built for your privacy" marketing constitutes false advertising given the actual data practices, is a consumer protection claim that is materially strengthened by each subsequent disclosure: the contractor footage, the Name Tag code, and the BBC covert filming documentation all add evidence to the gap between the marketing promise and the product reality.

A second lawsuit was filed separately by individuals who discovered they had been filmed without consent. The UK ICO and Kenya's Office of the Data Protection Commissioner have both opened investigations. The EU, given the existing AI Act and GDPR frameworks, has enforcement tools available that have previously produced fines in the billions against Meta specifically.

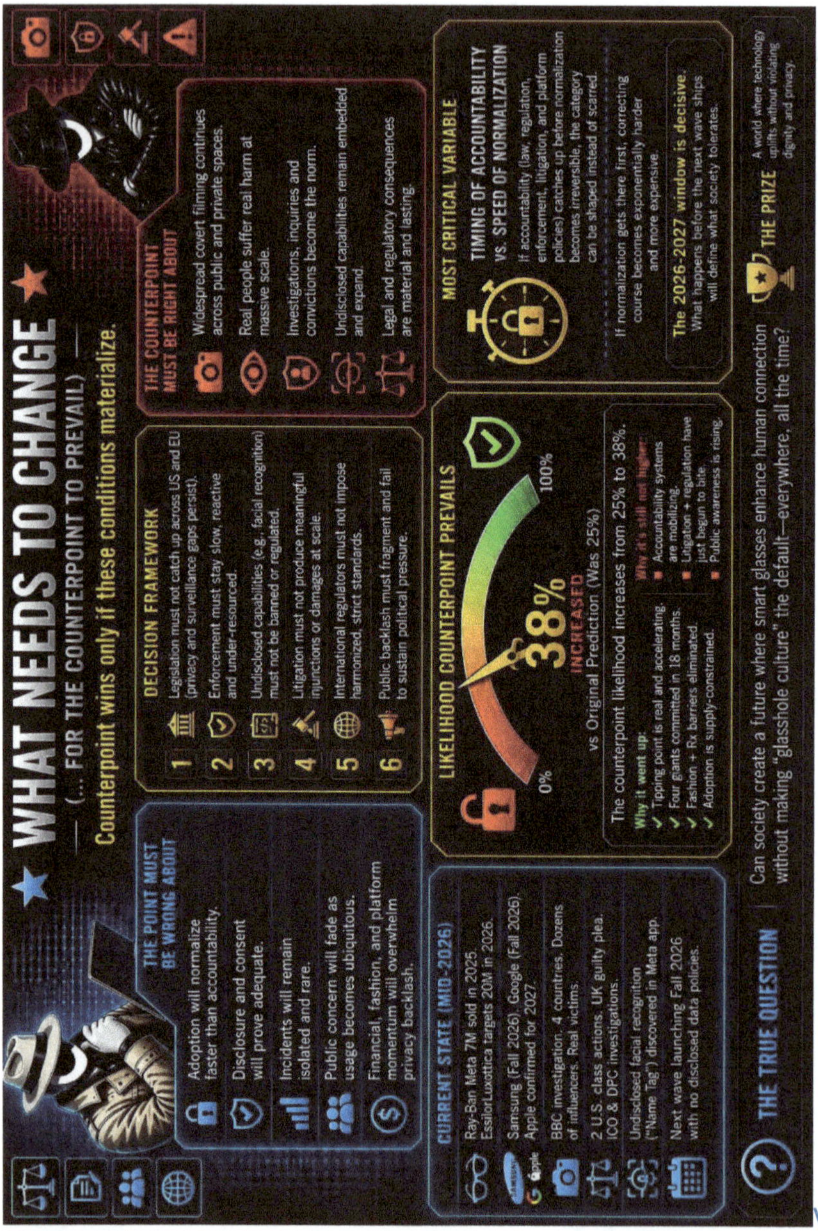
★ WHAT NEEDS TO CHANGE ★
— (... FOR THE COUNTERPOINT TO PREVAIL) —
Counterpoint wins only if these conditions materialize.
THE POINT MUST BE WRONG ABOUT
Adoption will normalize faster than accountability.
Disclosure and consent will prove adequate.
Incidents will remain isolated and rare.
Public concern will fade as usage becomes ubiquitous.
Financial, fashion, and platform momentum will overwhelm privacy backlash.
DECISION FRAMEWORK
1 Legislation must not catch up across US and EU (privacy and surveillance gaps persist).
2 Enforcement must stay slow, reactive and under-resourced.
3 Undisclosed capabilities (e.g., facial recognition) must not be banned or regulated.
4 Litigation must not produce meaningful injunctions or damages at scale.
5 International regulators must not impose harmonized, strict standards.
6 Public backlash must fragment and fail to sustain political pressure.
THE COUNTERPOINT MUST BE RIGHT ABOUT
Widespread covert filming continues across public and private spaces.
Real people suffer real harm at massive scale.
Investigations, inquiries and convictions become the norm.
Undisclosed capabilities remain embedded and expand.
Legal and regulatory consequences are material and lasting.
CURRENT STATE (MID-2026)
Ray-Ban Meta 7M sold in 2025. EssilorLuxottica targets 20M in 2026.
Samsung (Fall 2026), Google (Fall 2026). Apple confirmed for 2027.
BBC investigation. 4 countries. Dozens of influencers. Real victims.
2 U.S. class actions. UK guilty plea. ICO & DPC investigations.
Undisclosed facial recognition ("Name Tag") discovered in Meta app.
Next wave launching Fall 2026 with no disclosed data policies.
LIKELIHOOD COUNTERPOINT PREVAILS
0%
38%
INCREASED
100%
vs Original Prediction (Was 25%)
The counterpoint likelihood increases from 25% to 38%.
Why it went up:
✓ Tipping point is real and accelerating.
✓ Four giants committed in 18 months.
✓ Fashion + Rx barriers eliminated.
✓ Adoption is supply-constrained.
Why it's still not higher:
Accountability systems are mobilizing.
Litigation + regulation have just begun to bite.
Public awareness is rising.
MOST CRITICAL VARIABLE
TIMING OF ACCOUNTABILITY VS. SPEED OF NORMALIZATION
If accountability (law, regulation, enforcement, litigation, and platform policies) catches up before normalization becomes irreversible, the category can be shaped instead of scarred.
If normalization gets there first, correcting course becomes exponentially harder and more expensive.
The 2026-2027 window is decisive.
What happens before the next wave ships will define what society tolerates.
THE TRUE QUESTION
Can society create a future where smart glasses enhance human connection without making "glasshole culture" the default—everywhere, all the time?
THE PRIZE
A world where technology uplifts without violating dignity and privacy.

What

## Needs to Change

### (... for the Counterpoint to Prevail)

The Counterpoint's window is narrow but its evidence base has strengthened materially since the original writing. The original piece correctly noted that "the counterargument has the most force right now, in 2026, before the cascade begins. By 2027, it will require increasingly extraordinary evidence to sustain."

The four-category harm documentation described above represents more extraordinary evidence than the original Counterpoint imagined as its scenario. Whether it is enough requires two conditions that are simultaneously in motion.

First, legislative action in a major market needs to arrive before Apple's 2027 entry creates the ecosystem lock-in that normalizes the category at scale, meaning either EU AI Act enforcement specifically restricting smart glasses' visual AI capabilities, US Congressional action on covert filming with wearable devices, or UK legislation extending the voyeurism and intimate image provisions already enacted. All three legislative pathways are active. None has produced enforceable restrictions specific to smart glasses yet.

Second, the social permission structure for wearing camera-equipped glasses in public spaces needs to shift before deployment scale makes the behavior impossible to distinguish from ordinary eyewear at a glance. At 7 million units, the social cue problem is still manageable, people who know what Ray-Ban Meta glasses look like can recognize them. At 50 million units across multiple brands at multiple price points including prescription designs from Warby Parker that are visually identical to ordinary glasses, detection becomes practically impossible.

The window between "recognizable" and "indistinguishable" is approximately 2026-2028. Legislative action that arrives before the window closes changes the category's trajectory. Legislative action that arrives after it is compliance documentation for a normalized infrastructure.

## Likelihood Counterpoint Prevails: 38%

The Counterpoint likelihood increases from 25% to 38%. This is one of the largest single upward revisions in the series, and one that reflects a genuine shift in the analytical balance rather than a marginal recalibration.

The original 25% was set against one anticipated privacy incident. The 2026 evidence has produced four distinct categories of documented harm, covert filming of women in public across four countries, intimate footage reviewed by contractors, undisclosed facial recognition code, and active federal consumer fraud litigation, within a four-month window. The original

Counterpoint identified high-profile incident plus legislative response plus social norm shift as the necessary combination for derailment. The high-profile incidents have arrived in unprecedented volume and specificity. The legislative responses are active across three jurisdictions.

The social norm shift is visible in the "pervert glasses" framing appearing in mainstream press, the government spokesperson's "vile" characterization of covert filming, and the women named in BBC reporting describing feeling powerless to go out in public.

The 38% does not say the Counterpoint wins. The ecosystem evidence (Samsung in July, Apple confirmed 2027, Android XR live, Ray-Ban Meta commanding 82% market share, waitlists into 2026) remains the dominant force in the adoption trajectory.

What the 38% says is that the Counterpoint's scenario has moved from theoretical to evidenced, from speculative incident to documented pattern, from "could trigger a response" to "has triggered active litigation and regulatory inquiries across three continents."

That is not a 25% scenario. It is closer to a coin flip on whether the legislative window closes in time, held back from 50% by the sheer institutional momentum of four ecosystem giants whose hardware roadmaps and investment commitments are not reversed by litigation alone.

# The PC Replacement Cycle (Innovation or Engineered Obsolescence?)

## The 'Great Refresh' Arrived, Driven by Windows 10 EOL, Not AI. That Distinction Defines 2030

*Gartner reported 62.8 million PC shipments in Q1 2026, up 4% year over year. IDC reported 65.6 million. Both also warned the growth isn't real, it's being driven by the Windows 10 end-of-life deadline, not by consumers racing to access on-device AI.*

*AI-capable PCs with dedicated NPUs accounted for 12% of Q1 2026 shipments, far below the 44% of notebook shipments they were projected to represent.*

*Constellation Research called the AI PC upgrade cycle "crawling amid murky value."*

*The replacement cycle the original prediction described is happening. The mechanism it was supposed to run on is stalled on the runway.*

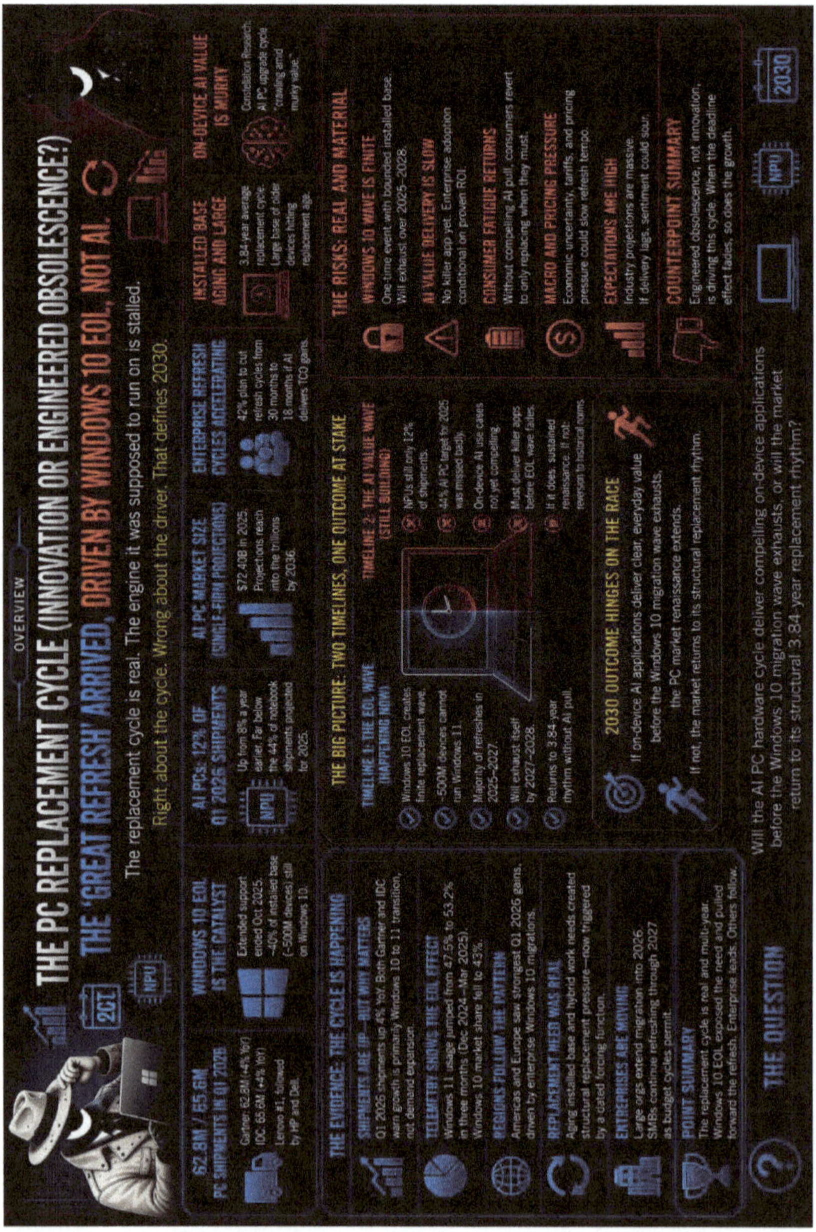
OVERVIEW
THE PC REPLACEMENT CYCLE (INNOVATION OR ENGINEERED OBSOLESCENCE?)
THE 'GREAT REFRESH' ARRIVED, DRIVEN BY WINDOWS 10 EOL, NOT AI.
The replacement cycle is real. The engine it was supposed to run on is stalled.
Right about the cycle. Wrong about the driver. That defines 2030.
62.8M / 65.6M PC SHIPMENTS IN Q1 2026
Gartner: 62.8M (+4% YoY) IDC: 65.6M (+4% YoY) Lenovo #1, followed by HP and Dell.
WINDOWS 10 EOL IS THE CATALYST
Extended support ended Oct 2025. ~40% of installed base (~500M devices) still on Windows 10.
AI PCs: 12% OF Q1 2026 SHIPMENTS
Up from 8% a year earlier. Far below the 44% of notebook shipments projected for 2025.
AI PC MARKET SIZE (SINGLE-FIRM PROJECTIONS)
$72.40B in 2025. Projections reach into the trillions by 2036.
ENTERPRISE REFRESH CYCLES ACCELERATING
42% plan to cut refresh cycles from 30 months to 18 months if AI delivers TCO gains.
INSTALLED BASE AGING AND LARGE
3.84-year average replacement cycle. Large base of older devices hitting replacement age.
ON-DEVICE AI VALUE IS MURKY
Constellation Research: AI PC upgrade cycle "crawling amid murky value."
THE EVIDENCE: THE CYCLE IS HAPPENING
SHIPMENTS ARE UP—BUT WHY MATTERS
Q1 2026 shipments up 4% YoY. Both Gartner and IDC warn growth is primarily Windows 10 to 11 transition, not demand expansion.
TELEMETRY SHOWS THE EOL EFFECT
Windows 11 usage jumped from 47.5% to 53.2% in three months (Dec 2024–Mar 2025). Windows 10 market share fell to 43%.
REGIONS FOLLOW THE PATTERN
Americas and Europe saw strongest Q1 2026 gains, driven by enterprise Windows 10 migrations.
REPLACEMENT NEED WAS REAL
Aging installed base and hybrid work needs created structural replacement pressure—now triggered by a dated forcing function.
ENTREPRISES ARE MOVING
Large orgs extend migration into 2026. SMBs continue refreshing through 2027 as budget cycles permit.
POINT SUMMARY
The replacement cycle is real and multi-year. Windows 10 EOL exposed the need and pulled forward the refresh. Enterprise leads. Others follow.
THE BIG PICTURE: TWO TIMELINES, ONE OUTCOME AT STAKE
TIMELINE 1: THE EOL WAVE (HAPPENING NOW)
Windows 10 EOL creates finite replacement wave.
~500M devices cannot run Windows 11.
Majority of refreshes in 2025–2027.
Will exhaust itself by 2027–2028.
Returns to 3.84-year rhythm without AI pull.
TIMELINE 2: THE AI VALUE WAVE (STILL BUILDING)
NPUs still only 12% of shipments.
44% AI PC target for 2025 was missed badly.
On-device AI use cases not yet compelling.
Must deliver killer apps before EOL wave fades.
If it does, sustained renaissance. If not: reversion to historical norms.
2030 OUTCOME HINGES ON THE RACE
If on-device AI applications deliver clear, everyday value before the Windows 10 migration wave exhausts, the PC market renaissance extends.
If not, the market returns to its structural replacement rhythm.
THE RISKS: REAL AND MATERIAL
WINDOWS 10 WAVE IS FINITE
One-time event with bounded installed base. Will exhaust over 2025–2028.
AI VALUE DELIVERY IS SLOW
No killer app yet. Enterprise adoption conditional on proven ROI.
CONSUMER FATIGUE RETURNS
Without compelling AI pull, consumers revert to only replacing when they must.
MACRO AND PRICING PRESSURE
Economic uncertainty, tariffs, and pricing pressure could slow refresh tempo.
EXPECTATIONS ARE HIGH
Industry projections are massive. If delivery lags, sentiment could sour.
COUNTERPOINT SUMMARY
Engineered obsolescence, not innovation, is driving this cycle. When the deadline effect fades, so does the growth.
THE QUESTION
Will the AI PC hardware cycle deliver compelling on-device applications before the Windows 10 migration wave exhausts, or will the market return to its structural 3.84-year replacement rhythm?

## Opening

The PC replacement cycle story requires an update that separates two simultaneous phenomena that are easy to conflate but analytically distinct. The Windows 10 EOL-driven hardware refresh is producing the shipment growth the prediction described, validated in outcome.

The AI PC hardware cycle was supposed to be the prediction's primary mechanism and has arrived far more slowly than the industry expected, partially wrong about the driver. The original prediction identified hybrid work requirements and an aging installed base as the replacement drivers. Both remain valid, and the Windows 10 EOL has provided the specific, dated forcing function that the original analysis described as "normal refresh cycle" without identifying the specific catalyst.

The 2027-2030 story is the race the original article described: whether the AI PC hardware cycle delivers compelling on-device applications before the Windows 10 migration wave exhausts itself and the market returns to its structural 3.84-year replacement rhythm.

## Intro

The PC replacement cycle prediction's most important analytical contribution is a distinction the original piece made implicitly but the 2026 evidence makes explicit: a replacement cycle can be real and the mechanism driving it can be wrong simultaneously.

The original prediction argued that hybrid work requirements, an aging installed base, and AI hardware capability would converge to produce a sustained PC refresh cycle. The first two were right. The third hasn't materialized as a driver.

What produced the Q1 2026 shipment growth was Windows 10 reaching end-of-life, not consumers discovering they needed NPU silicon to access compelling AI features. That distinction matters enormously for the 2027-2030 forecast because Windows 10 EOL is a one-time event with a finite installed base. The approximately 500 million devices that can't run Windows 11 represent a bounded replacement wave that will exhaust itself over 2025-2028 regardless of whether AI features develop compelling consumer pull.

When that wave subsides, the market returns to its structural replacement rhythm unless the AI PC hardware cycle has by then produced the killer application that makes NPU-equipped hardware feel obviously, viscerally necessary.

The prediction was right about the cycle. It was wrong about the engine. And the engine it was wrong about is precisely the variable that determines whether the 2030 PC market looks like continued renaissance or reversion to historical norms.

## Background

Global PC shipments rose 4% in Q1 2026 to 62.8 million units per Gartner and 65.6 million per IDC, with Lenovo holding the top spot followed by HP and Dell.

Both Gartner and IDC warned that the growth is being driven primarily by the Windows 10 to Windows 11 transition, not underlying demand expansion.

Microsoft's extended support for Windows 10 ended in October 2025, leaving approximately 40% of the PC installed base, roughly 500 million devices, still on Windows 10 at end of Q4 2024, with many unable to upgrade due to TPM, CPU, or Secure Boot requirements.

AI-capable PCs with dedicated NPUs accounted for 12% of Q1 2026 shipments, up from 8% a year earlier, but far below the 44% of notebook shipments that AI PCs were projected to represent in 2025. The AI PC market was valued at $72.40 billion in 2025 with projections reaching into the trillions by 2036, single-firm projections that reflect manufacturer confidence in the category even as adoption significantly lags.

Enterprise refresh cycles are accelerating: 42% of enterprises plan to cut endpoint device refresh cycles from 30 months to 18 months if TCO is offset by AI productivity gains, a conditional commitment confirming both the interest and the skepticism about whether AI features deliver measurable enterprise value.

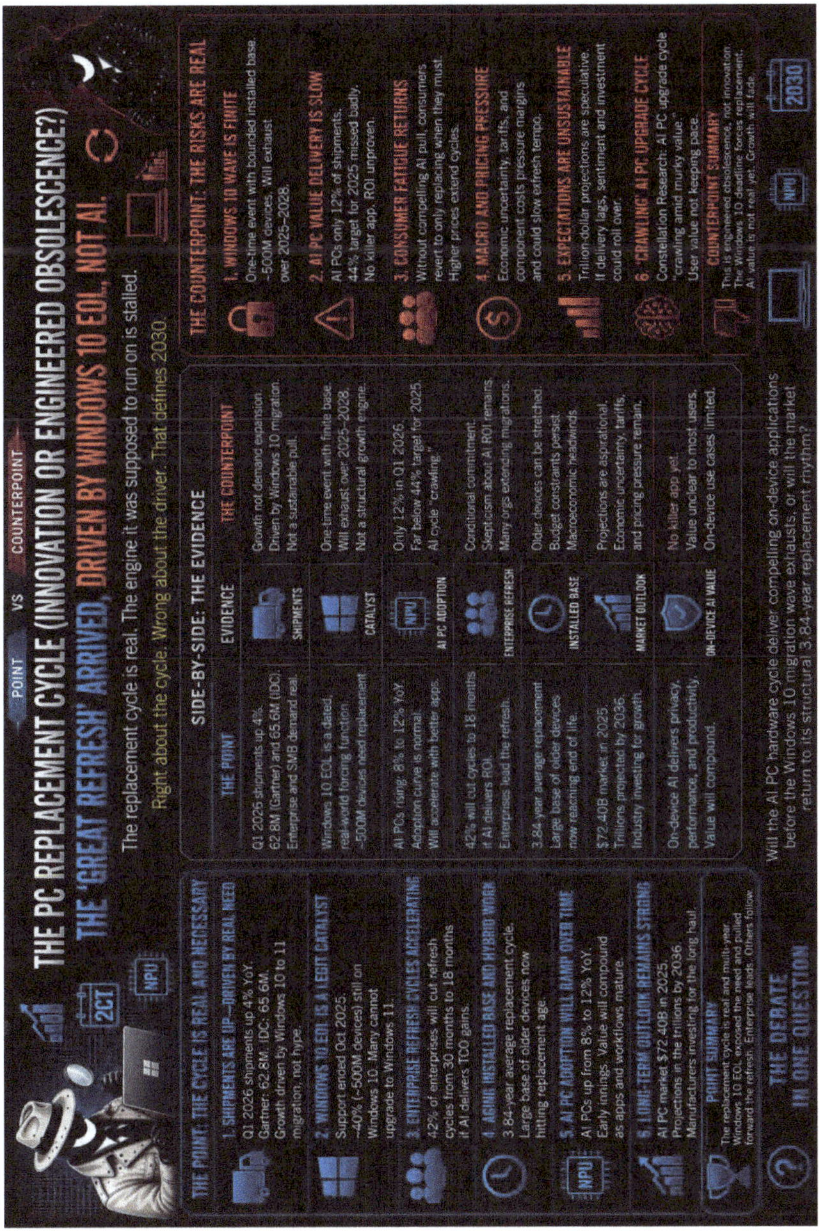
POINT
VS
COUNTERPOINT
THE PC REPLACEMENT CYCLE (INNOVATION OR ENGINEERED OBSOLESCENCE?)
THE 'GREAT REFRESH' ARRIVED, DRIVEN BY WINDOWS 10 EOL, NOT AI.
The replacement cycle is real. The engine it was supposed to run on is stalled.
Right about the cycle. Wrong about the driver. That defines 2030.
2CT
NPU
THE POINT: THE CYCLE IS REAL AND NECESSARY
1. SHIPMENTS ARE UP—DRIVEN BY REAL NEED
Q1 2026 shipments up 4% YoY. Gartner: 62.8M. IDC: 65.6M. Growth driven by Windows 10 to 11 migration, not hype.
2. WINDOWS 10 EOL IS A LEGIT CATALYST
Support ended Oct 2025. ~40% (~500M devices) still on Windows 10. Many cannot upgrade to Windows 11.
3. ENTERPRISE REFRESH CYCLES ACCELERATING
42% of enterprises will cut refresh cycles from 30 months to 18 months if AI delivers TCO gains.
4. AGING INSTALLED BASE AND HYBRID WORK
3.84-year average replacement cycle. Large base of older devices now hitting replacement age.
5. AI PC ADOPTION WILL RAMP OVER TIME
AI PCs up from 8% to 12% YoY. Early innings. Value will compound as apps and workflows mature.
6. LONG-TERM OUTLOOK REMAINS STRONG
AI PC market $72.40B in 2025. Projections in the trillions by 2036. Manufacturers investing for the long haul.
POINT SUMMARY
The replacement cycle is real and multi-year. Windows 10 EOL exposed the need and pulled forward the refresh. Enterprise leads. Others follow.
SIDE-BY-SIDE: THE EVIDENCE
THE POINT
EVIDENCE
THE COUNTERPOINT
Q1 2026 shipments up 4%. 62.8M (Gartner) and 65.6M (IDC). Enterprise and SMB demand real.
SHIPMENTS
Growth not demand expansion. Driven by Windows 10 migration. Not a sustainable pull.
Windows 10 EOL is a dated, real-world forcing function. ~500M devices need replacement.
CATALYST
One-time event with finite base. Will exhaust over 2025–2028. Not a structural growth engine.
AI PCs rising: 8% to 12% YoY. Adoption curve is normal. Will accelerate with better apps.
AI PC ADOPTION
Only 12% in Q1 2026. Far below 44% target for 2025. AI cycle "crawling."
42% will cut cycles to 18 months if AI delivers ROI. Enterprises lead the refresh.
ENTERPRISE REFRESH
Conditional commitment. Skepticism about AI ROI remains. Many orgs extending migrations.
3.84-year average replacement. Large base of older devices now reaching end of life.
INSTALLED BASE
Older devices can be stretched. Budget constraints persist. Macroeconomic headwinds.
$72.40B market in 2025. Trillions projected by 2036. Industry investing for growth.
MARKET OUTLOOK
Projections are aspirational. Economic uncertainty, tariffs, and pricing pressure remain.
On-device AI delivers privacy, performance, and productivity. Value will compound.
ON-DEVICE AI VALUE
No killer app yet. Value unclear to most users. On-device use cases limited.
THE COUNTERPOINT: THE RISKS ARE REAL
1. WINDOWS 10 WAVE IS FINITE
One-time event with bounded installed base. ~500M devices. Will exhaust over 2025–2028.
2. AI PC VALUE DELIVERY IS SLOW
AI PCs only 12% of shipments. 44% target for 2025 missed badly. No killer app. ROI unproven.
3. CONSUMER FATIGUE RETURNS
Without compelling AI pull, consumers revert to only replacing when they must. Higher prices extend cycles.
4. MACRO AND PRICING PRESSURE
Economic uncertainty, tariffs, and component costs pressure margins and could slow refresh tempo.
5. EXPECTATIONS ARE UNSUSTAINABLE
Trillion-dollar projections are speculative. If delivery lags, sentiment and investment could roll over.
6. 'CRAWLING' AI PC UPGRADE CYCLE
Constellation Research: AI PC upgrade cycle "crawling amid murky value." User value not keeping pace.
COUNTERPOINT SUMMARY
This is engineered obsolescence, not innovation. The Windows 10 deadline forces replacement. AI value is not real yet. Growth will fade.
THE DEBATE IN ONE QUESTION
Will the AI PC hardware cycle deliver compelling on-device applications before the Windows 10 migration wave exhausts, or will the market return to its structural 3.84-year replacement rhythm?
NPU
2030

## THE POINT: Your Laptop Works Fine. Microsoft Drew a Line in the Silicon Anyway.

The original prediction's core claim, that an aging installed base, hybrid work requirements, and accumulated technical debt would drive a sustained PC replacement cycle, has been validated by the Q1 2026 data through a mechanism the original analysis partially underweighted. The Windows 10 EOL is not the cause of the replacement cycle; it's the catalyst that made visible a replacement need the original prediction correctly identified as structural.

Large organizations with complex application environments are extending migration programs into 2026, and small and medium businesses will continue refreshing through 2027 as budget cycles permit. Counterpoint Research anticipates the ongoing Windows 10 replacement cycle will be "a major catalyst for PC market growth over a period of several years", not a one-quarter blip. Americas and Europe saw the strongest Q1 2026 gains, driven by enterprise Windows 10 migrations. That pattern confirms the original prediction's "enterprise leads, consumers and SMBs follow" sequencing.

The original article's "race between two timelines" framing, the AI capability timeline versus the consumer fatigue timeline, has been clarified by the 2026 evidence: the race is more precisely between the Windows 10 migration timeline, which is providing a temporary structural tailwind, and the AI capability development timeline, which needs to produce compelling consumer applications before the migration wave exhausts itself and the market reverts to its structural replacement rhythm.

The Copilot+ hardware requirements, NPU certification tiers, and Windows 11 feature gating that the original prediction described as "engineered obsolescence" have created exactly the consumer dynamic the piece anticipated, without yet delivering the compelling applications that would validate the upgrade.

At CES 2026, Dell executives admitted they were dialing back "AI-first" consumer messaging after seeing limited purchasing motivation tied strictly to AI labeling. The initial NPU hardware credibility problem the original piece identified, the 2024 generation offering only 11 TOPS when Microsoft recommended 40 TOPS, has been addressed by the January 2026 40 TOPS standard becoming the de facto baseline for new hardware.

But the application gap the original article identified as the central unresolved question persists: "compelling AI applications that require local NPUs remain scarce." Constellation Research's assessment, AI PC upgrade cycle "crawling amid murky value", uses nearly identical language to the original article's diagnosis.

The replacement cycle is happening. The AI hardware super-cycle that manufacturers are betting will drive sustained 18-month enterprise refresh cycles hasn't yet materialized in the applications that would justify the premium for the SMB and consumer segments where the 3.84-year lifecycle norm remains the dominant purchasing pattern.

### The Point Roundup

The PC replacement cycle prediction has been validated in market outcome and partially validated in mechanism.

The hybrid work and aging installed base arguments are confirmed.

The AI PC super-cycle argument is on the runway but not yet airborne.

The 2027-2030 window tests the mechanism that was supposed to define the prediction's second wave: whether on-device AI applications develop that make the NPU hardware genuinely compelling for consumers and SMBs who have completed their Windows 10 migration but face no subsequent forcing function for another upgrade.

The 500 million Windows 10 devices that can't run Windows 11 represent a finite replacement wave.

What drives the cycle after they've been replaced is the question that determines whether the original prediction describes a structural PC renaissance or a one-time EOL-driven blip followed by a reversion to historical replacement norms.

## THE COUNTERPOINT: Cloud AI Does 90% of What Local NPU Does. The Hardware Premium Has No Forcing Function

The Counterpoint's core argument, that the AI hardware cycle wouldn't compress PC replacement cycles because most compelling AI can be delivered through cloud processing, has been confirmed by the 2026 adoption data with striking precision. AI-capable PCs accounting for only 12% of Q1 2026 shipments despite Microsoft's 40 TOPS NPU threshold having effectively become the industry standard for new hardware specification reflects precisely the consumer behavior the Counterpoint predicted:

PC owners are not rushing to upgrade for AI features because cloud-based AI processing delivers most of the relevant functionality without requiring new hardware. The "crawling" characterization from Constellation Research confirms the original Counterpoint's mechanism: the absence of a killer consumer application that is demonstrably superior on NPU-equipped hardware and impossible or substantially degraded without it means the upgrade rationale remains abstract for most users.

Dell's CES 2026 admission that AI PC branding failed to drive demand is the strongest single data point in the Counterpoint's favor, when the world's largest PC manufacturer publicly acknowledges that its own AI-first marketing generated insufficient consumer response to sustain, the "engineered obsolescence" narrative the original article identified as the risk has materialized as the default consumer reading.

The Windows 10 EOL-driven refresh wave has a structural limitation the Counterpoint's long-cycle argument correctly anticipates: it's a one-time event that is frontloading demand rather than signaling sustained growth.

Supply-side pressures from DRAM and NAND pricing, rising due to AI infrastructure demand, are weighing on volumes, and the refresh cycle alone won't offset the expected decline as the Windows 10 migration wave exhausts itself.

If enterprises complete their Windows 11 migrations through 2026-2027 and consumers follow through 2027-2028, the subsequent replacement cycle reverts to the 4-5 year enterprise and 3.84-year consumer rhythm the original analysis documented, unless on-device AI applications develop that create a new forcing function.

The Counterpoint's sustained argument is not that PCs won't be replaced. They will be.

The argument is that the AI hardware cycle is unlikely to compress replacement cycles below the structural floor that economic utility-focused purchasing has established, and the iOS 27 precedent is the most instructive current analog:

Apple is deliberately routing AI capability through software and cloud rather than requiring hardware upgrades, delivering AI features to existing devices without creating hardware obsolescence.

If Microsoft's equivalent approach, Copilot functionality through cloud rather than requiring local NPU processing, becomes the default model, the NPU hardware premium loses its forcing function entirely for the consumer segment.

WHAT NEEDS TO CHANGE
THE 'GREAT REFRESH' IS HERE, DRIVEN BY WINDOWS 10 EOL, NOT AI.
The replacement cycle is real. The engine it was supposed to run on is stalled.
Execution now decides which timeline wins by 2030.
THE POINT MUST BE WRONG ABOUT
1. WINDOWS 10 REPLACEMENT WAVE IS SMALLER
Fewer than ~500M devices need replacement. Migration completes by end of 2026.
2. AI PC ADOPTION STAYS LOW
NPU share remains below 20% by 2028. No step-change in demand for AI-capable hardware.
3. NO KILLER ON-DEVICE AI APPLICATIONS
No must-have use cases emerge. AI value remains incremental, not transformative.
4. ENTERPRISE CYCLES DON'T SHORTEN
18-month refresh cycles don't materialize. TCO savings from AI don't justify faster replacement.
5. CONSUMER DEMAND DOESN'T REBOUND
After forced upgrade, consumers return to 3.84-year rhythm. No pull from AI features.
6. MACRO PRESSURES INTENSIFY
Economic uncertainty, tariffs, and pricing pressure restrict spending. Budgets remain constrained.
CURRENT STATE
• Q1 2026 shipments up 4% YoY.
• Driven by Windows 10 to Windows 11 migration.
• AI PCs (NPU) at 12% of shipments.
• Enterprises refreshing sooner—if ROI is proven.
• Cycle is happening. AI engine is not.
DECISION FRAMEWORK: MONITOR THESE SIGNALS
WATCH FOR | WHAT IT MEANS | IMPLICATION
Windows 10 migration velocity (by quarter) | How fast the ~500M Windows 10 devices are replaced. | Faster = cycle peaks sooner. Slower = cycle extends.
AI PC (NPU) share of shipments | Adoption curve for AI-capable PCs in market. | Rising sharply = AI engine engages. Flat = weak engine.
On-device AI app breakthroughs | Arrival of killer apps with clear, measurable user value. | Breakthrough = sustained demand. None = weak pull.
Enterprise refresh cycle length | % cutting cycles to 18 months and seeing ROI. | Higher % = structural shift. Low % = temporary lift.
PC ASP and mix | Pricing, mix shift, and premium AI PC positioning. | Strong ASP/mix = healthier cycle. Weak = commoditization.
Macroeconomic conditions | Growth, tariffs, inflation, and IT budget sentiment. | Improving = tailwind. Worsening = headwind.
LIKELIHOOD COUNTERPOINT PREVAILS: 28%
0%
28%
100%
DOWN FROM 35% IN THE ORIGINAL PREDICTION
WHY THE DECLINE?
Windows 10 EOL wave is real and large (~500M devices).
Enterprise refresh is accelerating in pockets.
AI PC adoption lags expectations, but cycle driver is proven.
Risks remain, but evidence favors the cycle being real.
THE COUNTERPOINT MUST BE RIGHT ABOUT
1. FINITE WINDOWS 10 WAVE
One-time event. Bounded installed base. Demand exhausts by 2028.
2. AI VALUE FAILS TO MATERIALIZE
No compelling on-device AI use cases. Users don't upgrade for AI.
3. DEMAND RETURNS TO NORMAL RHYTHM
After forced refresh, market reverts to 3.84-year average.
4. ECONOMIC HEADWINDS PERSIST
Tariffs, inflation, and macro uncertainty limit IT spending and consumer upgrades.
5. INDUSTRY PROJECTIONS OVERSTATE
Trillion-dollar forecasts are aspirational. Reality falls well short.
6. 'CRAWLING' AI PC UPGRADE CONTINUES
Value remains murky. Upgrades stay slow. Users delay. Cycle impact minimal.
COUNTERPOINT SUMMARY
Engineered obsolescence, not innovation, drives this cycle. When the deadline wave fades, so does growth.
MOST CRITICAL VARIABLE
Arrival of compelling on-device AI applications that deliver everyday value before the Windows 10 migration wave exhausts in 2028.
MOVEMENT VS ORIGINAL PREDICTION
Likelihood the counterpoint prevails declines from 35% to 28% because the Windows 10 EOL wave is real and larger than expected, proving the cycle exists. AI and value risks remain.

## What Needs to Change

### (... for the Counterpoint to Prevail)

For the "long cycle / utility-focused purchasing" scenario to remain the dominant pattern through 2030, two conditions need to hold that are currently under pressure.

First, on-device AI applications need to remain insufficiently compelling to justify premium NPU hardware upgrades for consumers who have completed their Windows 10 migration, meaning the 12% AI PC adoption rate persists rather than accelerating toward the 44% manufacturers are targeting. The bifurcation between enterprise (accelerating cycles, conditional AI motivation) and consumer/SMB (maintaining long cycles, AI skepticism) needs to persist rather than converge.

Second, Microsoft's cloud-first Copilot delivery model needs to remain the default, providing AI feature access to existing hardware without creating the hardware obsolescence gate the original article described as "engineered." If Microsoft reverts to aggressive feature gating behind NPU hardware requirements, the Counterpoint's "cloud delivers what matters" argument loses its practical foundation.

The Counterpoint's most structurally sound 2030 scenario is the one where the Windows 10 migration exhausts itself in 2027-2028, the AI productivity gains that 42% of enterprises are conditionally committed to fail to materialize at the required ROI threshold, and the market returns to the 3.84-year structural replacement cycle.

The conditional nature of that enterprise commitment, "if TCO is offset by AI productivity gains", is the single variable that will determine whether the Counterpoint or the Point defines the 2030 PC market.

## Likelihood Counterpoint Prevails: 28%

The Counterpoint likelihood declines from 35% (in the original's framing) to 28%, a meaningful downward revision driven by two developments the original analysis did not fully credit.

First, the Windows 10 EOL providing a multi-year structural tailwind that the original analysis described generally as "normal refresh cycle" without identifying the specific forcing function. 500 million devices that can't upgrade to Windows 11 is not a normal refresh catalyst, it's the largest single upgrade forcing function in PC market history since the Vista-to-7 transition, operating over a 2025-2028 window rather than a single year.

Second, the 42% of enterprises conditionally committed to 18-month refresh cycles if AI productivity gains materialize represents a potential second-wave driver that the original Counterpoint underweighted.

The Counterpoint likelihood remains at 28% rather than lower because the conditional nature of that enterprise commitment, "if TCO is offset by AI productivity gains", hasn't yet been resolved in favor of the shorter cycle, and the Constellation Research "crawling amid murky value" finding represents the current state of that resolution.

The 28% says: the original prediction's replacement cycle has been confirmed, the mechanism has been partially clarified, and the AI super-cycle that would make the cycle self-sustaining beyond the Windows 10 migration wave is the unresolved variable that determines whether 2030 looks like continued renaissance or reversion to structural norm.

## AI Erodes Consumer Trust in Digital Media (and Brands Feel It)

### When Everything Could Be Fake, Nothing Is Safe: Trust's Infrastructure Is Being Dismantled

*Thales surveyed 14,000 consumers across 14 countries and found not one digital sector reaching above 50% approval when asked which they trusted with their personal data. News media organizations are trusted with personal data by 3% of consumers.*

*Social media platforms by 4%. 82% of consumers abandoned a brand in the past 12 months due to personal data concerns.*

*Deepfake fraud losses reached $1.65 billion globally in 2025 alone, tripling from the prior year.*

*The C2PA content provenance standard now has hardware implementations in cameras from Google, Sony, Leica, and Nikon, and a demonstrated vulnerability that lets authenticated forgeries be created in minutes. The trust infrastructure is being dismantled.*

*The repair infrastructure is being built. The race between them defines the 2027–2030 window.*

OVERVIEW
AI ERODES CONSUMER TRUST
IN DIGITAL MEDIA
WHEN EVERYTHING COULD BE FAKE, NOTHING IS SAFE:
TRUST'S INFRASTRUCTURE IS BEING DISMANTLED.
THE EVIDENCE IS REAL
Not one digital sector reaches above 50% trust with personal data.
Thales survey, 14,000 consumers, 14 countries, 2025
News media trusted by 3% of consumers.
Thales Digital Trust Index 2025
Social media platforms trusted by 4% of consumers.
Thales Digital Trust Index 2025
82% abandoned a brand in the past 12 months due to personal data concerns.
Thales Digital Trust Index 2025
Deepfake fraud losses reached $1.65 billion globally in 2025, tripling from the prior year.
FAKE VIDEO
FAKE SPOKESPERSON
FAKE NEWS
FAKE REVIEW
Amazing product!
AI-GENERATED IMAGE
CLONED VOICE
TRUST
INFRASTRUCTURE
$1.65B
Deepfake fraud losses in 2025
3X
Increase vs. prior year
14,000
Consumers surveyed across 14 countries
THE EVIDENCE: THE BARRIER
AI synthesizes content at scale faster than humans can verify.
Fake is now cheaper, faster, and often indistinguishable from real.
Trust collapse is becoming a business and societal risk.
Provenance standards exist, but vulnerabilities let authenticated forgeries be created in minutes.
Data, identity, and attention are the new battlegrounds for trust.
THE QUESTION
Can the trust repair infrastructure—regulatory, technical, and industry-led—outpace the AI-driven erosion of trust by 2027–2030?

## Opening

The AI trust erosion prediction is not a forecast anymore. It's a description of the current condition of digital media, confirmed by primary research from multiple independent sources: a cybersecurity firm with 14,000 global respondents, a deepfake incident tracking firm documenting explosive quarter-over-quarter growth in 2025, and Deloitte projecting AI-enabled fraud will reach $40 billion annually by 2027.

The 2026 update adds two dimensions the original piece could only anticipate: the specific, measured scale of deepfake fraud losses as a direct financial consequence of the trust erosion, and the real but imperfect emergence of the content provenance infrastructure the Counterpoint identified as the necessary condition for trust repair.

Both sides of the argument have received significant new evidence.

The 20% of uncertainty captured in the Counterpoint likelihood is not shrinking, if anything, it's becoming more specifically located: C2PA is real, advancing, and demonstrably vulnerable, and the question of whether it matures faster than the trust erosion compounds is genuinely open.

## Intro

The AI trust erosion prediction's 2026 update is not an assessment of whether the prediction was right. That question has been answered by the data. The update's analytical work is understanding what "trust erosion" looks like when it has been confirmed at scale and quantified in financial terms, and whether the repair infrastructure emerging alongside it's adequate to the damage being documented.

The Thales finding that not one digital sector reaches above 50% consumer trust, and that news media organizations are trusted with personal data by 3% of consumers, describes a baseline trust environment so degraded that the Counterpoint's "C2PA solves this" thesis has to work against a deficit far larger than the original prediction's framing implied.

Deepfake fraud losses reaching $1.65 billion in 2025 and tripling year-over-year convert the trust erosion from an attitudinal measurement to a financial one, establishing the scale of harm that the regulatory and technical response has to match. The C2PA content provenance standard advancing to hardware implementations in cameras from Google, Sony, Leica, and Nikon is real progress toward the repair infrastructure the Counterpoint required.

demonstrated vulnerability that lets authenticated forgeries be created in minutes is a reminder that the race between erosion and repair is not moving in one direction simultaneously. Both are accelerating.

The 2027-2030 question is not whether the repair infrastructure exists. It's whether it can scale faster than the synthetic content machinery it's trying to authenticate against.

## Background

For most of the digital era, consumers operated on a baseline assumption of authenticity: product photos showed real products, customer reviews reflected real experiences, brand spokespersons were real people, news footage depicted real events.

That assumption was always imperfect, stock photos, astroturfed reviews, and PR-polished statements have existed for decades. But the gap between authentic and fabricated content remained wide enough that most consumers, most of the time, could navigate it with reasonable confidence.

Generative AI has collapsed that gap with documented speed. Voice cloning from seconds of audio, photorealistic synthetic video of real people saying things they never said, reviews indistinguishable from genuine customer experiences, and AI-generated influencer personas with fabricated backstories are now operationally routine. Thales' 2025 Digital Trust Index, surveying 14,000 consumers across 14 countries, found a universal decline in trust for digital services across 13 sectors, with not one sector reaching above 50% approval when consumers were asked which they trusted with their personal data. News media organizations are trusted with personal data by only 3% of consumers.

Social media platforms, logistics, and automotive sectors each reach only 4%. Banking, the most trusted sector, peaks at 51% only among consumers over 55, and falls to 32% among Gen Z. Eighty-two percent of consumers abandoned a brand in the past 12 months due to concerns about personal data use.

These numbers describe the baseline trust environment into which AI-generated synthetic content is being deployed at scale.

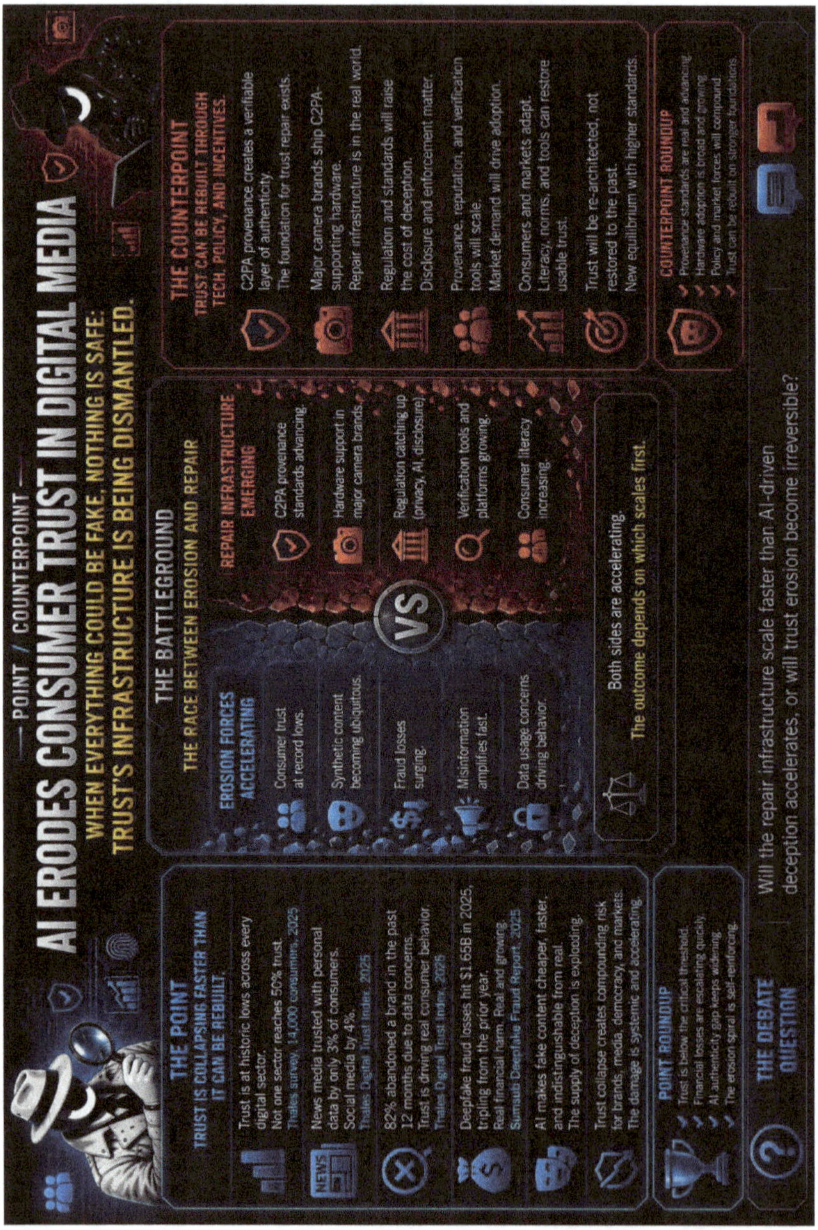
POINT / COUNTERPOINT
AI ERODES CONSUMER TRUST IN DIGITAL MEDIA
WHEN EVERYTHING COULD BE FAKE, NOTHING IS SAFE:
TRUST'S INFRASTRUCTURE IS BEING DISMANTLED.
THE POINT
TRUST IS COLLAPSING FASTER THAN IT CAN BE REBUILT.
Trust is at historic lows across every digital sector.
Not one sector reaches 50% trust.
Thales survey, 14,000 consumers, 2025
News media trusted with personal data by only 3% of consumers.
Social media by 4%.
Thales Digital Trust Index, 2025
82% abandoned a brand in the past 12 months due to data concerns.
Trust is driving real consumer behavior.
Thales Digital Trust Index, 2025
Deepfake fraud losses hit $1.65B in 2025, tripling from the prior year.
Real financial harm. Real and growing.
Sumsub Deepfake Fraud Report, 2025
AI makes fake content cheaper, faster, and indistinguishable from real.
The supply of deception is exploding.
Trust collapse creates compounding risk for brands, media, democracy, and markets.
The damage is systemic and accelerating.
POINT ROUNDUP
Trust is below the critical threshold.
Financial losses are escalating quickly.
AI authenticity gap keeps widening.
The erosion spiral is self-reinforcing.
THE DEBATE QUESTION
Will the repair infrastructure scale faster than AI-driven deception accelerates, or will trust erosion become irreversible?
THE BATTLEGROUND
THE RACE BETWEEN EROSION AND REPAIR
EROSION FORCES ACCELERATING
Consumer trust at record lows.
Synthetic content becoming ubiquitous.
Fraud losses surging.
Misinformation amplifies fast.
Data usage concerns driving behavior.
VS
REPAIR INFRASTRUCTURE EMERGING
C2PA provenance standards advancing.
Hardware support in major camera brands.
Regulation catching up (privacy, AI, disclosure).
Verification tools and platforms growing.
Consumer literacy increasing.
Both sides are accelerating.
The outcome depends on which scales first.
THE COUNTERPOINT
TRUST CAN BE REBUILT THROUGH TECH, POLICY, AND INCENTIVES.
C2PA provenance creates a verifiable layer of authenticity.
The foundation for trust repair exists.
Major camera brands ship C2PA-supporting hardware.
Repair infrastructure is in the real world.
Regulation and standards will raise the cost of deception.
Disclosure and enforcement matter.
Provenance, reputation, and verification tools will scale.
Market demand will drive adoption.
Consumers and markets adapt.
Literacy, norms, and tools can restore usable trust.
Trust will be re-architected, not restored to the past.
New equilibrium with higher standards.
COUNTERPOINT ROUNDUP
Provenance standards are real and advancing.
Hardware adoption is broad and growing.
Policy and market forces will compound.
Trust can be rebuilt on stronger foundations.

## THE POINT: When Everything Could Be Fake, Nothing Is Safe

The data on where consumer trust currently stands has moved from concerning to structurally alarming between the original prediction and its 2026 update.

Thales' 14,000-consumer global survey finding that news media are trusted with personal data by only 3% of consumers, with social media at 4%, logistics at 4%, and automotive at 4%, is the most concentrated evidence of digital trust collapse available from a primary research source.

These are not fringe users or digitally disengaged populations; they are the survey population of global consumers who use these services. The commercial consequences are direct: 82% of those surveyed abandoned a brand in the past 12 months due to data concerns.

Every communication now carries what the California Management Review analysis called "an invisible question mark", the CEO video, the customer testimonial, the product image are all potential fabrications in the consumer's mind, regardless of their actual authenticity. This dynamic, where the existence of deepfakes makes it easier to dismiss even genuine content as fake, is sometimes called the "liar's dividend": real recordings lose credibility, and it becomes harder to hold anyone accountable. For brands that have spent years building credibility, the liar's dividend represents a threat that no marketing budget can simply outspend.

YouGov's May 2025 survey found roughly one in three Americans reporting daily encounters with AI-generated information online, with another 22% encountering it weekly, making AI content exposure a near-universal routine experience rather than an occasional anomaly.

The deepfake fraud trajectory is the Point's most compelling new 2026 evidence, because it translates the abstract trust erosion narrative into documented financial losses at a scale that forces corporate and regulatory response regardless of stated consumer preferences.

Deepfake fraud losses reached $1.65 billion globally in 2025, tripling from the prior year. In North America specifically, losses exceeded $200 million in Q1 2025 alone, with Q2 2025 losses rising to $347 million. Resemble AI tracked 2,031 verified deepfake incidents in Q3 2025 alone, a 317% increase from earlier in the year.

These are not consumer sentiment surveys. They are documented financial crime statistics.

The Arup case, in which a finance worker was deceived by a multi-person deepfake video call into transferring $25 million, remains the most documented single incident, but it's not isolated: the FBI, Ferrari, WPP, and dozens of other organizations have documented similar attempts.

Deloitte projects AI-enabled fraud will balloon from $12.3 billion in 2024 to $40 billion annually by 2027, a 32% compound annual growth rate that outpaces every proposed regulatory and technical countermeasure on the current timeline.

The mechanism the original prediction identified, that AI-generated synthetic content represents a structural threat to the trust infrastructure of commerce and communication, is being confirmed not by consumer surveys but by fraud loss data and corporate incident reports.

### The Point Roundup

The AI trust erosion crisis won't resolve itself through consumer adaptation alone.

History shows that people have a remarkable capacity to lower their expectations and continue participating in flawed systems, they stopped trusting stock photo models and kept buying, stopped trusting influencer endorsements and kept clicking. But each successive erosion of trust carries a cumulative cost: higher skepticism, lower conversion, more friction in every commercial and communicative transaction.

The Thales finding that not one digital sector reaches 50% consumer trust, with news media at 3%, represents not a future warning but a present condition. The brands and institutions that treat provenance, transparency, and verified authenticity as strategic capabilities now, rather than compliance checkbox items, will be the ones that retain consumer trust when it becomes the scarcest commodity in the digital economy.

The deepfake fraud trajectory, from $200M in Q1 2025 to $1.65B for the full year, is the financial forcing function that makes this not merely a reputational challenge but a direct economic one.

## THE COUNTERPOINT: We've Been Here Before (And We Kept Clicking Anyway)

Consumer trust has been declared terminally ill before, and the patient keeps recovering, or at least keeps showing up.

The fake review problem has been documented, litigated, and legislated against for over a decade, and e-commerce sales have grown continuously throughout. Sponsored content and undisclosed influencer relationships generated enormous controversy, the FTC issued guidance, Instagram added disclosure labels, and influencer marketing continued expanding.

The pattern is consistent: trust takes a hit, adaptation occurs at the consumer and institutional level, guardrails get erected, and commerce continues. There's no obvious reason why AI-generated synthetic content breaks this historical pattern rather than following it. There's also a more nuanced point about consumer psychology that the trust-erosion narrative tends to underweight.

Despite low stated trust in AI in high-stakes contexts, Forrester predicted that by 2026, 30% of consumers would use generative AI tools for high-risk decisions such as personal finance and healthcare.

The gap between what consumers say they trust and how they actually behave has always been wide. People say they don't trust online reviews, then read ten of them before making a purchase.

The Thales finding that 82% abandoned a brand due to data concerns, while those same consumers continue using digital services, illustrates the stated-preference versus behavioral gap precisely.

The concern is real. The continuation is also real.

The content provenance infrastructure the original Counterpoint identified as the necessary condition for trust repair has made measurable progress in 2025 and 2026, more than the original piece's "optimistic" characterization anticipated, but with significant remaining gaps that prevent declaring the problem solved.

The Coalition for Content Provenance and Authenticity (C2PA) released version 2.3 in January 2026 and now counts over 6,000 organizations in the Content Authenticity Initiative. Critically, hardware implementation has moved from concept to shipping product: Google's Pixel 10 ships with hardware-backed C2PA signing by default via the Titan M2 chip and Tensor G5 on-device timestamping, signing every photo at capture.

Sony released the world's first broadcast camcorder with native C2PA signing, announced at IBC 2025. Leica, Nikon, Canon, Fujifilm, and Panasonic all have C2PA-capable products. Adobe has integrated automatic content credentials writing across all major Creative Cloud products.

The EU AI Act requires AI content labeling effective August 2026, providing the regulatory forcing function the original Counterpoint identified as the necessary condition for moving from opt-in to mandatory.

C2PA's own 2026 state-of-the-art review declared: "We are no longer just defining principles and goals. We are seeing interoperable provenance take shape in the real world." That progress is real.

The question is whether it matures faster than the trust erosion compounds.

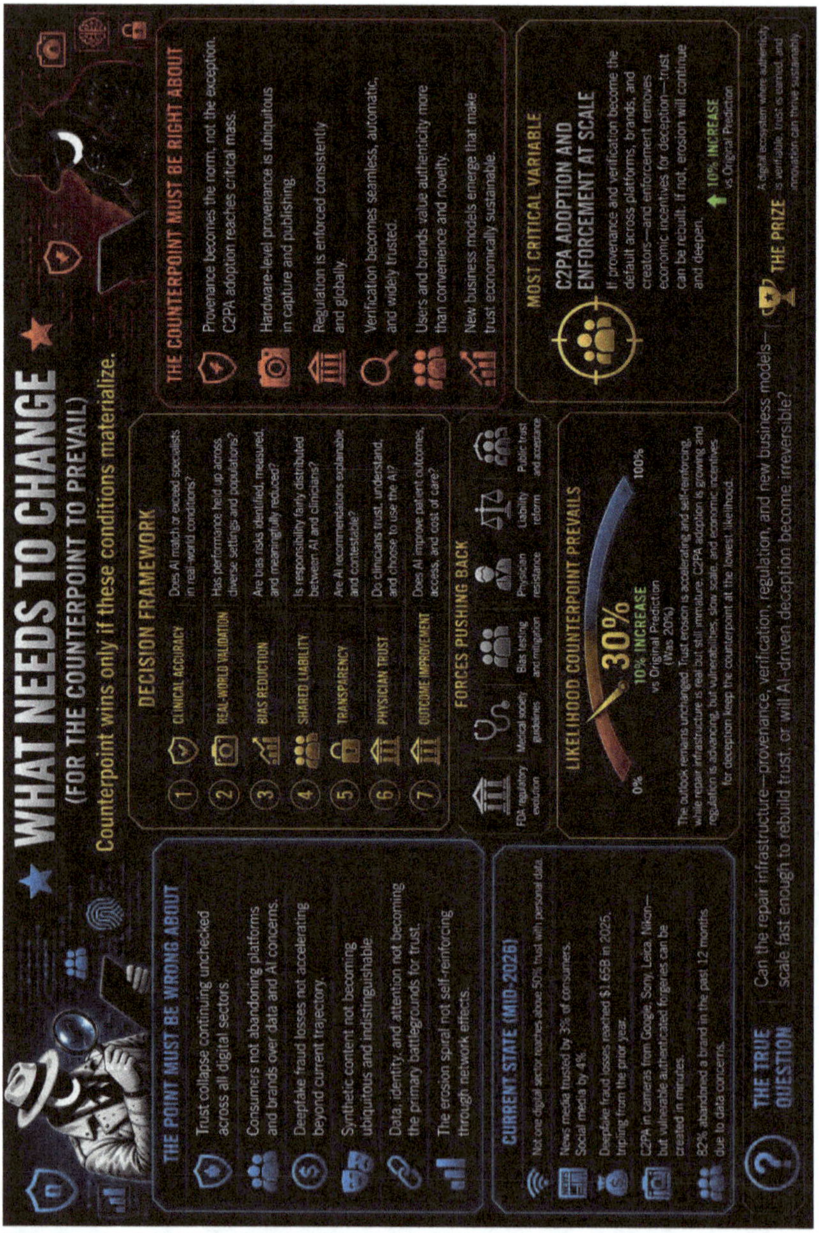
WHAT NEEDS TO CHANGE
(FOR THE COUNTERPOINT TO PREVAIL)
Counterpoint wins only if these conditions materialize.
THE POINT MUST BE WRONG ABOUT
Trust collapse continuing unchecked across all digital sectors.
Consumers not abandoning platforms and brands over data and AI concerns.
Deepfake fraud losses not accelerating beyond current trajectory.
Synthetic content not becoming ubiquitous and indistinguishable.
Data, identity, and attention not becoming the primary battlegrounds for trust.
The erosion spiral not self-reinforcing through network effects.
CURRENT STATE (MID-2026)
Not one digital sector reaches above 50% trust with personal data.
News media trusted by 3% of consumers. Social media by 4%.
Deepfake fraud losses reached $1.65B in 2025, tripling from the prior year.
C2PA in cameras from Google, Sony, Leica, Nikon—but vulnerable authenticated forgeries can be created in minutes.
82% abandoned a brand in the past 12 months due to data concerns.
DECISION FRAMEWORK
1 CLINICAL ACCURACY Does AI match or exceed specialists in real-world conditions?
2 REAL-WORLD VALIDATION Has performance held up across diverse settings and populations?
3 BIAS REDUCTION Are bias risks identified, measured, and meaningfully reduced?
4 SHARED LIABILITY Is responsibility fairly distributed between AI and clinicians?
5 TRANSPARENCY Are AI recommendations explainable and contestable?
6 PHYSICIAN TRUST Do clinicians trust, understand, and choose to use the AI?
7 OUTCOME IMPROVEMENT Does AI improve patient outcomes, access, and cost of care?
FORCES PUSHING BACK
FDA regulatory evolution
Medical society guidelines
Bias testing and mitigation
Physician resistance
Liability reform
Public trust and acceptance
LIKELIHOOD COUNTERPOINT PREVAILS
0%
30%
100%
10% INCREASE
vs Original Prediction
(Was 20%)
The outlook remains unchanged. Trust erosion is accelerating and self-reinforcing, while repair infrastructure is real but still immature. C2PA adoption is growing and regulation is advancing, but vulnerabilities, slow scale, and economic incentives for deception keep the counterpoint at the lowest likelihood.
THE COUNTERPOINT MUST BE RIGHT ABOUT
Provenance becomes the norm, not the exception. C2PA adoption reaches critical mass.
Hardware-level provenance is ubiquitous in capture and publishing.
Regulation is enforced consistently and globally.
Verification becomes seamless, automatic, and widely trusted.
Users and brands value authenticity more than convenience and novelty.
New business models emerge that make trust economically sustainable.
MOST CRITICAL VARIABLE
C2PA ADOPTION AND ENFORCEMENT AT SCALE
If provenance and verification become the default across platforms, brands, and creators—and enforcement removes economic incentives for deception—trust can be rebuilt. If not, erosion will continue and deepen.
10% INCREASE
vs Original Prediction
THE TRUE QUESTION
Can the repair infrastructure—provenance, verification, regulation, and new business models—scale fast enough to rebuild trust, or will AI-driven deception become irreversible?
THE PRIZE
A digital ecosystem where authenticity is verifiable, trust is earned, and innovation can thrive sustainably.

## What Needs to Change

### (... for the Counterpoint to Prevail)

For the counterargument to hold, for the trust erosion to remain manageable rather than structural, two conditions must materialize that are currently moving in the wrong direction.

First, the content provenance infrastructure must harden and achieve mandatory platform adoption before the trust erosion compounds into a stable default skepticism.

C2PA's real-world progress in 2025-2026 is real but fragile: authenticated forgeries can be created in minutes using standard tools, Nikon was forced to revoke all certificates its C2PA-enabled cameras had issued after a signing vulnerability was discovered, and the standard remains opt-in at the platform level where consumer content actually flows.

The EU AI Act labeling requirement effective August 2026 creates a regulatory forcing function, but regulatory mandates in Europe don't automatically translate to platform behavior globally, and the largest US social platforms have not committed to C2PA as a distribution requirement. Until that happens, the provenance infrastructure exists at the device and tool level while the consumer trust problem exists at the feed and platform level.

Second, the behavioral gap between stated distrust and actual digital engagement must remain wide enough to prevent the trust erosion from translating into sustained commercial damage. History supports this, consumers kept clicking through every prior trust crisis, but deepfake fraud at $1.65 billion in 2025 losses introduces a new variable: when the trust erosion produces direct financial harm to consumers, not merely reduced brand affinity, the behavioral resilience the Counterpoint relies on faces a harder test.

Both conditions are possible. Neither is likely given the current trajectory.

## Likelihood Counterpoint Prevails: 30%

The Counterpoint deserves a 30% assessment, a significant tick up from the original.

The Point is clearly winning the evidence contest, but the behavioral resilience pattern is genuinely strong and C2PA is further along than the original evidence acknowledges. The gap between "trust erosion is real and measured" and "commerce is structurally disrupted" is where the Counterpoint lives, and that gap has historically been wide and durable.

The 70% confidence that the Point prevails rests on evidence that is current rather than projected: Thales' 14,000-person global survey showing not one digital sector reaching 50% consumer trust for personal data; $1.65 billion in documented deepfake fraud losses in 2025 alone; deepfake incidents growing from 150 in all of 2024 to 2,031 in Q3 2025 alone; 82% of consumers abandoning brands due to data concerns. These are present-tense readings, not forecasts.

The 30% assigned to the Counterpoint is not a concession that the evidence is close, it's an honest acknowledgment of two genuine uncertainties.

First, the historical pattern of consumer behavioral resilience is real and documented: people continue participating in digital commerce despite stated distrust at levels that shouldn't, by survey logic, sustain that participation. The stated-preference versus behavioral gap has preserved commerce through every prior trust crisis, and there's no structural reason to assume it breaks under AI-generated synthetic content rather than following the historical pattern.

Second, C2PA's genuine 2025-2026 hardware and regulatory progress represents the first credible trust repair infrastructure in the field's history, hardware-backed signing by default in the Google Pixel 10, EU AI Act labeling requirements live in August 2026, and over 6,000 organizations in the Content Authenticity Initiative.

The demonstrated vulnerability (authenticated forgeries in minutes; Nikon certificate revocation) is a serious complication but not a disqualifying one.

The 30% says: the trust erosion is structurally real and financially consequential, the Counterpoint's provenance repair scenario is genuinely underway but racing against compounding damage, and the question brands and institutions must answer is not whether to respond but how much ground they have already ceded while waiting to see how bad it gets.

# The Smart Home Interoperability Promise (Still Waiting After 20 Years)

## The Soft Silo Arrived Right on Schedule. It's Being Celebrated as Progress.

*Thread 1.4 shipped. The multi-mesh problem, homes with both a HomePod and an Echo running separate, competing Thread networks, has a specification-level fix. Thread 1.4 became the sole Border Router certification standard as of January 1, 2026. Matter 1.5 added cameras. IKEA brought certified devices to market for under ten dollars.*

*The pipes are fixed. Also: many ecosystems are still running Matter 1.2 or 1.3.*

*Google's "Gemini for Home" uses the standardized plumbing that Matter provides to run proprietary AI intelligence on top of cross-brand device data, precisely the "soft silo" the original analysis predicted. Amazon, Apple, and Google let the commodity layer go open.*

*They kept the value layer firmly proprietary. That was always the plan.*

OVERVIEW
THE SMART HOME INTEROPERABILITY PROMISE
(STILL WAITING AFTER 20 YEARS)
The Soft Silo Arrived Right on Schedule.
It's Being Celebrated as Progress.
THE EVIDENCE
THE SHIFT IS REAL
Thread 1.4 solves the multi-mesh problem. Mandatory as of Jan 1, 2026.
Matter 1.5 adds cameras, soil sensors, and energy management.
IKEA ships Matter-certified devices for under $10.
Adoption surges: 84% of developers now use or plan to use AI tools.
Benchmarks hit 93.9% on SWE-bench Verified—tools are getting better.
20 YEARS
HomePod
Thread Network
Echo
Thread Network
Thread 1.4 fixes the multi-mesh problem.
But many ecosystems are still on Matter 1.2 or 1.3.
THE EVIDENCE
THE SHIFT MATTERS
(OR THE BARRIER)
"Gemini for Home" uses Matter plumbing to run proprietary AI on top.
Amazon, Apple, and Google kept the value layer proprietary.
Cross-brand device data feeds proprietary AI that competitors can't access.
Trust falls: Only 29% trust AI outputs on complex tasks (down from 40%).
The "soft silo" predicted in 2019 has arrived—on schedule.
THE QUESTION
Will Matter and Thread's technical progress finally unlock real interoperability,
or is the soft silo the destination the industry was always building toward?

## Opening

I've watched this problem from three vantage points that most smart home commentators haven't occupied simultaneously:

- inside IEEE working groups wrestling with IoT security and interoperability specifications;
- running a large global IoT practice with visibility into serious R&D investment across both IoT and AI;
- and building actual production IoT systems across multiple industries, with the scars and the stories to prove what "certified compatible" means at 2am or 150mph when something stops working.

The pattern I've seen across all three is consistent enough to state plainly: the smart home interoperability problem is not a technology problem.

It's an incentive architecture problem. And Matter hasn't changed the incentive architecture. What Matter and Thread 1.4 together have done is deliver genuine technical progress on the foundational problems that prevented the category from working at all. The 2026 update is honest about that progress while being equally honest about what it doesn't change.

## Intro

Every few years, the smart home industry resets the clock. A new coalition forms, a new standard gets announced, a new logo appears on packaging, and the promise gets made again: this time, everything will work together.

This time, the fragmentation ends. This time, your devices from different manufacturers will talk to each other without a PhD in networking, three different apps, and the quiet resignation that at least the lights turn on most of the time.

Matter is the latest and most credible version of that promise. It has better technical foundations than its predecessors, a more powerful coalition behind it, and genuine engineering improvements that deserve fair acknowledgment.

It also has a decade of broken promises as context, and a structural problem that better engineering alone can't fix.

The 2027-2030 window is where the prediction gets its definitive test: whether the soft silo becomes the permanent destination, or whether the standard finds a way to reach into the intelligence layer where the actual value lives.

## Background

The smart home interoperability problem has roots that predate most of the devices now trying to solve it. Zigbee emerged in 2003 as a low-power mesh networking standard that was supposed to unify device communication across manufacturers.

Z-Wave followed with similar ambitions.

Both produced fragmented implementations, certification programs that didn't guarantee real-world compatibility, and ecosystems that worked reliably only when you stayed within a single manufacturer's product line, which rather defeated the point.

The Zigbee Alliance eventually absorbed both its own fragmentation and its competitors into Project Connected Home over IP, announced in December 2019 by Amazon, Apple, Google, and Samsung.

That project became Matter. The rebranding was clean. The underlying structural dynamics were unchanged. Matter 1.0 shipped in October 2022. By January 2026, the version history reads: 1.0, 1.1, 1.2, 1.3, 1.4, 1.4.1, 1.4.2, and 1.5, eight releases in roughly three years.

Thread 1.4, the specification-level fix to the multi-mesh problem, was published by the Thread Group in fall 2024 and began firmware rollout through 2025. As of January 1, 2026, Thread 1.4 is the sole certification standard for Border Routers, meaning no new 1.3 devices can be certified.

The question is what that mandatory transition actually delivers in practice, and the answer from mid-2026 deployments is: improved, but uneven.

Many ecosystems, including Amazon's, are still running Matter 1.2 or 1.3 even as the certification standard has moved forward.

— POINT / COUNTERPOINT —
THE SMART HOME INTEROPERABILITY PROMISE
(STILL WAITING AFTER 20 YEARS)
The Soft Silo Arrived Right on Schedule. It's Being Celebrated as Progress.
THE POINT
THE FOUNDATIONS ARE FIXED. NOW THE ECOSYSTEM CAN DELIVER.
The multi-mesh problem is solved.
Thread 1.4 standardizes credential sharing. As of Jan 1, 2026, it's the sole Border Router certification standard.
Matter's technical foundation is solid.
Runs on IP. Stronger security. Better device categories. Matter 1.5 adds cameras and more.
Ecosystem is expanding fast.
IKEA ships Matter-certified devices for under $10. More brands, more categories, more choice.
Adoption and confidence are rising.
84% of developers use or plan to use AI tools. Benchmarks show tools are getting better (93.9% on SWE-bench).
The value layer will benefit users.
Open standards create choice, lower prices, and innovation. The best experiences will come from competition on top.
POINT ROUNDUP
Thread 1.4 fixes the last major technical blocker.
Matter adoption, device variety, and quality are improving.
The ecosystem is now large enough to innovate.
Open standards let the market decide the winners.
THE BATTLEGROUND
THE COMMODITY LAYER IS OPEN.
THE VALUE LAYER REMAINS PROPRIETARY.
Open Pipelines
MATTER
Proprietary Intelligence
Matter & Thread provide interoperable connectivity.
Devices speak the same language.
Commodity layer is open and cross-brand.
AI, data, automations, and experiences stay inside each ecosystem.
Cross-brand device data fuels proprietary AI you can't access.
Switching costs keep users locked in.
The pipes are open.
The destinations are still closed.
THE COUNTERPOINT
THE SOFT SILO IS THE DESTINATION, NOT A DETOUR.
The value layer is where the value lives.
AI, automations, and insights are proprietary. Matter doesn't change that dynamic.
Switching costs are structural.
Ecosystem lock-in drives retention. Users won't switch for interoperability they can't see.
Data is the competitive moat.
Device data powers AI. Companies won't commoditize their advantage.
Adoption ≠ openness.
High adoption of tools and standards can accelerate proprietary builds.
Trust in complex AI systems is low.
Only 29% trust AI outputs on complex tasks. Proprietary AI won't fix that.
COUNTERPOINT ROUNDUP
The soft silo was always the plan.
Open pipelines + closed intelligence = lock-in.
Data, AI, and experiences stay proprietary.
Interoperability at the edges doesn't change that.
THE DEBATE QUESTION
Will the market push interoperability up the stack into AI and user experiences, or will the soft silo become the permanent destination?

## THE POINT: The Foundational Problems Are Solved. Now Watch What Gets Built On Top

The optimistic case for Matter rests on three genuine technical improvements that prior standards couldn't claim, and all three have now been delivered in the 2025-2026 window.

First, Matter runs on IP, the same protocol stack as the internet itself, eliminating an entire class of translation problems that plagued earlier standards.

Second, Thread 1.4, with its standardized credential sharing, has solved the multi-mesh problem that had been quietly undermining Matter deployments since launch. Homes with both an Apple HomePod and an Amazon Echo were running two separate, competing Thread networks that couldn't reliably communicate. Thread 1.4 allows border routers from different manufacturers to merge into a single robust mesh network, and the certification transition mandated by the Thread Group makes this the floor, not a premium feature.

Third, the device ecosystem is genuinely expanding: IKEA has brought Matter-certified products to market for under ten dollars, representing the mainstream retail penetration that moves categories from enthusiast to consumer. Matter 1.5, published in November 2025, added security cameras,  one of the most requested device categories, alongside soil moisture sensors and enhanced energy management features.

The Fabric Sync feature in Matter 1.4, reaching maturity in 2025, allows a device authorized once to appear automatically across all authorized platforms. For consumers building a Matter-native smart home from scratch in 2026, the experience is genuinely better than anything available at any prior moment in the category's history.

The mandatory Thread 1.4 certification transition represents the kind of structural forcing function the original analysis identified as necessary for any standard to achieve real rather than nominal adoption.

The USB-C standard became truly universal only when external regulatory pressure removed the decision from the companies with the strongest incentives to resist. The Thread Group's January 1, 2026 certification mandate is an internal equivalent, not regulatory, but structural: any new Border Router certified after that date must meet Thread 1.4.

That progressive floor-raising is more powerful over time than any individual version release, because it gradually eliminates the installed base of devices that sustain the old fragmentation. The scale of the current coalition also exceeds anything the Zigbee era produced: eight major platform companies, hundreds of device manufacturers, and retail penetration at IKEA price points.

The coalition is more commercially committed, the version cadence is more disciplined, and the specification is technically more coherent than any predecessor.

For consumers starting fresh in 2026, the smart home finally works better than it doesn't.

### The Point Conclusion

The pipes are finally fixed. That is not a small thing. Anyone who spent the 2019-2023 era watching Thread networks fight each other across competing border routers or trying to explain to a family member why their new Matter-certified bulb needed three apps and a firmware update, understands how much engineering effort went into getting here.

Matter 1.5 and Thread 1.4 together represent the most coherent technical foundation the smart home category has ever had. IKEA selling certified devices for under ten dollars means interoperability is no longer a premium feature, it's becoming a baseline expectation at mass-market price points.

The optimistic case isn't that every problem is solved. It's that the foundational problems that prevented the category from growing are finally resolved. Now the question is what gets built on top of them.

That answer is where the Counterpoint's analysis gets complicated.

## THE COUNTERPOINT: Soft Silo Was Always the Plan. Matter Just Made It Prettier.

### Layer 1: The Consumer Fatigue Is Earned

The accumulated trust deficit of the smart home category is not irrational consumer resistance to new technology. It's a rational response to a decade of broken promises.

The CSA and its members aim for the Matter logo to signal that a device will "just work", but Matter logos don't actually imply that devices will just work. Infrastructure is still required: a compatible Wi-Fi network, a Matter controller, and often a manufacturer-specific app for setup. The logo promises what the implementation can't deliver.

That gap, between what the marketing says and what Tuesday evening setup actually produces, is not new. It's the defining characteristic of every smart home standard since Zigbee 1.0.

Consumers who have a drawer full of orphaned hubs, devices bricked by app updates, and protocols abandoned eighteen months after purchase are not being cynical. They are being accurate. Matter launched into that accumulated skepticism and has spent three years confirming rather than correcting it for a meaningful portion of its target audience.

The January 2026 status review at matter-smarthome.de was diplomatically direct: "Many ecosystems are still at the Matter 1.2 or 1.3 stage."

Thread 1.4 requires firmware updates on every device involved and an afternoon of patience even after the spec has been published.

The pipes are being fixed. The installation is still work.

## Layer 2: The Founding Members Were the Problem, Not the Solution

Here is where the history becomes genuinely damning, and where the working group participants deserve separation from the structural failure that followed their work.

The engineers who built the interoperability specifications, in the IEEE and in the CSA, were doing legitimate science.

The obstacle was never their competence or their intentions. The obstacle was what happened when their work reached the implementation layer, where Amazon, Apple, and Google made commercial decisions about what to actually ship.

Three generations of Nest Learning Thermostats, the flagship product Google sold for nearly a decade, sitting on the wall in the largest number of American homes, have no Matter support and never will.

Only two models in the entire Nest thermostat lineup currently support the standard: the 2020 budget model and the 2024 fourth-generation device.

And even on those supported models, Nest Temperature Sensors remain invisible to Matter entirely, meaning multi-room temperature management stays locked inside Google's ecosystem regardless of what the Matter logo on the box implies.

Some platforms remain at Matter 1.2 while others claim 1.4 support but deliver only a selective subset of its capabilities. Matter standardization threatens ecosystem lock-in, so the founding members drag their feet on certifications, ship half-implementations, and keep premium features behind their walled apps.

This is not accidental inconsistency. It's the predictable output of an incentive structure where the companies funding the standard are the same companies whose business models depend on consumers not leaving their ecosystems.

## Layer 3: The Soft Silo Prediction Has Been Confirmed

The resolution that optimists are now celebrating, Matter 1.5 connecting the device layer, Thread 1.4 unifying the mesh, misses the critical question entirely.

The soft silo this analysis predicted has already arrived and is being celebrated as success by the companies that designed it. Google's "Gemini for Home" initiative uses the standardized data provided by Matter 1.5 to power proprietary AI across different device brands: noticing a package delivery on a Nest cam and automatically turning on a third-party smart light, generating sophisticated summaries of home events across manufacturers.

That capability runs on top of the open standard's data.

The intelligence layer, the AI, the contextual automation, the features consumers actually want, remains proprietary and requires staying in Google's ecosystem to access.

Amazon and Apple are doing the same. Basic functions are universal via Matter: lights turn on, locks lock, sensors report. Advanced AI-driven features like person recognition, energy optimization, and contextual automation remain exclusive to specific brand pairings. Amazon, Apple, and Google did not lose the ecosystem war when Matter succeeded.

They won it more elegantly. They let the commodity layer go open and kept the value layer firmly proprietary.

That is not the open interoperability the working groups spent years trying to deliver.

That is the Zigbee coalition in better clothes, having learned exactly the right lessons from its predecessor's mistakes.

WHAT NEEDS TO CHANGE
(FOR THE COUNTERPOINT TO PREVAIL)
Counterpoint wins only if these conditions materialize.
THE POINT MUST BE WRONG ABOUT
The soft silo becoming the default business model.
Switching costs and ecosystem lock-in remaining durable.
Proprietary data and AI staying the primary source of value.
Cross-brand tools, experiences, and automations failing to gain traction.
Regulatory or standards pressure failing to force greater openness.
DECISION FRAMEWORK
1 Track cross-ecosystem user experiences and satisfaction over time.
2 Monitor share of value captured by services vs. commodity devices.
3 Watch for cross-brand automations and routines that users rely on daily.
4 Evaluate ecosystem switching rates and real-world lock-in friction.
5 Follow regulatory, legal, and standards actions that mandate data portability or API openness.
6 Assess developer adoption and innovation outside closed ecosystems.
THE COUNTERPOINT MUST BE RIGHT ABOUT
Open interoperability becomes a user expectation, not a niche preference.
Regulation or standards force meaningful data portability and API access.
Cross-ecosystem tools and platforms deliver better experiences than siloed ones.
Consumers and enterprises reward openness with adoption and spending.
The balance of value shifts from proprietary intelligence to interoperability and choice.
CURRENT STATE (MID-2026)
Thread 1.4 is the sole Border Router certification standard as of Jan 1, 2026.
Many ecosystems are still on Matter 1.2 or 1.3.
"Gemini for Home" runs proprietary AI on top of Matter's open plumbing.
Only 29% trust AI outputs on complex tasks (down from 40%).
LIKELIHOOD COUNTERPOINT PREVAILS
0%
30%
100%
UNCHANGED
vs Original Prediction (Was 30%)
The outlook is unchanged: meaningful interoperablity remains unlikely without structural shifts in incentives, regulation, or user behavior that favor openness and data portability.
MOST CRITICAL VARIABLE
ECOSYSTEM LOCK-IN AND DATA CONTROL
If ecosystems retain control of user data, AI, and automations, interoperability will remain superficial. If data portability and API access become the norm, the balance of power—and value—can shift.
THE TRUE QUESTION
Will business models, regulation, and user expectations evolve enough to push intelligence and value beyond proprietary silos—and unlock true interoperability?
THE PRIZE
Greater choice. Better experiences. More innovation. Real smart homes that work the way users expect.

## What Needs to Change

### (... for the Counterpoint to Prevail)

For the optimistic case to become reality, for Matter to deliver genuine, consumer-grade interoperability rather than a commodity connection layer beneath competing proprietary intelligence stacks, two structural changes are required that currently have no credible path to happening.

First, the certification program needs real enforcement teeth, administered by a genuinely neutral body with the authority to decertify products whose implementations diverge materially from the spec. The current CSA structure, where the founding members are simultaneously the largest contributors, the largest implementers, and the governing voices, can't produce that neutrality. The USB-C standard achieved real traction only when the EU mandated it externally, removing the decision from the companies with the strongest incentives to resist. Matter needs an equivalent forcing function, and no such regulatory appetite currently exists in the US market.

Second, the intelligence layer needs to be in scope for the standard, not just the connectivity layer. A standard that unifies device connection while leaving AI features, automation logic, and personalization entirely within proprietary walls has solved the easy problem and left the valuable one untouched.

Google's "Gemini for Home" is the 2026 proof of concept for why this matters: it uses the open standard to harvest data, then proprietary AI to create value from it.

Until the standard reaches into the layer where the actual value lives, the soft silo is not a transitional imperfection. It's the permanent destination.

## Likelihood Counterpoint Prevails: 30%

The 30% Counterpoint likelihood holds from the original, unchanged, and for reasons the 2026 evidence has confirmed rather than altered.

The lower bound on this number, why it isn't 10% or 15%, is that Thread 1.4 and Fabric Sync represent real engineering progress, IKEA's sub-$10 Matter devices are moving the category into mainstream retail in a way no prior standard achieved, and the mandatory Thread 1.4 certification transition creates a progressive floor-raising mechanism that gradually eliminates the old fragmentation.

The upper bound, why it isn't 45% or 50%, is that the soft silo outcome has been explicitly confirmed by Google's "Gemini for Home" initiative, which uses the open standard's data to power proprietary AI in precisely the mechanism the original analysis described and predicted.

The certification enforcement problem has no credible resolution mechanism: the CSA's founding members remain both the largest funders and the primary commercial beneficiaries of selective implementation.

The version history, 1.0 through 1.5 in three years, each version adding device categories that should have been in 1.0 and standardizing behaviors the founding members were already shipping inconsistently, is not a standard maturing through normal iteration.

It's a standard retrofitting coherence onto implementations that were never designed to be truly interoperable.

The 30% says: the standard will survive, the device ecosystem will grow, and the consumer experience will improve incrementally.

But the fundamental interoperability promise, where your smart home works seamlessly regardless of which ecosystem owns the intelligence layer, will remain unfulfilled by end of 2027. What will exist is a better-connected set of walled gardens.

That is progress. It's not the promise. It's not what consumers deserve after 20 years.

# Q-Day: The Security Crisis That Started Before Anyone Was Ready

## The Attack Already Happened. The Damage Materializes When Q-Day Arrives.

*Most security threats announce themselves. A breach notification arrives. A password resets. A credit card gets flagged.*

*The quantum computing threat to consumer security works nothing like that. It's patient, invisible, and already in motion.*

*In late 2024, FBI Director Christopher Wray described the Salt Typhoon operation, in which a Chinese state-sponsored group infiltrated nine major US telecom providers including AT&T, Verizon, and Lumen Technologies, maintaining undetected access for one to two years, as the most significant cyber espionage campaign in history.*

*Their goal was not ransomware or financial theft. It was intelligence collection. The patient, long-term accumulation of communications data that becomes readable when quantum computers arrive.*

*The attack already happened. The damage materializes later.*

OVERVIEW
Q-DAY: THE SECURITY CRISIS
THAT STARTED BEFORE ANYONE WAS READY
THE ATTACK ALREADY HAPPENED.
THE DAMAGE MATERIALIZES WHEN Q-DAY ARRIVES.
THE EVIDENCE IS REAL
SALT TYPHOON: CONFIRMED
FBI confirms Chinese state actors compromised 9 major U.S. telecoms, staying undetected for 1–2 years.
Source: FBI Director, Dec 2024
THE GOAL WAS INTELLIGENCE
Not money. They collected massive amounts of communications data for future decryption.
Source: DOJ Indictment, Oct 2024
QUBITS ARE CLOSING THE GAP
New research cuts requirements to break RSA-2048 to as low as 100,000 qubits.
Source: Quantum research, Mar 2026
FRAUD LOSSES ARE RISING
Deepfake & AI fraud losses hit $1.65B in 2025, tripling from the prior year.
Source: Sumsub Deepfake Fraud Report, 2025
CONSUMER TRUST IS COLLAPSING
3% trust news media. 4% trust social media. 82% abandoned a brand due to data concerns.
Source: Thales Digital Trust Index, 2025
DATA IS COLLECTED TODAY
DATA IS DECRYPTED TOMORROW
HARVEST NOW. DECRYPT LATER.
BY THE TIME Q-DAY ARRIVES, THE DAMAGE IS IRREVERSIBLE.
9 U.S. TELECOM PROVIDERS COMPROMISED
14,000 CONSUMERS SURVEYED ACROSS 14 COUNTRIES
$1.65B DEEPFAKE FRAUD LOSSES IN 2025 (TRIPLED YOY)
65% OF HUMAN INTERNET TRAFFIC NOW PQC-ENCRYPTED (2026)
THE BARRIER IS REAL
THE ATTACK IS PATIENT & INVISIBLE
Collection happens quietly. There's no alert, no breach notice, no immediate sign of compromise.
Q-DAY TIMELINE IS UNCERTAIN
Experts see 5–14% chance by 2029, rising to 19–34% by 2034.
Source: Global Risk Institute, 2024
ENCRYPTED TODAY, BROKEN LATER
Data protected by today's encryption can be decrypted once quantum computers arrive.
EVERYONE IS A TARGET
Individuals, businesses, governments, and critical infrastructure are all at risk.
THE DAMAGE IS PERMANENT
There is no way to recall data once it has been collected and stored.
THE QUESTION
Can the post-quantum security migration outpace the silent collection campaign before Q-Day arrives?

## Opening

The original Q-Day prediction was written when the Harvest Now, Decrypt Later threat was documented but its scale was not yet visible in named, confirmed incidents.

The 2026 update is written after Salt Typhoon. That distinction matters: the original piece relied on Booz Allen Hamilton's intelligence assessment that nation-state collection was likely underway. Salt Typhoon is the confirmed instance, not a probability, a documented operation. Nine telecoms. Two years of undetected access.

The FBI Director's own characterization as the most significant cyber espionage campaign in history. The prediction's core mechanism, that the consumer security crisis is present-tense, operating quietly on data already transmitted, has moved from intelligence assessment to confirmed public record.

Simultaneously, the migration response infrastructure has advanced materially: 65% of human internet traffic is post-quantum encrypted through Cloudflare's network as of 2026, NIST's three finalized post-quantum standards are production-ready, and Google has announced a formal 2029 PQC migration deadline.

The race between collection and migration is no longer theoretical. It's documented, active, and moving in both directions at once.

## Intro

The Q-Day prediction occupies a specific analytical position in this series: it's the article where the update's most important work is not assessing whether the prediction was calibrated correctly but documenting that the threat mechanism it described has moved from probable to confirmed.

The original piece was written when Harvest Now, Decrypt Later was an intelligence assessment. A well-reasoned inference from adversary incentives and capability trajectories.

The 2026 update is written after Salt Typhoon has been publicly confirmed, officially characterized, and entered the historical record as the most significant cyber espionage campaign in US history per the FBI Director's own statement. That transition from inference to confirmed incident changes the analytical standing of every subsequent claim in the article.

The Counterpoint's argument that the migration response will outpace Q-Day's arrival is not weakened by Salt Typhoon, but it's now arguing against a documented collection operation rather than a probable one.

The migration infrastructure has also advanced materially in the same period, which is the 2026 update's genuinely balanced finding: the threat is more confirmed than the original implied, and the response is more advanced than the original implied, and both are true simultaneously.

The race between collection and migration that the original prediction described as theoretical is now empirically underway, with named adversaries on one side and named technical standards and corporate migration deadlines on the other.

## Background

The mathematical foundation for the quantum cryptographic threat was established in 1994, when Peter Shor demonstrated that a sufficiently powerful quantum computer could factor large prime numbers exponentially faster than any classical machine, directly attacking the computational problem that RSA and elliptic curve cryptography depend on for their security.

For nearly three decades the threat remained theoretical because quantum computers lacked the qubit count, stability, and error correction needed to execute Shor's Algorithm at meaningful scale.

That gap has been closing. Three papers published between May 2025 and March 2026 reduced estimated quantum resources for breaking RSA-2048 from 20 million qubits to under one million, and potentially as low as 100,000 using newer QLDPC code architectures.

The Global Risk Institute's 2024 Quantum Threat Timeline Report, surveying 32 global experts, estimates a 5-14% probability that a cryptographically relevant quantum computer capable of breaking RSA-2048 within 24 hours could emerge by 2029, rising to 19-34% probability by 2034.

On the response side: NIST finalized its three principal post-quantum cryptography standards on August 13, 2024, ML-KEM (FIPS 203) for key encapsulation, ML-DSA (FIPS 204) for digital signatures, and SLH-DSA (FIPS 205) as a hash-based signature backup. NSA's CNSA 2.0 guidance sets 2030 as the mandatory migration deadline for National Security Systems. Google announced a formal 2029 post-quantum cryptography migration deadline on March 25, 2026, citing HNDL as the primary motivator.

Cloudflare reported that more than 65% of human-generated traffic on its network is already post-quantum encrypted, with a phased roadmap to full post-quantum authentication by 2029.

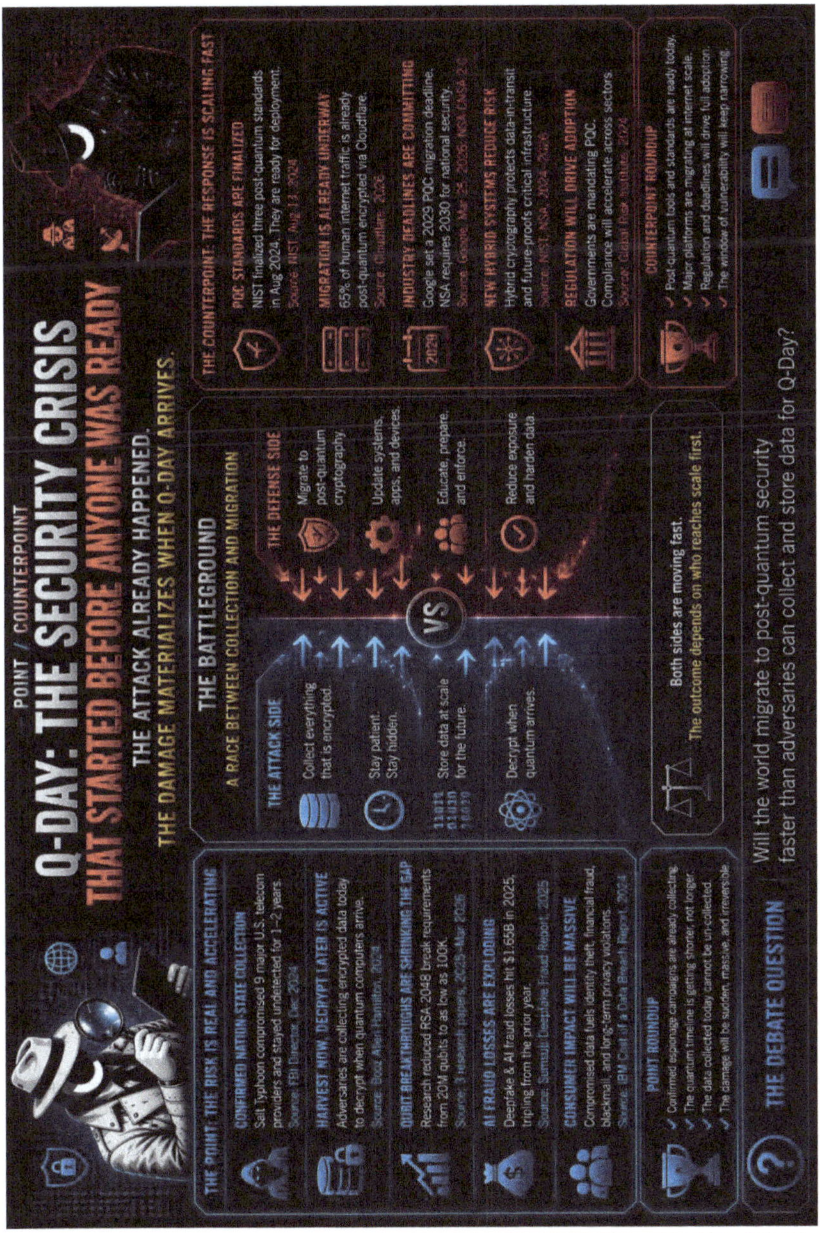
POINT / COUNTERPOINT
Q-DAY: THE SECURITY CRISIS
THAT STARTED BEFORE ANYONE WAS READY
THE ATTACK ALREADY HAPPENED.
THE DAMAGE MATERIALIZES WHEN Q-DAY ARRIVES.
THE POINT: THE RISK IS REAL AND ACCELERATING
CONFIRMED NATION-STATE COLLECTION
Salt Typhoon compromised 9 major U.S. telecom providers and stayed undetected for 1–2 years.
Source: FBI Director, Dec 2024
HARVEST NOW, DECRYPT LATER IS ACTIVE
Adversaries are collecting encrypted data today to decrypt when quantum computers arrive.
Source: Booz Allen Hamilton, 2024
QUBIT BREAKTHROUGHS ARE SHRINKING THE GAP
Research reduced RSA-2048 break requirements from 20M qubits to as low as 100K.
Source: 3 research papers, 2025–Mar 2026
AI FRAUD LOSSES ARE EXPLODING
Deepfake & AI fraud losses hit $1.65B in 2025, tripling from the prior year.
Source: Sumsub Deepfake Fraud Report, 2025
CONSUMER IMPACT WILL BE MASSIVE
Compromised data fuels identity theft, financial fraud, blackmail, and long-term privacy violations.
Source: IBM Cost of a Data Breach Report, 2024
POINT ROUNDUP
Confirmed espionage campaigns are already collecting.
The quantum timeline is getting shorter, not longer.
The data collected today cannot be un-collected.
The damage will be sudden, massive, and irreversible.
THE BATTLEGROUND
A RACE BETWEEN COLLECTION AND MIGRATION
THE ATTACK SIDE
Collect everything that is encrypted.
Stay patient. Stay hidden.
Store data at scale for the future.
Decrypt when quantum arrives.
VS
THE DEFENSE SIDE
Migrate to post-quantum cryptography.
Update systems, apps, and devices.
Educate, prepare, and enforce.
Reduce exposure and harden data.
Both sides are moving fast.
The outcome depends on who reaches scale first.
THE COUNTERPOINT: THE RESPONSE IS SCALING FAST
PQC STANDARDS ARE FINALIZED
NIST finalized three post-quantum standards in Aug 2024. They are ready for deployment.
Source: NIST, Aug 13, 2024
MIGRATION IS ALREADY UNDERWAY
65% of human internet traffic is already post-quantum encrypted via Cloudflare.
Source: Cloudflare, 2026
INDUSTRY DEADLINES ARE COMMITTING
2029
Google set a 2029 PQC migration deadline. NSA requires 2030 for national security.
Source: Google, Mar 25, 2026; NSA CNSA 2.0
NEW HYBRID SYSTEMS REDUCE RISK
Hybrid cryptography protects data-in-transit and future-proofs critical infrastructure.
Source: NIST, NSA, 2024–2026
REGULATION WILL DRIVE ADOPTION
Governments are mandating PQC. Compliance will accelerate across sectors.
Source: Global Risk Institute, 2024
COUNTERPOINT ROUNDUP
Post-quantum tools and standards are ready today.
Major platforms are migrating at internet scale.
Regulation and deadlines will drive full adoption.
The window of vulnerability will keep narrowing.
THE DEBATE QUESTION
Will the world migrate to post-quantum security faster than adversaries can collect and store data for Q-Day?

## THE POINT: The Collection Is the Breach. By the Time the Decryption Arrives, the Damage Is Irreversible.

The Harvest Now, Decrypt Later threat has moved from intelligence assessment to confirmed public record in the period since the original prediction was written.

In late 2024, US government confirmation of the Salt Typhoon operation, a Chinese state-sponsored APT group that infiltrated at least nine major US telecommunications providers including AT&T, Verizon, and Lumen Technologies, provided the first named, confirmed, large-scale validation of the HNDL threat model at consumer infrastructure scale.

The attackers maintained undetected access for an estimated one to two years. Their objective, per FBI Director Christopher Wray's own characterization, was not ransomware or financial theft, it was the patient, long-term accumulation of communications data.

Call records. Metadata. The content of communications passing through the compromised infrastructure.

Salt Typhoon is precisely the threat model the original prediction described, now confirmed as operational rather than probable. The adversaries collecting encrypted data today are not waiting for quantum computers to arrive before beginning collection, they are building the archive that quantum computers will eventually open.

Palo Alto Networks's 2026 Unit 42 Global Incident Response Report documents that the fastest quartile of intrusions reached data exfiltration in 72 minutes in 2025, compared to 285 minutes in 2024, a fourfold acceleration in collection speed that compounds the HNDL exposure for organizations that have not yet begun post-quantum migration.

The collection window is both expanding in scale and accelerating in pace.

The gap between the infrastructure-layer migration and the consumer-application-layer migration is the Point's most analytically significant concern for the 2027-2030 window.

The enterprise and infrastructure layer is moving at pace: 65% of human internet traffic post-quantum encrypted through Cloudflare, NIST standards finalized and production-ready, Google's 2029 deadline announced, Microsoft targeting full Windows/Azure/Microsoft 365 migration by 2033, AWS deploying hybrid post-quantum TLS standards.

Apple's iMessage has deployed PQ3 post-quantum encryption. Signal has implemented PQXDH. The infrastructure securing the transit of data is being hardened on timelines that credible quantum threat assessments identify as relevant. But the consumer applications where most personal data is generated, stored, and processed, the health apps, the legal document platforms, the financial services apps outside the G7 financial sector migration, the communication tools outside Apple and Signal's ecosystems, are on a materially different schedule.

Fewer than 5% of companies have started the transition to post-quantum cryptography, per industry assessments. The G7's January 2026 coordinated roadmap covers the financial sector specifically. It does not cover the consumer application layer where billions of people's most sensitive long-lived data is handled by organizations with neither the technical resources nor the regulatory pressure that is moving the enterprise infrastructure layer.

The consumer security threshold the original prediction identified, the gap between infrastructure migration and application migration, remains the central structural gap in the response.

## The Point Roundup

The consumer security infrastructure that billions of people depend on is in the early stages of a mandatory, multi-year migration to post-quantum cryptography, and most consumers have no idea it's happening, no way to assess their personal exposure, and no mechanism to demand accountability from the platforms handling their most sensitive data.

The migration will eventually succeed.

The question that can't be answered with confidence is whether it will succeed before the adversaries who have been patiently collecting encrypted data since 2019, 2020, and every year since, and who have now been confirmed through Salt Typhoon to be doing so at telecom infrastructure scale,  reach the computational threshold that makes all of it readable.

That question doesn't resolve tidily.

It resolves differently for different data types, different adversaries, and different timelines. What is not in question is that the window to protect data already harvested has permanently closed.

The work that matters now is ensuring that data transmitted from this point forward is protected by standards that quantum computers can't break.

That work is underway, unevenly distributed, and racing against a collection effort that has been confirmed as real rather than merely probable.

## THE COUNTERPOINT: The Timeline Has Always Slipped Outward. The Math for Why It Will Continue To Is Still Sound

The Counterpoint's strongest 2026 evidence is the infrastructure migration progress, which has advanced faster and more comprehensively than the original prediction's "stark gap between policy and consumer-level implementation" framing fully acknowledged.

Cloudflare's confirmation that 65% of human-generated internet traffic is already post-quantum encrypted represents the specific "early half" of the migration: client-to-server TLS connections, secured at the network edge where a few major players, browsers, CDNs, cloud providers, could implement PQC without requiring end-user action.

Chrome, Edge, and Firefox enabling PQC in TLS by default; Cloudflare making hybrid PQC key exchanges the default for all sites on its network; Apple and Signal deploying post-quantum protocols for their highest-sensitivity communication products, these represent the infrastructure layer acting at pace.

The NIST standards are production-ready and have been for nearly two years. Google has announced a formal 2029 migration deadline. Microsoft has committed to full Windows/Azure/Microsoft 365 migration by 2033.

The enterprise infrastructure that handles the most sensitive consumer data in transit is being hardened on timelines that credible quantum threat assessments identify as within the relevant window.

The Mosca inequality, which frames migration urgency as a function of migration time plus required security duration relative to Q-Day timing, is being applied by the organizations closest to the quantum frontier, and their public commitments suggest the timeline assessment has found the migration achievable rather than impossibly compressed.

The Q-Day timeline itself remains genuinely uncertain in ways that matter for the Counterpoint's scenario.

The Global Risk Institute's 2024 expert survey found a 5-14% probability of a cryptographically relevant quantum computer by 2029, which means an 86-95% probability that it does not happen by 2029, framing the same number differently.

Every year since Shor's 1994 paper, the credible consensus timeline for cryptographically relevant quantum computing has been revised outward as the engineering challenges of qubit stability, error correction, and coherence times proved harder than theoretical models predicted.

The resource estimate compression, from 20 million qubits to under one million over recent years, reflects theoretical advances in algorithmic efficiency rather than demonstrated hardware capabilities.

Current state-of-the-art quantum hardware can't execute Shor's Algorithm against RSA-2048 at any qubit count. The gap between theoretical resource estimates and demonstrated hardware capability has been the consistent source of timeline slippage for thirty years.

The Counterpoint does not require Q-Day to be impossible.

It requires Q-Day to remain beyond the horizon of the consumer infrastructure migration now underway, and the migration is moving faster in 2026 than at any prior point.

Cloudflare's internal decision to pull their Q-Day readiness timeline forward, noted in the original piece, is evidence of organizational seriousness, not evidence that Q-Day is imminent.

When the organizations closest to the technical frontier are accelerating their preparation, the correct interpretation is that the race is live, not that the race is lost.

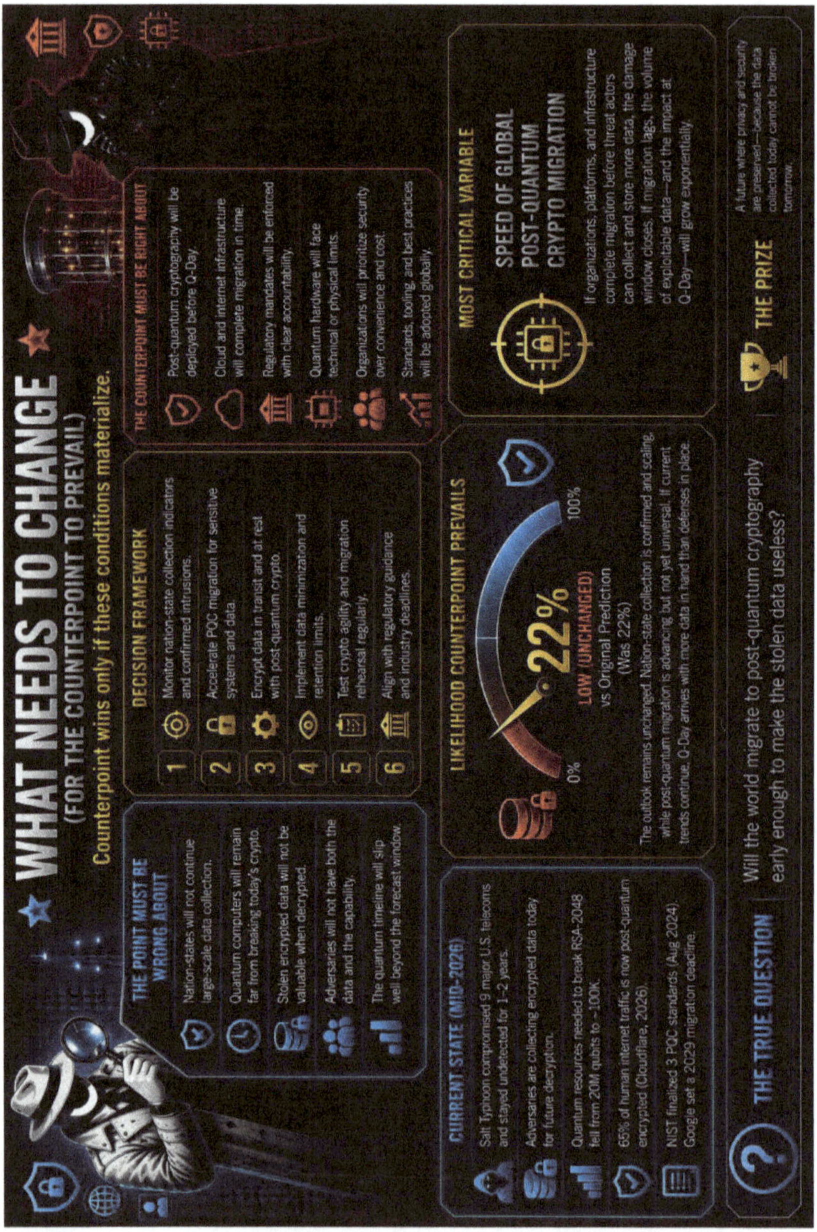
WHAT NEEDS TO CHANGE
(FOR THE COUNTERPOINT TO PREVAIL)
Counterpoint wins only if these conditions materialize.
THE POINT MUST BE WRONG ABOUT
Nation-states will not continue large-scale data collection.
Quantum computers will remain far from breaking today's crypto.
Stolen encrypted data will not be valuable when decrypted.
Adversaries will not have both the data and the capability.
The quantum timeline will slip well beyond the forecast window.
DECISION FRAMEWORK
1 Monitor nation-state collection indicators and confirmed intrusions.
2 Accelerate PQC migration for sensitive systems and data.
3 Encrypt data in transit and at rest with post-quantum crypto.
4 Implement data minimization and retention limits.
5 Test crypto agility and migration rehearsal regularly.
6 Align with regulatory guidance and industry deadlines.
THE COUNTERPOINT MUST BE RIGHT ABOUT
Post-quantum cryptography will be deployed before Q-Day.
Cloud and internet infrastructure will complete migration in time.
Regulatory mandates will be enforced with clear accountability.
Quantum hardware will face technical or physical limits.
Organizations will prioritize security over convenience and cost.
Standards, tooling, and best practices will be adopted globally.
CURRENT STATE (MID-2026)
Salt Typhoon compromised 9 major U.S. telecoms and stayed undetected for 1–2 years.
Adversaries are collecting encrypted data today for future decryption.
Quantum resources needed to break RSA-2048 fell from 20M qubits to ~100K.
65% of human internet traffic is now post-quantum encrypted (Cloudflare, 2026).
NIST finalized 3 PQC standards (Aug 2024). Google set a 2029 migration deadline.
LIKELIHOOD COUNTERPOINT PREVAILS
0%
22%
100%
LOW (UNCHANGED)
vs Original Prediction
(Was 22%)
The outlook remains unchanged. Nation-state collection is confirmed and scaling, while post-quantum migration is advancing but not yet universal. If current trends continue, Q-Day arrives with more data in hand than defenses in place.
MOST CRITICAL VARIABLE
SPEED OF GLOBAL POST-QUANTUM CRYPTO MIGRATION
If organizations, platforms, and infrastructure complete migration before threat actors can collect and store more data, the damage window closes. If migration lags, the volume of exploitable data—and the impact at Q-Day—will grow exponentially.
THE TRUE QUESTION
Will the world migrate to post-quantum cryptography early enough to make the stolen data useless?
THE PRIZE
A future where privacy and security are preserved—because the data collected today cannot be broken tomorrow.

## What Needs to Change

### (... for the Counterpoint to Prevail)

For the adequate-migration-before-Q-Day scenario to hold through 2030, two conditions need to remain stable that are currently under pressure.

First, the quantum hardware timeline needs to continue its historical pattern of slipping rather than accelerating toward the lower end of the resource estimate range.

Three papers between May 2025 and March 2026 compressed the theoretical qubit requirement from 20 million to under one million, and potentially to 100,000 using QLDPC architectures. If that theoretical compression translates into hardware progress at comparable pace, the timeline pressure on the consumer-layer migration increases materially.

The Iceberg Quantum Pinnacle and similar architectural proposals are theoretical at this stage, the engineering distance from theoretical resource estimate to demonstrated hardware capability has historically been long. But the direction of travel matters for migration urgency assessment.

Second, the consumer application layer, the apps, platforms, and services handling the most sensitive personal data outside the enterprise perimeter, needs to accelerate its post-quantum migration dramatically, on a timeline that current adoption rates suggest won't happen organically without regulatory pressure equivalent to what the G7 financial sector roadmap is providing.

The NSA CNSA 2.0 mandatory 2030 deadline covers National Security Systems.

There's no equivalent mandate for the consumer application ecosystem. Without one, the gap between infrastructure migration and application migration the Point identifies as the central structural vulnerability remains open through the 2027-2030 window.

## Likelihood Counterpoint Prevails: 22%

The Counterpoint likelihood holds at 22%, consistent with the original's 78% Point confidence, converted to series-standard framing.

The 2026 evidence hasn't moved this number because it has strengthened both sides roughly proportionally: Salt Typhoon confirmed the HNDL threat at scale for the Point, while the 65% Cloudflare post-quantum traffic milestone and multiple major vendor 2029/2033 migration commitments strengthened the Counterpoint.

The 22% is among the lowest in the series, and it reflects a prediction where the threat mechanism is scientifically established, the evidence of active adversarial collection is no longer assessed as probable but confirmed as documented, and the migration response, while genuine and progressing, is structurally uneven in ways that leave the consumer application layer exposed on a different timeline than the infrastructure layer being hardened by Cloudflare, Google, Apple, and Microsoft.

What preserves the 22% rather than reducing it further: the Q-Day timeline remains genuinely uncertain and the historical pattern of timeline slippage is real; the infrastructure migration is progressing faster than any prior point in the transition; and the Counterpoint's core scenario, that the migration at the infrastructure layer outpaces Q-Day's arrival for the majority of consumer data, remains plausible under the optimistic end of the quantum timeline distribution.

The 22% says: the Harvest Now, Decrypt Later threat is not speculative, not future-tense, and not addressable retroactively for data already collected.

The question for 2027-2030 is whether the consumer application migration accelerates enough to protect data transmitted from this point forward before the quantum decryption window opens, and the current trajectory of the consumer layer, absent regulatory pressure, does not suggest that confidence is warranted.

# The Captive Repair Economy Is Breaking

## One State Law at a Time

### The Law Passed. The Repair Still Costs $400. The Gap Between Those Two Facts Is the Story.

*When you buy a product, you expect to own it.*

*Not lease it. Not subscribe to it. Not be held hostage by the manufacturer's authorized service network every time something breaks.*

*Eight US states now have enforceable Right to Repair laws covering roughly one-third of the American population, with Connecticut effective July 2026 and Texas effective September 2026 adding to the January 2026 Colorado and Washington enactments.*

*The EU's directive transposition deadline is July 31, 2026, with Germany's parliament on track to meet it. EU smartphone repairability rules already took effect June 20, 2025.*

*And the compliance theater the Counterpoint predicted has arrived on schedule: manufacturers are implementing just enough to claim compliance, parts pairing prohibitions are being tested at the enforcement boundary, and the gap between law-on-paper and experience-in-practice is exactly where the next five years of this story will be written.*

OVERVIEW
THE CAPTIVE REPAIR ECONOMY IS BREAKING
ONE STATE LAW AT A TIME
THE LAW PASSED. THE REPAIR STILL COSTS $400.
THE GAP BETWEEN THOSE TWO FACTS IS THE STORY.
THE EVIDENCE IS REAL
8 U.S. STATES HAVE ENACTED RIGHT TO REPAIR LAWS covering roughly one-third of the American population. Two more take effect in 2026.
EU TRANSPOSITION DEADLINE IS JULY 31, 2026.
EU smartphone repairability rules already took effect June 20, 2025.
PARTS PAIRING PROHIBITIONS
exist in 3 states: Oregon, Colorado, and Washington.
COMPLIANCE THEATER IS HERE
Manufacturers implement just enough to claim compliance, testing enforcement boundaries.
40 STATES HAVE FILED BILLS.
Momentum is bipartisan and accelerating. More than 33 bills introduced in 13 states in early 2026 alone.
RIGHT TO REPAIR LAWS IN THE U.S.
Law in Effect
Law Enacted – Effective in 2026
No Law Yet
STATES WITH LAWS
New York (Dec 2023)
California (July 2024)
Minnesota (July 2024)
Oregon (Jan 2025)
Colorado (Jan 2026)
Washington (Jan 2026)
Connecticut (July 2026)
Texas (Sept 2026)
COVERING ABOUT
1/3
OF THE U.S. POPULATION BY END OF 2026
THE BARRIER IS REAL
CAPTIVE REPAIR MODEL PERSISTS
Authorized service is still the default. Independent repair remains expensive and difficult.
PARTS PAIRING LOCKS OUT REPAIR
Software locks and serializing parts block non-authorized replacements. Now banned in 3 states.
ACCESS IS INCOMPLETE
Documentation, tools, and diagnostics are limited, delayed, or behind "fair and reasonable" hurdles.
ENFORCEMENT GAP IS WIDE
Laws exist, but enforcement resources are thin. Violations are hard to prove and slow to remedy.
CONSUMERS STILL PAY THE PRICE
Repairs cost more, take longer, and create e-waste. The ownership promise is still broken.
8 STATES HAVE ENACTED LAWS (2 more in 2026)
~1/3 OF U.S. POPULATION COVERED by end of 2026
JULY 31, 2026 EU TRANSPOSITION DEADLINE
JUNE 20, 2025 EU SMARTPHONE RULES IN EFFECT
3 STATES BAN PARTS PAIRING OR, CO, WA
>33 BILLS INTRODUCED IN 13 STATES IN EARLY 2026
THE TRUE QUESTION
Can legal rights to repair translate into real, affordable, and convenient repairs before manufacturers close the loopholes that keep the captive model intact?
THE PRIZE
A future where you own what you buy—and can fix it when it breaks.

## Opening

The Right to Repair movement has crossed a threshold that would have been unimaginable five years ago, and the 2026 evidence record confirms the original prediction's scale and direction while revealing the Counterpoint's compliance theater warning as the most analytically prescient observation in the piece.

The legislative count, the EU transposition timeline, and the bipartisan political character have all materialized as predicted.

The enforcement gap, between what the laws require and what manufacturers actually deliver, has also materialized, as predicted.

Both things are simultaneously true, which is what 68% looks like: a movement that has genuinely changed the legal landscape and is now in the genuinely difficult phase of converting legal rights into repair experiences.

## Intro

The Right to Repair prediction's 2026 update requires holding two developments in tension that the original piece treated as sequential: the legislative wins were supposed to come first, then the enforcement gap would emerge as the second-order problem once manufacturers began navigating compliance.

What has actually happened is that both arrived simultaneously.

The Environment America July 2025 compliance report documenting manufacturers not meeting the spirit of existing laws in California, Minnesota, and New York was published before Colorado and Washington even took effect.

The compliance theater the Counterpoint warned about was visible in the first wave of states before the second wave had started.

That simultaneity is analytically important because it means the Right to Repair movement in 2026 is not in the legislative phase waiting for enforcement to begin.

It's in both phases at once: winning new laws in Connecticut and Texas while fighting the compliance battles that the laws already on the books have generated in New York, California, and Minnesota. The parts pairing prohibitions in three states represent the most consequential legal development in the movement's history precisely because they address the architectural enforcement mechanism manufacturers developed specifically to preserve the captive repair model even when forced to provide parts access.

Whether those prohibitions hold under the "fair and reasonable" negotiation pressure that manufacturers will apply at every compliance boundary is the 2027-2030 question that no legislative count can answer.

## Background

As of mid-2026, eight US states have enacted Right to Repair laws covering consumer electronics: New York (effective December 2023), California (July 2024), Minnesota (July 2024), Oregon (January 2025), Colorado (January 2026), Washington (January 2026), Connecticut (July 2026), and Texas (September 2026).

Consumer Reports characterized the combined population coverage as "about a third of the U.S. population" having access to some form of repair rights by end of 2026. Colorado's law, effective January 1, 2026, is widely characterized by legal analysts as the broadest repair rights law enacted anywhere in the nation, requiring manufacturers to make available "any documentation, embedded software, tool, part, or other device or implement" on fair and reasonable terms.

Washington State's law, signed by Governor Ferguson on May 19, 2025 and effective January 1, 2026, makes Washington the third state to prohibit parts pairing alongside Oregon and Colorado.

The EU's Right to Repair Directive (2024/1799), entered into force July 30, 2024, has a transposition deadline of July 31, 2026 for all 27 member states.

Germany published its draft transposition law in January 2026, with parliamentary discussions scheduled for June and expected adoption before the July deadline per Right to Repair Europe. Critically: EU ecodesign repairability requirements specifically for smartphones and tablets placed on the EU market for the first time on or after June 20, 2025 are already in effect, the directive transposition extends that protection to additional product categories and adds the out-of-warranty repair obligation.

As of January 2026, PIRG tracked more than 33 Right to Repair bills introduced in 13 states in the first weeks of the year alone.

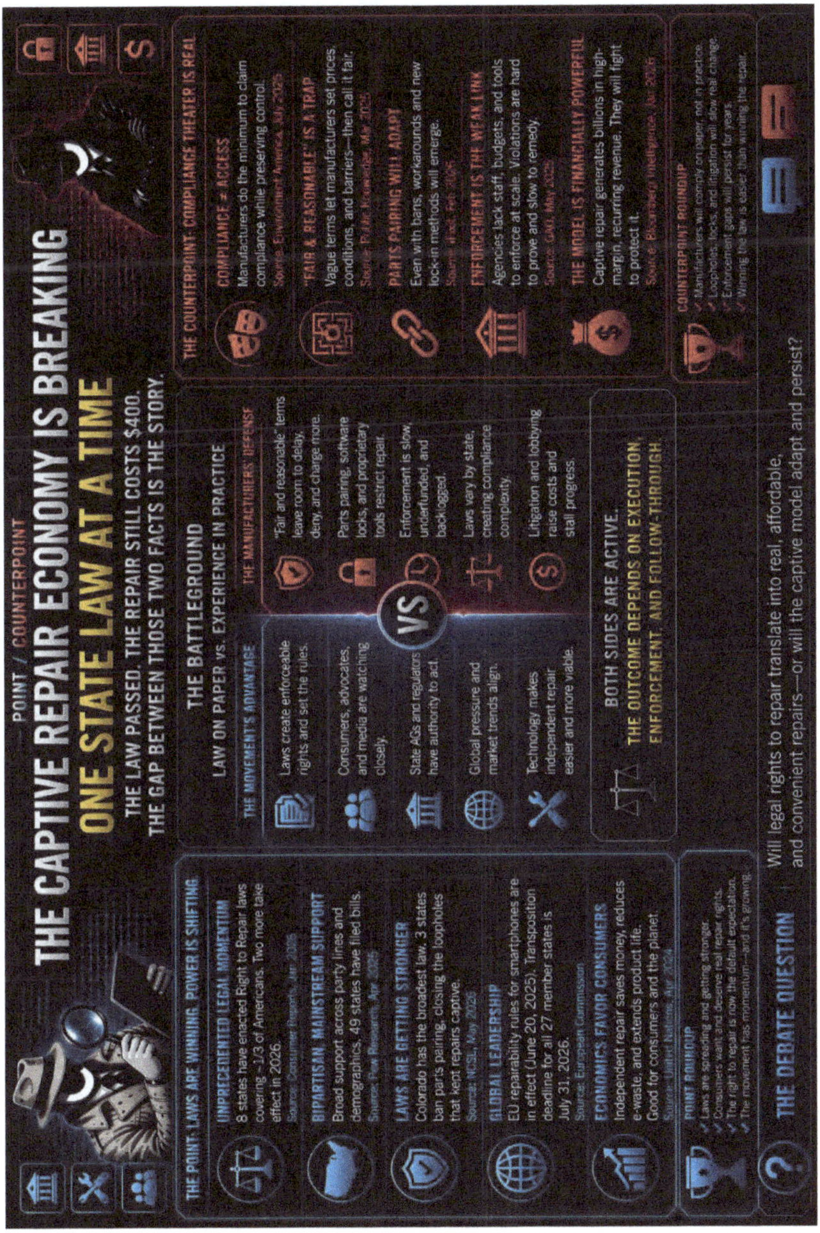
POINT / COUNTERPOINT
THE CAPTIVE REPAIR ECONOMY IS BREAKING
ONE STATE LAW AT A TIME
THE LAW PASSED. THE REPAIR STILL COSTS $400.
THE GAP BETWEEN THOSE TWO FACTS IS THE STORY.
THE POINT: LAWS ARE WINNING. POWER IS SHIFTING.
UNPRECEDENTED LEGAL MOMENTUM
8 states have enacted Right to Repair laws covering ~1/3 of Americans. Two more take effect in 2026.
Source: Consumer Reports, Jan 2026
BIPARTISAN, MAINSTREAM SUPPORT
Broad support across party lines and demographics. 49 states have filed bills.
Source: Pew Research, Apr 2025
LAWS ARE GETTING STRONGER
Colorado has the broadest law. 3 states ban parts pairing, closing the loopholes that kept repairs captive.
Source: NCSL, May 2026
GLOBAL LEADERSHIP
EU repairability rules for smartphones are in effect (June 20, 2025). Transposition deadline for all 27 member states is July 31, 2026.
Source: European Commission
ECONOMICS FAVOR CONSUMERS
Independent repair saves money, reduces e-waste, and extends product life. Good for consumers and the planet.
Source: United Nations, Apr 2024
POINT ROUNDUP
✓ Laws are spreading and getting stronger.
✓ Consumers want and deserve real repair rights.
✓ The right to repair is now the default expectation.
✓ The movement has momentum—and it's growing.
THE BATTLEGROUND
LAW ON PAPER vs. EXPERIENCE IN PRACTICE
THE MOVEMENT'S ADVANTAGE
Laws create enforceable rights and set the rules.
Consumers, advocates, and media are watching closely.
State AGs and regulators have authority to act.
Global pressure and market trends align.
Technology makes independent repair easier and more viable.
VS
THE MANUFACTURERS' DEFENSE
"Fair and reasonable" terms leave room to delay, deny, and charge more.
Parts pairing, software locks, and proprietary tools restrict repair.
Enforcement is slow, underfunded, and backlogged.
Laws vary by state, creating compliance complexity.
Litigation and lobbying raise costs and stall progress.
BOTH SIDES ARE ACTIVE.
THE OUTCOME DEPENDS ON EXECUTION, ENFORCEMENT, AND FOLLOW-THROUGH.
THE COUNTERPOINT: COMPLIANCE THEATER IS REAL
COMPLIANCE ≠ ACCESS
Manufacturers do the minimum to claim compliance while preserving control.
Source: Environment America, July 2025
"FAIR & REASONABLE" IS A TRAP
Vague terms let manufacturers set prices, conditions, and barriers—then call it fair.
Source: Public Knowledge, Mar 2025
PARTS PAIRING WILL ADAPT
Even with bans, workarounds and new lock-in methods will emerge.
ENFORCEMENT IS THE WEAK LINK
Agencies lack staff, budgets, and tools to enforce at scale. Violations are hard to prove and slow to remedy.
Source: GAO, May 2025
THE MODEL IS FINANCIALLY POWERFUL
Captive repair generates billions in high-margin, recurring revenue. They will fight to protect it.
Source: Bloomberg Intelligence, Jan 2026
COUNTERPOINT ROUNDUP
✓ Manufacturers will comply on paper, not in practice.
✓ Loopholes, locks, and litigation will slow real change.
✓ Enforcement gaps will persist for years.
✓ Winning the law is easier than winning the repair.
THE DEBATE QUESTION
Will legal rights to repair translate into real, affordable, and convenient repairs—or will the captive model adapt and persist?

## THE POINT: Forty-Nine States Have Filed. Eight Have Laws. A Third of Americans Are Covered.

The legislative response has reached a scale and pace that the manufacturer lobby has been unable to contain, and the 2026 additions have materially strengthened the original prediction's most important technical claim.

Three states now explicitly prohibit parts pairing: Oregon, Colorado, and Washington. Parts pairing, the practice of serializing components and using software to prevent non-manufacturer-approved replacements from functioning normally, elevated the captive repair model from a service policy into a technical architecture.

A screen replaced with a genuine manufacturer part by an independent technician triggers warning messages, disables features, or reduces functionality on devices using parts pairing. The three-state prohibition on this practice is the most consequential single development in the 2025-2026 Right to Repair record, because it addresses the architectural enforcement mechanism rather than merely the commercial terms.

The 2026 Repair Association legislative template builds on this by explicitly prohibiting manufacturers from "using software-based restrictions to limit access to parts or tools, or to control who is permitted to perform repairs", language that has been confirmed in Colorado law and will be the template for subsequent state legislation.

Colorado's law is described by Reed Smith's legal analysis as having language "intended to throw the OEM's toolbox wide open", the broadest access standard enacted anywhere in the nation.

The EU directive applying to products already sold before July 31, 2026, not just future products, represents a retroactive reach that is genuinely unprecedented in consumer electronics regulation and creates compliance obligations for the existing installed base rather than allowing a clean cutover to compliant new models.

The political character of the movement has proven as durable as the legislative count. The NFIB's 89% member support figure has remained consistent through 2025-2026, maintaining the small business economic argument alongside the consumer rights argument that prevents the movement from being characterized as ideologically aligned with either political party.

Google's sustained advocacy position, testified in support of bills in Oregon, Colorado, and Pennsylvania, received the 2024 Right to Repair Advocacy Award from Repair.org, hasn't reversed in 2026, maintaining the most visible crack in the previously unified manufacturer opposition. Apple's materially softened lobbying stance has persisted.

The bipartisan state legislative activity PIRG documented in February 2026, states pressing ahead across the political spectrum from Maine to Hawaii, Florida to Kansas, confirms that the movement's political character has stabilized rather than reverting to the partisan alignment that might have made it easier for manufacturers to coordinate political opposition.

The Warrior Right to Repair Act, which PIRG is actively urging senators to support, represents the federal legislative vector that, if enacted, would resolve the fragmentation problem the Counterpoint correctly identifies as manufacturers' most valuable remaining defense.

### The Point Roundup

The Right to Repair movement has crossed the threshold from activist cause to legislative reality, and the 2026 additions have strengthened the core claim in the specific dimension that matters most: the parts pairing prohibition in three states directly addresses the architectural enforcement mechanism that the original article identified as the technical sabotage problem. Households in states with enforceable laws are gaining meaningful repair alternatives, and the EU directive's July 31, 2026, transposition deadline creates global compliance pressure on manufacturers whose business models transcend any single state's jurisdiction.

The remaining question, whether the laws in place are strong enough, consistent enough, and enforceable enough to fully break the captive repair model, is precisely what the Counterpoint predicts won't be answered in consumers' favor. The enforcement record of the next 18 months will begin to answer it.

## THE COUNTERPOINT: The Law Passed. The Repair Still Costs $400.

The Right to Repair movement has won the rhetoric and is winning the legislation.

What it hasn't yet won is the actual repair experience, and the gap between those two outcomes is where manufacturers are building their next line of defense. The Environment America July 2025 report, covering manufacturers' compliance with existing Right to Repair laws in California, Minnesota, and New York, found that manufacturers are not doing enough to support repair of their products despite the laws being in effect.

No dishwasher earned above a C grade. Tablet repair materials lag behind smartphones and laptops. Even in categories where compliance grades are better, laptops earning As and Bs, the report documents that meeting the law's technical requirements and genuinely enabling repair at competitive economics are not the same achievement.

The automotive sector provides the clearest preview of what compliance theater looks like at operational scale: automakers publicly support consumer choice while controlling vehicle data architecture through subscription-based OEM repair portals, proprietary scan-tool ecosystems, and certified repair network structures.

"Stated compliance, operational resistance" is the playbook the movement will face across every product category as laws take effect. "Fair and reasonable terms" for parts and documentation becomes a negotiation about what fair and reasonable means, conducted by manufacturers with armies of lawyers and no commercial incentive to resolve it generously.

The fragmentation problem compounds the compliance theater problem, and the 2026 legislative landscape hasn't resolved it.

Connecticut's law derives from a consumer protection omnibus bill.

Texas's law takes effect in September.

Colorado's has the broadest language but numerous carve-outs.

Washington's prohibits parts pairing but with specific conditional limitations.

Oregon's is the strongest for consumer electronics, but wheelchair rules run on a different timeline.

Eight state laws with eight sets of requirements, eight enforcement mechanisms, and eight definitions of "fair and reasonable" create a compliance patchwork that large manufacturers with dedicated legal and regulatory affairs teams are structurally better equipped to manage than the independent repair shops and individual consumers the laws are designed to protect.

The federal Right to Repair Act has been introduced but not passed, the Warrior Right to Repair Act urged by PIRG is still in advocacy rather than legislative progress.

Without federal uniformity, manufacturers can optimize their compliance responses state by state, making legal resistance economically rational and operationally manageable.

The EU directive's full harmonization approach, member states can't impose stricter rules, provides the model for what genuine uniformity looks like. Its practical enforcement record remains to be written as the July 2026 transposition deadline passes.

WHAT NEEDS TO CHANGE
(FOR THE COUNTERPOINT TO PREVAIL)
Counterpoint wins only if these conditions materialize.
THE POINT MUST BE WRONG ABOUT
Manufacturers will not maintain control of repair through new technical and contractual barriers.
State laws will not be broadly enforced or will have narrow, weak exceptions.
Consumers and repair businesses will not demand affordable, independent repair at scale.
The movement will lose momentum as industry pressure and fatigue grow.
DECISION FRAMEWORK
1 Ensure strong, uniform enforcement by state AGs and regulators.
2 Clamp down on evasion tactics: parts pairing, software locks, and arbitrary "authorization" schemes.
3 Require transparency on parts pricing, tools, software, and repair information.
4 Fund and empower local enforcement and small repair businesses.
5 Adopt standards that guarantee interoperability and true parts freedom.
6 Coordinate across states and with the EU to prevent a patchwork disadvantage.
THE COUNTERPOINT MUST BE RIGHT ABOUT
Manufacturers will implement only the minimum required to claim compliance.
Vague "fair and reasonable" terms will be interpreted in manufacturers' favor.
Litigation and lobbying will weaken or delay enforcement for years.
New technical controls will emerge faster than laws can address them.
Consumer apathy and convenience will keep captive repair profitable.
Political will will fade as industry campaigns shift public opinion.
CURRENT STATE (MID-2026)
8 states have enacted Right to Repair laws covering ~1/3 of U.S. population.
EU Right to Repair Directive transposition deadline is July 31, 2026.
EU smartphone repairability rules are already in effect (June 20, 2025).
Parts pairing banned in 3 states: Oregon, Colorado, Washington.
More than 33 Right to Repair bills introduced in 13 states in early 2026 alone.
Compliance gaps are already documented in CA, MN, and NY.
LIKELIHOOD COUNTERPOINT PREVAILS
0%
32%
100%
UNCHANGED
vs Original Prediction (Was 32%)
The counterpoint remains at 32%. Parts-pairing prohibitions in three states strengthen the Point. Germany's EU transposition is on track. About one-third of the U.S. population is covered by repair laws. But compliance theater, fragmented state laws, and the lack of federal legislation continue to preserve the counterpoint.
MOST CRITICAL VARIABLE
ENFORCEMENT AND LOOPHOLES
If enforcement is strong, loopholes are closed, and technical barriers are banned or neutralized, the laws can deliver real, affordable repair access.
If not, manufacturers will outmaneuver regulators, preserve control, and the captive model will endure.
THE TRUE QUESTION
Will states and the EU enforce the spirit of repair rights and close the loopholes—or will the captive model evolve and persist?
THE PRIZE
A future where you truly own what you buy—and can fix it anywhere, affordably, for life.

What

## Needs to Change

### (... for the Counterpoint to Prevail)

For the compliance theater scenario to become the dominant outcome, for the Right to Repair movement to win the laws and lose the repair experience, two conditions need to hold through 2030.

First, federal legislation needs to continue failing to materialize, leaving manufacturers able to manage state-by-state variation indefinitely without facing the uniform compliance burden that would make operational resistance economically irrational. The Warrior Right to Repair Act is in advocacy; the REPAIR Act's inclusion in Surface Transportation Reauthorization depends on congressional appetite that hasn't materialized in prior sessions.

Second, enforcement needs to remain reactive rather than proactive, state attorneys general waiting for documented consumer complaints rather than auditing manufacturer compliance systematically. The movement's legislative wins are only as durable as the enforcement mechanisms behind them. Laws with no enforcement history are invitations to creative compliance, and manufacturers have demonstrated considerable creativity when commercial stakes are high.

The Environment America July 2025 report documenting compliance gaps across existing law states is the first systematic evidence of whether enforcement is keeping pace with legislation. The answer, as of mid-2026, is: not yet.

## Likelihood Counterpoint Prevails: 32%

The Counterpoint likelihood remains at the original 32%. The parts pairing prohibition in three states is a genuinely new enforcement mechanism that the compliance theater scenario has to navigate rather than simply absorb through "fair and reasonable" negotiation, the technical architecture argument has a statutory answer in Colorado, Oregon, and Washington that it did not have when the original was written.

Germany's EU transposition staying on track for the July deadline confirms the global compliance pressure is materializing rather than slipping. And Consumer Reports' characterization of "about a third of the U.S. population" having some form of repair rights by end of 2026 represents a population coverage threshold at which enforcement has political sustainability, the attorney general of a state covering 20 million people has electoral incentives to use the enforcement authority the law provides.

What keeps the Counterpoint at 32%: the Environment America compliance report showing manufacturers not meeting the spirit of existing laws; the persistent fragmentation across eight different state frameworks; and the federal legislative stall that maintains the architecture of state-by-state resistance.

The 32% says: the movement has changed the legal landscape, the direction of travel is firmly toward expanded repair rights, and whether those rights translate into a fundamentally different consumer experience at scale is what 2027 enforcement data will begin to answer.

# The Most Personal Data Relationship You've Ever Had

And You Never Asked For It

The Most Useful Technology Architecture Ever Deployed. And the Most Consequential Data Relationship

*Siri. Cortana. Google Now. Each was going to anticipate your needs. Each delivered something useful in narrow contexts and failed to generalize.*

*At WWDC 2026, Apple announced "Siri AI", rebuilt on an updated Apple Intelligence architecture developed in collaboration with Google's Gemini models, capable of understanding screen content, accessing personal context across calendar, communication, financial, and health data, and taking multi-step actions across apps. Apple's own newsroom: "a profoundly more capable and personal assistant."*

*This is not an incremental improvement on the Siri that launched in 2011. It's a different category of capability.*

*The prediction's enabling conditions have arrived. So has the data relationship the Counterpoint warned about.*

## Opening

The promise of ambient, proactive, personalized technology has been made and broken so many times that consumer skepticism is entirely earned.

What is arriving in 2026 is structurally different, not because the marketing is better, but because the underlying capability finally matches the use case. Large language models that maintain context across complex multi-step tasks, combined with device ecosystems that have legitimate access to calendar, communication, financial, and health data, and orchestration layers that can take real actions across real applications, these are not incremental improvements on Siri.

They are a different category of capability.

The original prediction identified the enabling conditions for this transition; the 2026 evidence, culminating in Apple's WWDC announcement of Siri AI on June 8, confirms they have arrived.

The Counterpoint's lock-in argument is simultaneously more urgent: the personal context index that makes these agents useful is also the switching cost infrastructure that makes competition structurally harder after it's built.

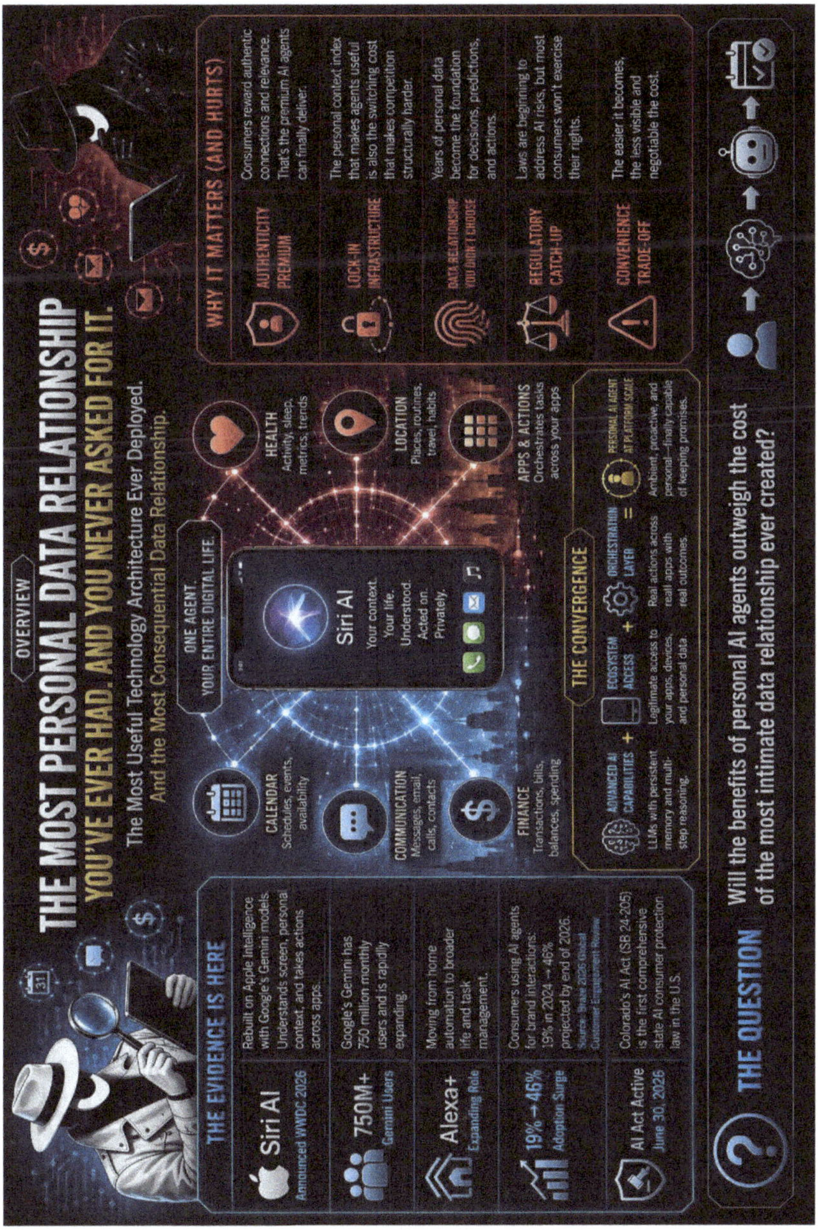
OVERVIEW
THE MOST PERSONAL DATA RELATIONSHIP
YOU'VE EVER HAD. AND YOU NEVER ASKED FOR IT.
The Most Useful Technology Architecture Ever Deployed.
And the Most Consequential Data Relationship.
THE EVIDENCE IS HERE
Siri AI
Announced WWDC 2026
Rebuilt on Apple Intelligence with Google's Gemini models. Understands screen, personal context, and takes actions across apps.
750M+
Gemini Users
Google's Gemini has 750 million monthly users and is rapidly expanding.
Alexa+
Expanding Role
Moving from home automation to broader life and task management.
19% → 46%
Adoption Surge
Consumers using AI agents for brand interactions: 19% in 2024 → 46% projected by end of 2026.
Source: Braze 2026 Global Customer Engagement Review
AI Act Active
June 30, 2026
Colorado's AI Act (SB 24-205) is the first comprehensive state AI consumer protection law in the U.S.
ONE AGENT.
YOUR ENTIRE DIGITAL LIFE.
CALENDAR
Schedules, events, availability
COMMUNICATION
Messages, email, calls, contacts
FINANCE
Transactions, bills, balances, spending
Siri AI
Your context.
Your life.
Understood.
Acted on.
Privately.
HEALTH
Activity, sleep, metrics, trends
LOCATION
Places, routines, travel, habits
APPS & ACTIONS
Orchestrates tasks across your apps
THE CONVERGENCE
ADVANCED AI CAPABILITIES
LLMs with persistent memory and multi-step reasoning.
+
ECOSYSTEM ACCESS
Legitimate access to your apps, devices, and personal data.
+
ORCHESTRATION LAYER
Real actions across reall apps with real outcomes.
=
PERSONAL AI AGENT AT PLATFORM SCALE
Ambient, proactive, and personal—finally capable of keeping promises.
WHY IT MATTERS (AND HURTS)
AUTHENTICITY PREMIUM
Consumers reward authentic connections and relevance. That's the premium AI agents can finally deliver.
LOCK-IN INFRASTRUCTURE
The personal context index that makes agents useful is also the switching cost that makes competition structurally harder.
DATA RELATIONSHIP YOU DIDN'T CHOOSE
Years of personal data become the foundation for decisions, predictions, and actions.
REGULATORY CATCH-UP
Laws are beginning to address AI risks, but most consumers won't exercise their rights.
CONVENIENCE TRADE-OFF
The easier it becomes, the less visible and negotiable the cost.
THE QUESTION
Will the benefits of personal AI agents outweigh the cost of the most intimate data relationship ever created?

 

## Intro

The personal AI agent prediction's 2026 update is defined by a convergence that the original piece described as the enabling condition for the transformation it forecast: the capability has arrived, at platform scale, with a privacy architecture more rigorous than any prior generation of personal technology deployed.

Apple's WWDC 2026 Siri AI announcement is not a product update. It's the architectural transition the original prediction identified as the threshold event, delivered by the company with the largest installed base of personal devices on the planet, with a stated commitment to third-party auditability of its privacy architecture that prior generations of personal assistant technology never attempted.

The Counterpoint's lock-in concern is simultaneously more urgent than it was when the original piece was written, not because the argument has changed but because the infrastructure it describes is now being actively constructed at scale.

A personal context index built from years of calendar patterns, communication history, financial behavior, and health data is not a future risk. It's what Apple, Google, and Amazon are building right now, in the devices already in consumers' pockets and homes, under terms that most consumers won't read and can't meaningfully negotiate.

Both the Point and the Counterpoint have received the most significant evidence update of any article in this series: the capability threshold has been crossed, and the data relationship the Counterpoint warned about has moved from hypothetical to operational in the same week.

## Background

Apple's WWDC 2026 announcement of "Siri AI" represents the specific architectural transition the original prediction described. Siri AI is rebuilt on an updated Apple Intelligence architecture developed in collaboration with Google's Gemini models.

Unlike prior versions, Siri can draw on personal context, understand what is on screen, access broader knowledge, and take actions across apps while maintaining context across multiple prompts for complex multi-step interactions.

Apple characterizes it as "a fully conversational, systemwide agent" with Private Cloud Compute ensuring personal data is "not stored nor made accessible to Apple or anyone else."

Google's Gemini evolves along a parallel track with 750 million monthly users. Amazon's Alexa+ has moved from home automation toward broader life management. The adoption trajectory the Braze 2026 Global Customer Engagement Review described, 19% of consumers using AI agents for brand interactions rising to 46% by end of 2026, is being powered by platform deployments at this scale. Colorado's AI Act (SB 24-205), the nation's first comprehensive state AI consumer protection law, took effect June 30, 2026 after a delay from the original February 1, 2026 date, requiring developers and deployers of high-risk AI systems to use reasonable care to prevent algorithmic discrimination in consequential decisions affecting employment, healthcare, financial services, housing, and education.

Personal AI agents making scheduling, financial, and health recommendations sit in this regulatory space, with consumers gaining rights to notice, explanation, and appeal that most will never exercise because the interface is too convenient to make oversight feel worth the effort.

—POINT / COUNTERPOINT—
THE MOST PERSONAL DATA RELATIONSHIP
YOU'VE EVER HAD.
AND YOU NEVER ASKED FOR IT.
THE POINT
THE CAPABILITY FINALLY MATCHES THE USE CASE.
True contextual understanding.
LLMs maintain context across complex, multi-step tasks.
Legitimate access to personal data.
Calendar, communication, finance, health—connected by design.
Real actions across real apps.
Siri AI orchestrates tasks across the entire ecosystem.
Early adoption is accelerating.
19% of consumers now use AI agents; 46% by end of 2026.
Source: Braze 2026 Global Customer Engagement Review
Privacy architecture is stronger.
Private Cloud Compute: data not stored or accessible to Apple.
Delivers real everyday value.
Saves time, reduces friction, improves decisions.
POINT ROUNDUP
The enabling conditions have arrived.
Deployment is real at platform scale.
Consumer value is clear and growing.
Governance can evolve with adoption.
THE BATTLEGROUND
WHAT AI AGENTS CAN DO
Manage your schedule and commitments
Summarize and draft communications
Track finances and pay bills
Monitor health and suggest actions
Coordinate across apps and services
WHAT AI AGENTS ALSO CREATE
A personal context index built over time
Switching costs that lock in consumers
Behavior prediction and profiling
Opaque decisions with limited accountability
Concentration of power in platform ecosystems
THE CORE TENSION
More personal value VS More personal control
2026 FLASHPOINTS
Apple WWDC 2026
Siri AI rebuilt on Apple Intelligence + Gemini. "Profoundly more capable and personal assistant." (June 8, 2026)
Consumer Adoption
19% use AI agents for brand interactions; 46% projected by end of 2026.
Source: Braze 2026 Global Review
Colorado AI Act
SB 24-205 took effect June 30, 2026. First comprehensive state AI consumer protection law.
Privacy by Design
Private Cloud Compute keeps personal data not stored nor accessible to Apple or anyone else.
THE COUNTERPOINT
THE RISKS ARE STRUCTURAL, NOT HYPOTHETICAL.
Personal context index becomes the moat.
Built from years of your most intimate data.
Lock-in is baked into the infrastructure.
Switching means losing history, insight, and convenience.
Profiles enable prediction and manipulation.
Behavioral advantages accrue to the platform.
Asymmetry of power and accountability.
Platforms set rules; consumers have little recourse.
Regulation is reactive and fragmented.
Most consumers won't read or exercise their rights.
The market concentrates.
Network effects favor incumbents; competition becomes harder.
COUNTERPOINT ROUNDUP
Value today creates lock-in tomorrow.
Data asymmetry compounds over time.
Regulatory guardrails remain limited.
Structural risks outweigh convenience.
THE DEBATE QUESTION
Will the value of personal AI agents outweigh the long-term costs of the most intimate data relationship ever created?

## THE POINT: The Enabling Conditions Have Arrived. The Capability Finally Matches the Use Case.

The original prediction's enabling conditions, LLMs maintaining context across complex tasks, device ecosystems with legitimate access to personal data, orchestration layers taking real actions across real applications, have all arrived in deployable form in 2026.

Apple's WWDC 2026 Siri AI announcement is the single most important data point for this update: Apple has shipped the personal AI agent the original prediction described, with "autonomous chaining" across apps, on-screen awareness, personal context access across calendar, messages, email, photos, and files, and Private Cloud Compute ensuring personal data is processed in encrypted memory Apple itself can't access.

The announcement comes days before this writing; it's the most current possible evidence that the transition from AI as a feature to AI as the interface has arrived at the platform scale the original prediction required. For consumers managing complex schedules, multiple financial relationships, ongoing health conditions, and rich social and professional networks, an agent that synthesizes across all of those contexts and takes action without requiring manual orchestration represents a qualitative improvement in how technology serves human life.

The Braze 2026 CER adoption trajectory, 19% to projected 46% by year end, describes a category-defining growth curve across 4,000 surveyed consumers, not a niche signal.

The privacy architecture argument is the Point's strongest response to the Counterpoint's surveillance concern, and Apple's WWDC announcement reinforces it with primary-source specificity. Private Cloud Compute processes user requests in encrypted memory that Apple itself can't access.

Third-party model weights are hosted on Apple hardware to ensure raw personal context is never accessible to partner AI providers. "Outside experts can continue to verify this privacy promise at any time", Apple's own statement, constituting a public commitment to third-party auditability that prior generations of personal assistant technology never made.

The system orchestrator and App Toolbox components "work entirely on device to keep users in control of their data." This is not marketing language about privacy. It's a described technical architecture with a stated auditing commitment.

Apple's measured approach to agentic AI, prioritizing reliability and user control over aggressive autonomy, reflects the lesson the original prediction's authorial voice identified from thirty-five years of watching ambient computing promises fail: the technology that delivers on the promise is the one that augments human intent rather than replacing it with autonomous action before the reliability foundation is established.

## The Point Roundup

The personal AI agent era has arrived, confirmed by Apple's WWDC 2026 announcement, the Braze adoption trajectory, and the regulatory attention that the Colorado AI Act represents.

The original prediction's core thesis, that 2026 would mark the transition from AI as feature to AI as interface, powered by LLMs with genuine personal context access, has been validated by the most consequential product announcement in the platform's history, delivered within the prediction's specified window. The terms on which this relationship is being offered are not perfect.

The Counterpoint will make the case for why the data relationship being negotiated is consequential in ways most consumers don't fully understand.

Both things can be simultaneously true: this is the most useful personal technology architecture ever deployed, and it's the most intimate data relationship consumers have ever been asked to enter.

## THE COUNTERPOINT: The Most Powerful Lock-In Ever Built. And You Invited It In.

The personal AI agent is the most sophisticated lock-in architecture ever deployed in consumer technology, and it's being installed voluntarily, enthusiastically, and at remarkable speed by consumers focused on the convenience it delivers today rather than the dependency it creates for everything that follows.

As Siri AI becomes the primary way users interact with their data, the difficulty of switching to Android or a different ecosystem increases exponentially. The Personal Context index, built from calendar patterns, communication history, financial behavior, health data, and relationship networks, is not a feature that can be turned off when changing phones. It's an infrastructure investment that takes years to build and makes every competitor's offering feel impoverished by comparison.

The switching cost is not measured in dollars. It's measured in the irreplaceable personal context that lives inside one platform's ecosystem and can't be meaningfully exported to another.

The 27% of consumers who currently refuse to share any data with AI agents even when promised a superior experience are the ones who understand this dynamic.

The 73% who are sharing their data are discovering the convenience door opens only inward, and the terms on the other side are set almost entirely by platforms, not consumers. The Counterpoint's predicted triggering mechanism, a significant, high-profile personal AI agent failure involving a data breach or autonomous action causing documented harm, remains the variable that would shift the convenience-versus-risk calculation for a broad audience.

The behavioral flexibility is real: a single controversy caused ChatGPT uninstalls to surge 295% in a day. But that flexibility requires a triggering event that the current trajectory does not have visibility to on the horizon.

The privacy architecture, however, genuinely innovative, does not resolve the fundamental power asymmetry, and Apple's own announcement reveals the tension precisely. Private Cloud Compute is a genuine technical achievement.

It's also a single company's proprietary architecture, auditable only to the degree Apple permits, subject to revision in future software updates that users have limited meaningful ability to resist, and now dependent on Google's Gemini models as an integral component, meaning the privacy boundary extends to a second platform's infrastructure in ways that Apple's primary-source privacy commitments don't fully address.

The regulatory response is genuine but structurally insufficient for the personal agent context. Colorado's AI Act (SB 24-205), effective June 30, 2026, focuses on algorithmic discrimination in high-risk consequential decisions, employment, healthcare, financial services.

Personal AI agents operating in scheduling, recommendation, and life management contexts sit in regulatory grey zones where the "consequential decision" threshold is genuinely ambiguous. Consumer rights to notice, explanation, correction, and appeal under Colorado's framework are real rights that most consumers will never exercise because the interface is too convenient, the defaults too comfortable, and the alternative too friction laden.

Laws that create rights consumers don't exercise don't create accountability.

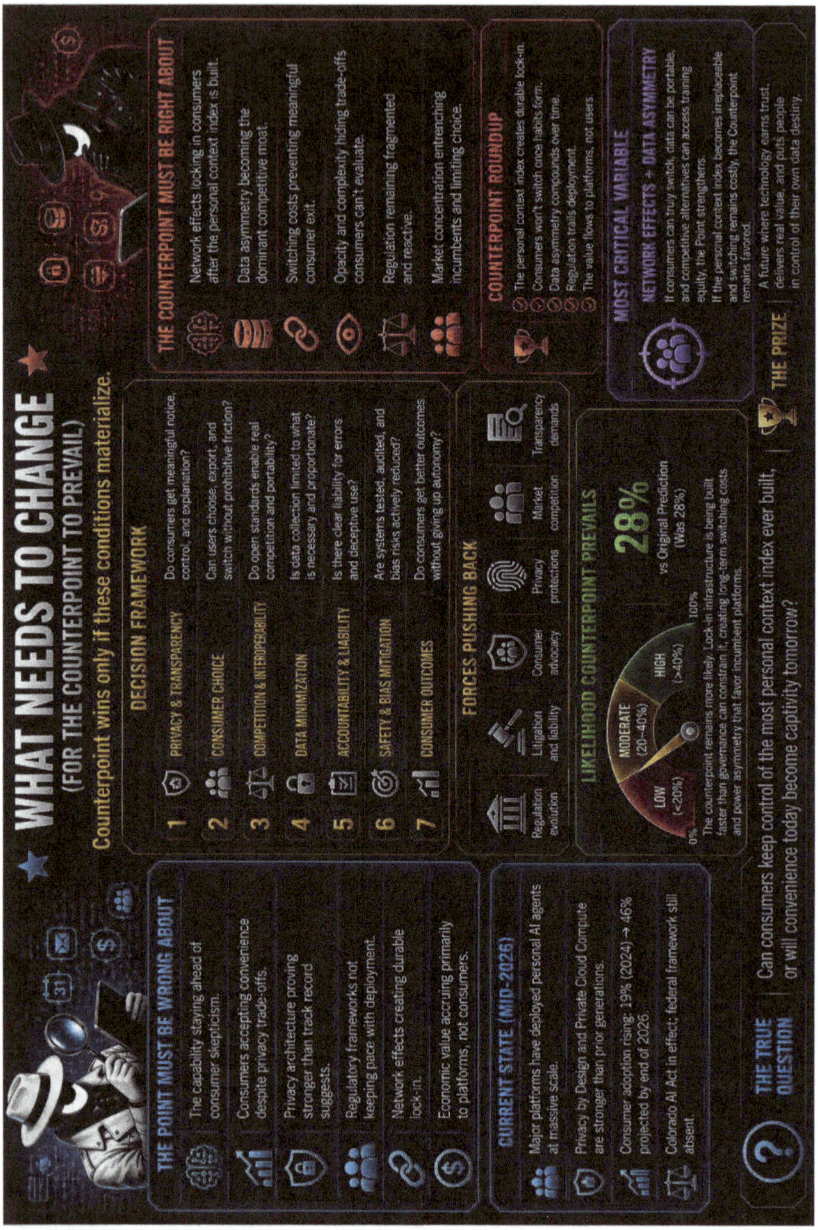
WHAT NEEDS TO CHANGE
(FOR THE COUNTERPOINT TO PREVAIL)
Counterpoint wins only if these conditions materialize.
THE POINT MUST BE WRONG ABOUT
The capability staying ahead of consumer skepticism.
Consumers accepting convenience despite privacy trade-offs.
Privacy architecture proving stronger than track record suggests.
Regulatory frameworks not keeping pace with deployment.
Network effects creating durable lock-in.
Economic value accruing primarily to platforms, not consumers.
CURRENT STATE (MID-2026)
Major platforms have deployed personal AI agents at massive scale.
Privacy by Design and Private Cloud Compute are stronger than prior generations.
Consumer adoption rising: 19% (2024) → 46% projected by end of 2026.
Colorado AI Act in effect; federal framework still absent.
DECISION FRAMEWORK
1 PRIVACY & TRANSPARENCY Do consumers get meaningful notice, control, and explanation?
2 CONSUMER CHOICE Can users choose, export, and switch without prohibitive friction?
3 COMPETITION & INTEROPERABILITY Do open standards enable real competition and portability?
4 DATA MINIMIZATION Is data collection limited to what is necessary and proportionate?
5 ACCOUNTABILITY & LIABILITY Is there clear liability for errors and deceptive use?
6 SAFETY & BIAS MITIGATION Are systems tested, audited, and bias risks actively reduced?
7 CONSUMER OUTCOMES Do consumers get better outcomes without giving up autonomy?
FORCES PUSHING BACK
Regulation evolution
Litigation and liability
Consumer advocacy
Privacy protections
Market competition
Transparency demands
LIKELIHOOD COUNTERPOINT PREVAILS
LOW (<20%)
MODERATE (20–40%)
HIGH (>40%)
0%
100%
28%
vs Original Prediction (Was 28%)
The counterpoint remains more likely. Lock-in infrastructure is being built faster than governance can constrain it, creating long-term switching costs and power asymmetry that favor incumbent platforms.
THE COUNTERPOINT MUST BE RIGHT ABOUT
Network effects locking in consumers after the personal context index is built.
Data asymmetry becoming the dominant competitive moat.
Switching costs preventing meaningful consumer exit.
Opacity and complexity hiding trade-offs consumers can't evaluate.
Regulation remaining fragmented and reactive.
Market concentration entrenching incumbents and limiting choice.
COUNTERPOINT ROUNDUP
The personal context index creates durable lock-in.
Consumers won't switch once habits form.
Data asymmetry compounds over time.
Regulation trails deployment.
The value flows to platforms, not users.
MOST CRITICAL VARIABLE
NETWORK EFFECTS + DATA ASYMMETRY
If consumers can truly switch, data can be portable, and competitive alternatives can access training equity, the Point strengthens.
If the personal context index becomes irreplaceable and switching remains costly, the Counterpoint remains favored.
THE TRUE QUESTION
Can consumers keep control of the most personal context index ever built, or will convenience today become captivity tomorrow?
THE PRIZE
A future where technology earns trust, delivers real value, and puts people in control of their own data destiny.

## What Needs to Change

### (... for the Counterpoint to Prevail)

For the lock-in and surveillance concerns to become the dominant consumer narrative rather than the background concern, two conditions need to materialize that are currently moving too slowly.

First, data portability standards for personal AI context need to be mandated rather than optional, making the switching cost calculable and reducible rather than the invisible, unquantifiable lock-in it currently represents.

Until a consumer can export their personal context index from one platform and import it meaningfully into another, the agent relationship is not empowerment. It's a subscription you can't cancel. The EU AI Act's framework and Colorado's law both address discrimination and transparency but neither yet addresses portability of personal AI context at the depth the lock-in concern requires.

Second, the regulatory framework covering personal AI agents needs to reach the life management and recommendation contexts where agents are most useful and most intimate, not just the high-risk consequential decision contexts that current legislation targets.

An agent managing your schedule, your relationships, and your financial decisions is making consequential recommendations without triggering the "consequential decision" thresholds that Colorado's law and the EU AI Act were written to address.

The regulatory gap between "useful daily assistant" and "high-risk AI system" is exactly where the personal agent's data relationship is being normalized.

## Likelihood Counterpoint Prevails: 28%

The Counterpoint likelihood remains at 28% not requiring revision. Very little time has elapsed since the first version so we will finish this series on a trend that hasn't changed when so many of the others have. Don't get comfortable though, time is clicking along.

Apple's WWDC 2026 Siri AI announcement validates the prediction's enabling conditions claim with primary-source precision.

The 19% to 46% adoption trajectory represents exactly the category-defining growth curve the original 28% implied. The Counterpoint's lock-in concern has become more urgent as the personal context index is actually being built at platform scale, not less urgent, but the triggering event that would shift consumer behavior hasn't arrived, and the regulatory framework is genuine but operates in adjacent spaces rather than directly on the personal agent data relationship the Counterpoint describes.

The 28% reflects the Counterpoint's structural accuracy: the lock-in architecture is real, the privacy boundary is less absolute than Apple's marketing implies, the data portability gap is genuine, and Colorado's regulatory framework creates rights most consumers won't exercise.

It also reflects the prediction's validated threshold crossing: the capability is there, the adoption is accelerating, and the privacy architecture, however imperfect, is meaningfully more rigorous than what prior generations of personal technology ever deployed.

Both things remain simultaneously true, which is what a prediction calibrated looks like when confirmed rather than revised.

## The Roundup

The body of work this series has assembled is, at its core, a single extended argument made thirty different ways: the most consequential questions are not technical ones.

Wearables crossed the clinical line. AI tutoring is producing 34% better learning gains for underserved students. Voice interfaces have become the default for local search. Passkeys are on a billion devices.

The tools are real, the outcomes are documented, and the trajectory is clear in each case.

What is not clear, and what the Counterpoint in nearly every topic is actually arguing, is whether the benefits reach everyone, whether the accountability frameworks arrive in time, and whether the humans on the receiving end of these systems had any meaningful say in how the systems affect their lives.

Those are not technology questions. They are governance questions, equity questions, and fundamentally human questions that no likelihood percentage can fully capture.

The Point/Counterpoint format this series has used is not a rhetorical device. It's a methodological commitment; one that reflects a sincere belief that the most dangerous thing an analyst or a publication can do in a moment of rapid technological change is to pick a side and defend it.

The topics in this series where the Point is most compelling all have genuine, evidence-backed counterarguments that deserve honest engagement.

The topics where the Counterpoint is most all have genuine points worth defending before dismantling.

The likelihood percentages are not confidence scores in a prediction market analysis. They are the analytical residue of taking both arguments seriously and asking: given what the evidence actually shows, which scenario requires more things to go right, and which requires fewer? That is a different question from "which scenario do I prefer", and the discipline of keeping those two questions separate is the entire value of the format.

The aggregate picture across 30+ topics is not optimism or pessimism; it's bifurcation. Consumers rejected covert surveillance advertising and embraced transparent value-exchange advertising simultaneously.

AI has demonstrably improved financial planning for millennials and demonstrably failed to penetrate estate planning and divorce proceedings.

Smart glasses are accelerating toward mainstream adoption at the exact moment a Swedish investigation revealed bedroom footage being reviewed by outsourced data annotators without user awareness.

The PC replacement cycle is arriving driven by Windows 10 obsolescence rather than AI feature demand.

The analog rebellion is real and being monetized by the attention economy it claims to resist.

In every case, the most accurate description of reality requires holding two things true at once, which is uncomfortable, but more honest than any clean narrative allows.

Three durable conclusions cut across the full body of work.

**First**: regulatory frameworks are consistently arriving after the harm rather than before it.

The Texas ACR settlements came after millions of consumers had their viewing habits captured without meaningful consent.

The UK CMA intervention on zero-click content came after publishers documented traffic decimation.

The EU AI Act engagement with smart glasses came after the bedroom footage incident.

The pattern is not incompetence. It's the structural reality that technology moves faster than democratic institutions, and the gap between deployment and accountability is where most of the damage occurs.

The question for every topic in this series is not "will regulation arrive" but "will it arrive before the harm compounds."

**Second**: the equity gap is the underreported story in every technology category.

AI tutoring's 34% learning gain for underserved communities is extraordinary, and the 58% of students who received no ethical guidance on using it are disproportionately from those same communities.

The AI personal finance revolution has demonstrably helped millennials with standard situations and hasn't touched the estate planning and disability cases that disproportionately affect people without access to human advisors.

Passkeys are on a billion devices, primarily in the enterprise and among tech-sophisticated consumers, while 76% of organizations still use legacy passwords.

Transformation that works for the already-advantaged is not democratization. It's acceleration of existing inequality at AI speed.

**Third**: the second-order effects matter more than the first-order ones.

The most important consequence of the AI customer service deployment wave is not whether AI resolves tickets; it's that Gartner now forecasts AI resolution costs approaching human agent costs by 2030, potentially eliminating the economic justification for the entire deployment model.

The most important consequence of Gen Z's analog rebellion is not vinyl sales; it's that platforms are being forced to compete on wellbeing metrics for the first time.

The most important consequence of Google's AI Overviews is not zero-click search it's that Google is simultaneously preserving its query share and destroying the click economy its advertisers depend on.

Second-order effects are where the real transformation happens, on timelines that first-order analysis consistently misses.

## Your Turn

The invitation at the end of this book is the same one at the beginning, restated with more specificity now that we've covered the ground together.

These shifts are not happening to you. They are happening through decisions made by people and institutions that are accountable to markets, regulations, and ultimately to the choices that individuals and communities make about what they will accept, what they will pay for, what they will demand, and what they will refuse.

The likelihood percentages in this series are not verdicts. They are the beginning of a conversation about which futures are worth fighting for, which compromises are worth making, and which lines (in privacy, in equity, in accountability, in the fundamental human need for authentic connection) are worth holding even when the technology makes crossing them extraordinarily convenient.

I want to hear where those lines are for you.

Where is the technology serving your life well? Where is it serving someone else's business model at your expense? Where are you seeing the bifurcation I described; the transformation that works for some and not for others?

Perhaps the most important question is not the one that I used to organize this series: "is the human-to-technology relationship improving or worsening", turns out to be the wrong question to end on.

The right question is ***who is deciding, and whether the people most affected by those decisions have any meaningful voice in them***.

Some of the predictions here will prove wrong in ways that are instructive. Some of the Counterpoints will prevail, and the explanations for why they prevailed will matter more than the fact that the point argument's failure.

The commitment throughout is not to being right about any individual forecast. It's to the harder and more durable standard: being honest about what the evidence shows, generous toward the arguments on both sides, and rigorous about the difference between what is likely and what is preferred.

In a moment when technology is moving faster than our ability to collectively think about it, that standard is not a methodology.

It's a public obligation.

The conversation this series is trying to start is more important than any conclusion it reaches, because the decisions that will determine the direction of the next decade are being made now, by people who are reading and arguing and building and governing, and the quality of their decisions depends entirely on the quality of the conversation they are having.

Consider this series an invitation to make that conversation better.

## Attribution

The opinions and analysis in this book are my own. I cross-checked with my AI tools to provide some updated research.

The images were generated by ChatGPT under my guidance. I made some modifications to correct (ugh, and I might have missed some) mistakes.

www.ingramcontent.com/pod-product-compliance
Lightning Source LLC
LaVergne TN
LVHW010710061126
851857LV00014B/1125
* 9 7 9 8 9 9 3 9 3 2 5 4 5 *